Springer Tracts in Modern Physics
Volume 189

Available online at
SpringerLink.com

Starting with Volume 165, Springer Tracts in Modern Physics is part of the [SpringerLink] service. For all customers with standing orders for Springer Tracts in Modern Physics we offer the full text in electronic form via [SpringerLink] free of charge. Please contact your librarian who can receive a password for free access to the full articles by registration at:

www.springerlink.com

If you do not have a standing order you can nevertheless browse online through the table of contents of the volumes and the abstracts of each article and perform a full text search.

There you will also find more information about the series.

Springer
Berlin
Heidelberg
New York
Hong Kong
London
Milan
Paris
Tokyo

Physics and Astronomy ONLINE LIBRARY

http://www.springer.de

Springer Tracts in Modern Physics

Springer Tracts in Modern Physics provides comprehensive and critical reviews of topics of current interest in physics. The following fields are emphasized: elementary particle physics, solid-state physics, complex systems, and fundamental astrophysics.
Suitable reviews of other fields can also be accepted. The editors encourage prospective authors to correspond with them in advance of submitting an article. For reviews of topics belonging to the above mentioned fields, they should address the responsible editor, otherwise the managing editor.
See also www.springer.de

Edeltraud Gehrig Ortwin Hess

Spatio-Temporal Dynamics and Quantum Fluctuations in Semiconductor Lasers

With 91 Figures

Springer

Dr. Edeltraud Gehrig
Professor Ortwin Hess

Advanced Technology Institute
School of Electronics and Physical Sciences
University of Surrey
Guildford, Surrey
GU2 7XH, United Kingdom

E-mail: e.gehrig@surrey.ac.uk
o.hess@surrey.ac.uk

Library of Congress Cataloging-in-Publication Data

Hess, Ortwin, 1966-
Spatio-temporal dynamics and quantum fluctuations in semiconductor lasers /
Ortwin Hess, Edeltraud Gehrig.
p. cm. -- (Springer tracts in modern physics, ISSN 0081-3869 ; v. 189)
Includes bibliographical references and index.

1. Semiconductor lasers. I. Gehrig, Edeltraud, 1969- II. Title. III. Series.

TA1700.H47 2003
621.36'6--dc21 2003050427

Physics and Astronomy Classification Scheme (PACS):
42.55.Px, 07.05.Tp, 42.65.Sf, 42.50.Lc, 42.65.Re

ISSN print edition: 0081-3869
ISSN electronic edition: 1615-0430
ISBN 978-3-642-05638-3 e-ISBN 978-3-540-36558-7

Springer-Verlag Berlin Heidelberg New York
a member of BertelsmannSpringer Science+Business Media GmbH

http://www.springer.de

Printed in Germany

Cover concept: eStudio Calamar Steinen
Cover production: *design & production* GmbH, Heidelberg

To Jens
and Rita Marie

Preface

The semiconductor laser is a part of everyday life. It is at the heart of today's global telecommunication networks and hidden in many sophisticated consumer products like DVD-players or personal computers. Indeed, due to their small size and unrivaled efficiency in transforming electrical current into coherent light semiconductor lasers have become key elements in information networks, display technology, data storage, medical and biological applications and spectroscopy. Recent improvements in material processing and laser technology allow the realization of customized novel lasers as well as large-area high-power lasers with high spatial and spectral purity. Yet the exploration of new lasers opens up both new questions and possibilities. The physical processes involved in semiconductor lasers may be as fast as the femtosecond carrier dynamics and span the whole range up to the "slow" dynamics of the propagating light fields (picoseconds) and of the transport of carriers in the nanosecond regime. The interplay of temporal and spatial dynamics leads to complex transversely and longitudinally varying structures and filaments within the active area and thus strongly determines the overall behavior of semiconductor lasers. It is thus not surprising that the highly nonlinear character of these processes may frequently lead to chaotic spatio-temporal instabilities. Clearly, it is not only the microscopically coupled carrier and light field dynamics which determines the spatio-temporal dynamics in such a laser. The external constraints given by the carrier injection, the specific material properites and structuring of the active layer (e.g. wave guiding properties, carrier confinement in quantum wells or quantum dots) as well as the geometry of the active layer are other important factors that influence the laser-internal dynamics and consequently the emission properties.

In this monograph we discuss the theoretical framework for and give an overview of the spatio-temporal dynamics of spatially extended semiconductor lasers. After a brief introduction to semiconductor lasers in Chap. 1 a hierarchy of microscopic and phenomenological models is discussed in Chap. 2 along with the various approximations that lead from fundamental microscopic descriptions (Maxwell–Bloch equations, explicitly taking into account the quantum-kinetic processes in the active semiconductor laser) to simplified rate equations.

In Chap. 3 we analyze the dynamic physical processes in the active charge carrier plasma of a semiconductor laser on the basis of the Maxwell–Bloch equations that take into account the various spatio-temporal optical, kinetic and thermal interactions. In particular, we discuss spatio-spectral hole burning, carrier heating and relaxation. Extending this approach, the Maxwell–Bloch Langevin equations consider, in particular, the interplay of spontaneous and stimulated emission as well as light and dipole fluctuations relevant for e.g. high-power semiconductor lasers and ultrashort-time dynamics.

Chapter 4 gives an impression of the of the high-speed dynamics of edge-emitting semiconductor lasers. The simulations are based on multi-mode Maxwell–Bloch equations that allow an efficient calculation of temporal emission characteristics and of current-modulated lasers in view of a direct comparison with experimental results. In the high-speed modulation dynamics of segmented contact semiconductor lasers the existence of a high-frequency modulation response is predicted.

Simulations of the polarization dynamics and a theory for the fluctuations of vertical-cavity surface emitting lasers are discussed in Chap. 5. Chapter 6 discusses the spatio-temporal dynamics of a broad-area semiconductor laser amplifier with injection of one or more ultrashort light pulses based on the theory presented and discussed in Chap. 3. Chapter 7 gives an overview of simulations of the spatio-temporal dynamics of various semiconductor laser amplifier configurations. In particular, we discuss the dynamics of the carrier distributions and the interband polarization that are self-consistently coupled to the macroscopic constraints imposed by the geometry, electronic contacts, epitaxial structure and facet reflectivities of the semiconductor laser.

In Chap. 8 we discuss a mesoscopic theory for quantum dot lasers and explore ultrashort-time and spatio-temporal dynamics of quantum dot lasers and amplifiers. In Chap. 9 the spatio-temporal dynamics of laser-arrays is analyzed on the basis of a phenomenological model. Inclusion of spatially resolved delayed optical feedback into the phenomenological theory represents a spatial extension of the famous Lang–Kobayashi description to spatially extended semiconductor lasers. In particular, we apply delayed optical feedback to achieve suppression of spatio-temporal instabilities, e.g. the stabilization of stationary operation in the fundamental transverse mode from an originally spatio-temporally chaotic state. Chapter 10 discusses the theoretical basis for describing spatially inhomogeneous luminescence (spontaneous and amplified spontaneous emission) in semiconductor lasers.

This monograph cannot draw but a very rough picture of the richness of phenomena that unravels when looking through the magnifying glass provided by theory and simulation to nano-electronics and photonics: the spatio-temporal dynamics of light and matter on mesoscopic and ultrafast scales. It is our hope that theory and simulations in close cooperation with state-of-the-art semiconductor material technology and experimental characterization may provide the basis for the exploration of this vast nano-space as well as the conception and realization of innovative information technologies.

We would sincerely like to thank all members of the Computational Quantum Electronics group. Their dedication, superb contributions and excellent work have laid the foundations for this monograph. In particular, we would like to acknowledge the central roles of K. Böhringer, J. Hamm, C. Hermann, Dr. H. F. Hofmann, A. Klädtke, Dr. M. Münkel, Dr. D. Preißer, and Dr. C. Simmendinger.

We would gladly like to acknowledge fruitful discussion and cooperation with Prof. Dr. W. Elsäßer and Prof. Dr. F. Kaiser (Darmstadt), Prof. Dr. H. Lamela (Madrid), Prof. Dr. D. Lenstra (Amsterdam), Prof. Dr. G. Mahler (Stuttgart), Prof. Dr. M. Pessa (Tampere), Prof. Dr. R. Wallenstein (Kaiserslautern) and their research groups as well as many others whom we cannot mention all of by name. It is our pleasure to further acknowledge fruitful cooperation with OSRAM Opto-Semiconductors (Regensburg) and Infineon (Munich).

We are very grateful to the Institute of Technical Physics of the Deutsches Zentrum für Luft- und Raumfahrt in Stuttgart that has been our scientific home for many years. In particular, we would like to thank Prof. Dr. H. Opower and Prof. Dr. W. Bohn for their support and encouragement.

O. H. would like to gratefully acknowledge the funding agencies that have supported our research: the Deutsche Forschungsgemeinschaft (DFG), Bundesministerium für Bildung und Forschung (BMBF), the Stiftung Volkswagenwerk as well as the European Commission within the network FALCON (ERB-FMRX-CT-98-0223) and project DOTCOM (IST-2000-28713).

Guildford,
July 2003

Edeltraud Gehrig
Ortwin Hess

Contents

1 Introduction to Semiconductor Lasers

Every laser consists of or involves – in principle – two vital parts or mechanisms: an active medium that allows light amplification and some form of feedback, e.g. realized in the form of an optical resonator. In a semiconductor laser, the fundamental principle of the active medium is the recombination of electron–hole pairs in a suitable layer of semiconductor medium embedded in a chip forming an optical cavity for optical fields propagating within. Inversion of this medium may be achieved by optical or electronic pumping via an optical pump beam or directly via electrical contacts by applying a current. Once the pump beam or the current exceeds a characteristic threshold intensity spontaneous emission processes are exceeded by stimulated emission and lasing starts. In the following we will give a brief overview of typical semiconductor active media and resonator structures employed in modern semiconductor lasers. For details please see e.g. [1; 2; 3; 4].

1.1 Semiconductor Gain Media

To achieve gain in semiconductor media a characteristic density of electron–hole pairs has to be appropriately spatially localized. With moderate current densities this may be realized near the junction of a pn-diode. Today the most common type of semiconductor (diode) lasers is based on III-V semiconductors. The active region of a typical system is based on GaAs and $Ga_{1-x}Al_xAs$ where the subscript x indicates the fraction of Ga atoms in GaAs that is replaced by Al. Semiconductor lasers based on $Ga_{1-x}Al_xAs$ emit – depending on composition and doping – in the spectral region from about 750 nm to 880 nm. Although the theoretical treatment of semiconductor lasers in this book is largely independent of the particular material system, in the following we will center our discussions around this most common material type. Depending on the geometry and thickness of the active layer one differentiates (independent on the various basis materials) between bulk (heterostructure, three-dimensional "3D") or quantum well (two-dimensional "2D"), quantum wire (one-dimensional "1D"), and quantum dot (zero-dimensional "0D") nanostructures. Thereby the dimensionality marks the number of dimensions the charge carriers effectively "see".

1.1.1 Heterostructure Lasers

Semiconductor (double) heterostructure lasers consist of at least three semiconductor layers sandwiched on top of each other [5; 6]. The middle layer, the active zone, has a smaller bandgap than the two outer ones, the cladding layers, and the layers are such that a pn-junction is formed. The thin active region is usually undoped or doped slightly p-type, while one of the adjacent layers is doped heavily p-type and the other one n-type. When a positive bias current is applied to the diode, electrons are injected from the n-type layer into the active layer. Correspondingly, holes are created in the p-side. The charge carriers that are injected are prevented from diffusing out of the active region into the p- or n-type layer, due to the potential barriers resulting from the energy difference between the energy gaps of the different compound semiconductors (e.g. GaAs and $Al_xGa_{1-x}As$).

As the individual layers are differently doped each laser has a different index of refraction and forms a dielectric waveguide for the optical field in the laser structure. In our example, the GaAs layer has a higher index of refraction than the $Al_xGa_{1-x}As$ so that most of the energy of the optical mode is confined to the central active layer.

As the double heterostructure offers double confinement, i.e. confinement of the injected carriers and the optical radiation in the vertical direction in the active region, the structure lasers have been a significant step in the development of semiconductor lasers by allowing continuous-wave laser operation at room temperature. Many of today's mass-produced semiconductor lasers still rely on this principle.

1.1.2 Quantum Well Semiconductor Lasers

Today advanced semiconductor growth techniques allow realization of a semiconductor structure with a precision down to a single atomic layer. This means in turn that quantum wells can now be realized in semiconductor quantum structures with a precision of a tenth of a nanometer.

Quantum well formation. If a given semiconductor material with a small energy gap is sandwiched between energy barriers made of a semiconductor material with a larger energy gap, a quantum well is formed between the barriers. Typical layer thicknesses are just a limited number of atomic layers, say 1 to 10 nm. Once an electron is captured in this well, the probability of escaping from the well is limited. Moreover, the restriction on the movement of the electron in this plane forms a "two-dimensional world" for the carriers, affecting also their energy as compared to a "free" electron in the three-dimensional case. These so-called quantization effects result in allowed energy bands whose energy positions are dependent on the height and width of the barrier and can be calculated by means of fundamental quantum mechanics.

Bandgap engineering. A large number of semiconductor materials can be used to manufacture quantum structures. In addition, one can "mix" different semiconductors with favorable properties in an alloy. This variation of input parameters is usually referred to as bandgap engineering. Bandgap engineering opens fascinating possibilities, in particular for fabrication of novel laser structures [6]. One can select semiconductor materials in such a way that the laser emits at an almost arbitrary wavelength.

The best controlled III-V quantum system is the GaAs/AlGaAs structure with emission in the red range [6]. An attractive material, which is expected to have significant advantages over the conventional long wavelength structures, is the quaternary alloy GaInAsN, which was first proposed by Kondow et al. [7]. Single or multiple quantum well structures with widths less than 10 nm can be obtained by combining GaInAsN with wide gap materials such as GaAs or $Ga_{1-x}Al_xAs$. Providing that GaInAsN is compressively strained, quantum wells with type-I band line up can be formed and can be utilized in light emitting devices. In such structures, due to the large electronegativity of N atoms the conduction band offset can be as large as 500 meV and hence very strong carrier confinement can be maintained even at high temperatures. Therefore, the poor temperature characteristics of the conventional long wavelength sources can be overcome in this material system. In addition to improved temperature performance, GaInAsN allows the use of well established high refractive index contrast GaAs/AlAs DBRs in VCSEL fabrication. Good temperature characteristics and simpler fabrication makes GaInAsN VCSELs very attractive for applications in high-speed optical networks [8; 9; 10].

1.1.3 Quantum Dot Semiconductor Lasers

Semiconductor quantum dots (QDs) have unique electronic and optical properties due to their discrete, atom-like energy level structure, which results from confinement of electronic wavefunctions in all three spatial dimensions. The rapid progress made in recent years in the epitaxial fabrication of self-assembled III/V QDs has triggered tremendous efforts to use them as a gain medium in semiconductor lasers [11]. It has been theoretically predicted that QD lasers should have advantageous properties, such as low and almost temperature-independent threshold current densities and high material gain [12]. Indeed, QDs lasers have been realized that exhibit threshold current densities similar to good quantum well lasers and promising characteristic temperatures T_0. Recently, the potential of QDs for being used in lasers with high-power outputs and extended wavelength ranges has attracted additional interest. In particular, the possibility of fabricating QD lasers emitting in the 1.3 μm wavelength region that are grown on inexpensive GaAs substrates and integratable with existing III/V technology appears extremely interesting for telecommunication applications in the spectral window of minimum dispersion in glass fibers. QD vertical-cavity surface-emitting laser (VCSEL) structures emitting at 1.3 μm have recently been demonstrated and have also been predicted to be suitable for applications in wavelength division multiplexing.

1.2 Laser Cavities

Concerning the geometry of the laser resonator, one generally diffentiates between the families of edge-emitters or in-plane lasers and vertical-cavity lasers or surface-emitters.

1.2.1 In-Plane Edge-Emitting Lasers

In the *edge-emitting* laser or amplifier (Fig. 1.1) carriers are injected via the contacts on the top and light travels in the longitudinal (z) direction inside the cavity formed by the two mirrors at the front and back. To assure

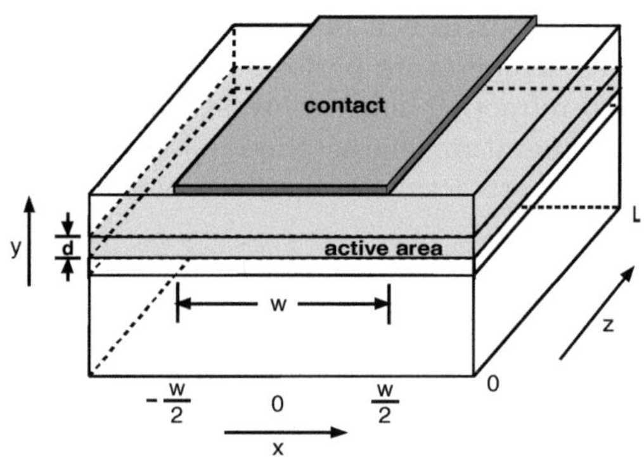

Fig. 1.1. Schematic of the geometry of an edge-emitting semiconductor laser. Charge carriers injected through the contact region at the top of the device (*hatched*) recombine in the active zone. The active layers of the edge-emitter are located between the cladding layers. Light, generated by stimulated emission and amplification propagates in the longitudinal (z) direction

sufficient gain during the counterpropagation (induced by reflection at the mirrors) the resonator length L measures at least 100 times the wavelength (e.g. for GaAs with $\lambda \approx 815$ nm the resonator length typically ranges from 300 µm to 2000 µm). The comparatively small thickness (about 0.1 µm) of the active layer in the vertical (y) direction is a result of the semiconductor epitaxial layer structure and assures a vertical guiding of the optical waves. This is in strong contrast to the transverse (x) direction that may be considerably larger (about 3–5 µm for a single-mode laser and 50–200 µm in case of a multi-mode high-power broad-area laser). In an edge-emitting laser the typical dimensions of the active area strongly suppress one of the polarization directions and thus lead to the emission of linear polarized light. As a consequence no differentiation between σ^+ and σ^- has to be made in the theory and modelling of edge-emitting devices. The respective Bloch equation describing the dipole dynamics is therefore restricted to averaged dipoles without consideration of polarization properties (see Chap. 2 and Chap. 3).

1.2.2 Vertical-Cavity Surface-Emitting Lasers

In the class of *vertical-cavity* or *surface-emitting* lasers (Fig. 1.2), the geometry of the cavity is completely different. Most notably, the length of the resonator now only measures about one wavelength. Consequently, only a single longitudinal mode will be dynamically relevant and propagation effects may be disregarded. At the same time, both transverse (x and y) directions are equally large (typically 3–30 μm). To assure sufficient gain the mirror reflectivities have to be sufficiently high and this is technologically realized by dielectric multilayers. For the VCSEL we thus will have to consider two transverse dimensions. In the active area of the VCSEL, the recombination

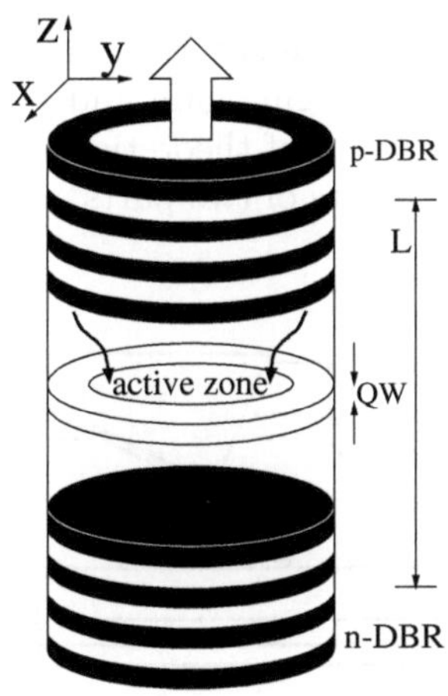

Fig. 1.2. Schematic of the geometry of a vertical-cavity surface-emitting laser. The active layers of the VCSEL are located between distributed Bragg reflector (DBR) layers

of an electron–hole pair leads with equal probability to the two possible polarization directions. It is therefore, in particular, the specific design of the laser resonator or the epitaxial design that determines the polarization properties of the emitted radiation. As the laser resonator of a VCSEL has a highly symmetric geometry around the axis of laser light emission the polarization of the emitted light is highly sensitive to the microscopic carrier and light field dynamics, anistropies in the crystal structure or strain and optical anistropies in the mirrors. Consequently, VCSELs may exhibit polarization instabilities in the input-output characteristics which are the limiting factor in polarization-sensitive applications. The theoretical description of VCSELs consequently has to consider the dynamics of microscopic dipoles for each polarization. This can be done in the frame of the semiconductor Bloch equations (see Chap. 2 and Chap. 5). Also the wave equations for the light fields have to be solved for the two possible polarizations.

1.2.3 High-Power Laser Amplifiers

The need for high output power and good spatial and spectral purity often required by applications in e.g. nonlinear optics or communication has lead in recent years to the realization of laser amplifier configurations with improved beam quality [13; 14; 15; 16; 17; 18; 19; 20; 21; 22; 23]. A coherent light signal (e.g. a single-stripe laser) is injected into the active region of an antireflection-coated large-area semiconductor laser and is – via the induced recombination in the inverted medium – amplified. Thereby it basically maintains its spatial and spectral properties [13]. Up to now various amplifier systems have been realized (broad-area amplifiers in single-pass or double-pass configuration, amplifiers with tapered geometry). In particular the tapered amplifier (Fig. 1.3) has due to its small signal gain and good wave-guiding properties in recent years been in the focus of theoretical [24; 25; 26] and experimental [18; 27] investigations. It consists of two parts, a single-stripe waveguide with

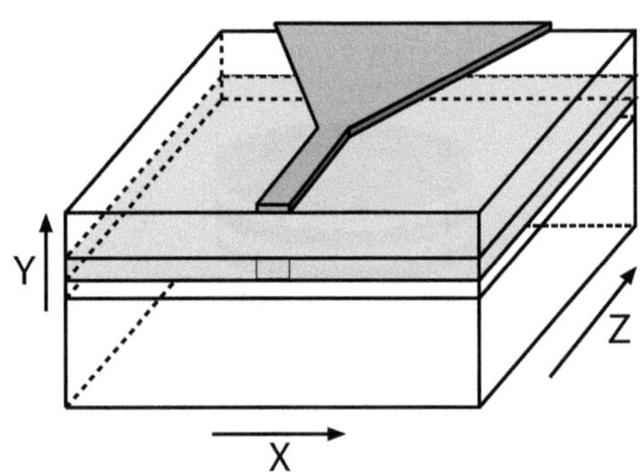

Fig. 1.3. Scheme of a large-area semiconductor laser with tapered geometry

a width of ≈ 3 to 5 µm and a tapered section in which the active area enlarges in the propagation direction so that the intensity at the output facet is kept below the threshold value for catastropic optical mirror damage (COMD). The facets of the tapered amplifiers are antireflection coated. For good beam quality, this facet reflectivity should be less than 10^{-4} [28]. Alternatively, the wave propagation in the resonator should be off-axis, i.e. the facets should be angled with respect to the resonator axis. The small transverse dimension of the waveguide at the input facet of the active area leads to a high small-signal gain allowing efficient saturation of the inversion within the active layer for very moderate input powers of a few mW. Typical lengths of the small waveguide are a few 100 μm at a total length of 1 to 3 mm of the device.

1.2.4 Optically Pumped Lasers

The emission properties of semiconductor lasers with large extension of the active area typically show a complex filamentation behavior that arises from the spatio-temporal coupling and interplay of light propagation, diffraction, carrier diffusion and microscopic carrier scattering processes. Recently the realization of optically pumped semiconductor lasers (typically a vertical-cavity surface-emitting laser in external resonator configuration) has attracted attention. Thereby the transfer of the concept of optical pumping that has so far been successfully applied to solid state lasers to the semiconductor laser allows combination of the power scaling involved in the high gain of semiconductor laser devices and the high beam quality provided by direct optical excitation. Optically pumped VCSELs thus combine the advantages of VCSELs and large-area laser amplifiers. First experimental investigation prove that this concept will be very promising for future laser technologies and applications. The spatio-temporal dynamics and the emission properties of optically pumped lasers are determined by the dynamic interplay of laser-internal light matter interactions with external conditions given by the optical pump laser (with specific frequency, beam profile and duration of the excitation) or a specific external resonator design. In the situation of optically pumped semiconductor laser devices the injection of a light field high above the bandgap generates a (partially incoherent) excited electron–hole plasma which then leads to a hierarchy of carrier relaxation processes in the bands followed by (low-momentum) radiative carrier recombination that can be observed as (photo) luminescence.

1.3 Microlasers

In recent years increasing miniaturisation and high-speed application has lead to a growing influence of quantum properties of the light fields on the spatio-temporal dynamics and emission properties of modern semiconductor laser structures. A theoretical analysis of quantum effects thus is of high relevance for a profound understanding of the physical properties in advanced semiconductor lasers. The underlying physical mechanisms that are responsible for the generation of light affect the interplay between spontaneous and induced radiation and consequently the spatial and spectral emission as well as noise properties and loss mechanisms.

1.3.1 Optical Microcavities

By embedding semiconductor quantum wells or semiconductor quantum dots in microcavities we can – to a certain degree – engineer the photonic emission characteristics of a semiconductor laser. This results from the fact that a microcavity is of the order of the wavelength of the emitted light and thus allows to a higher degree than e.g. an edge-emitting laser influencing the emission pattern and even the local spontaneous emission rate. In combination with semiconductor quantum dots microcavities have been investigated with respect to an enhancement of spontaneous emission in microposts [29] or by exploiting whispering-gallery modes in microdisks [30]. Figure 1.4 shows an example of a disc-shaped microcavity laser together with the op-

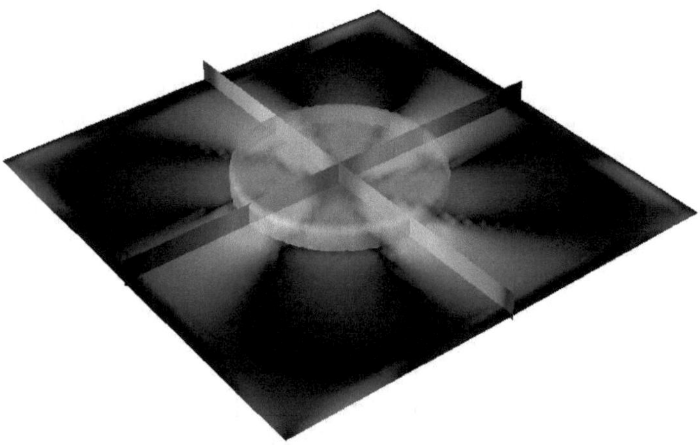

Fig. 1.4. Electric field within and emitted from a microcavity laser. The diameter ($d = 210$ nm) of the microcavity (transparent disc) is of the order of the wavelength ($\lambda = 235$ nm) of the laser; the hight of the disc is 30 nm.

tical field distribution within as well as the one emitted from the laser. The diameter ($d = 210$ nm) of the microcavity is of the order of the wavelength ($\lambda = 235$ nm) of the laser; the hight of the disc is 30 nm.

If we imagine a microcavity small enough to contain only one mode, the combination of a semiconductor quantum dot with a photonic dot formed by the microcavity resonator results in a dot-in-a-dot system that should be capable of controlled, possibly single photon emission – an important basis e.g. for many quantum communication and cryptography schemes [31].

Microcavity lasers are also of great interest for future low power applications and, due to their fast response to external pumping, potentially in high-speed optical communications.

1.3.2 Photonic Bandgap Laser

Photonic crystals are three-dimensional periodic composites of dielectric materials with a lattice spacing of the order of the wavelength of light. For an overview of this fascinating field see e.g. [32; 33]. Arranging the periodic holes in an appropriate way – in the form of a hexagon (insert of Fig. 1.5) – and controlled inclusion of defects in photonic crystal structures offers the possibility of designing high-Q cavities and waveguides on scales of the wavelength. Figure 1.5 (insert) shows an example of the geometry of a photonic crystal laser cavity. The dynamic loss of photonic energy stored in the cavity (Fig. 1.5) is a measure of the quality factor of the cavity.

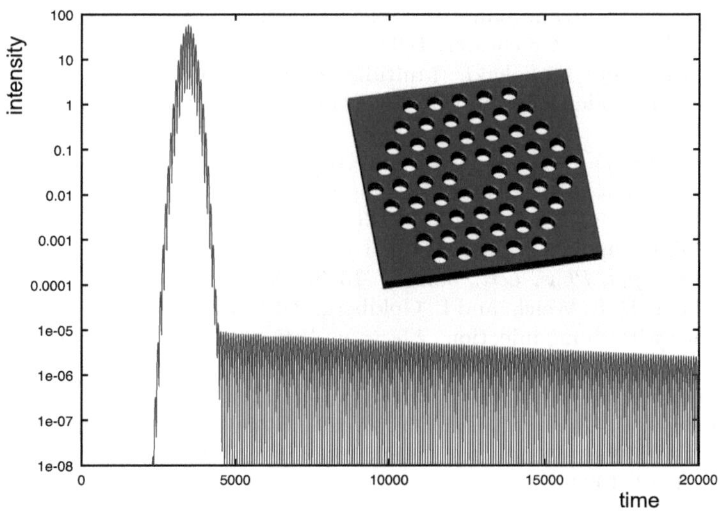

Fig. 1.5. Decay of the electromagnetic intensity stored in a photonic crystal defect cavity (*insert*). The exponentional decay of the cavity mode sets in after the initial excitation pulse

References

1. L. A. Coldren and S. W. Corzine. *Diode Lasers and Photonic Integrated Circuits.* John Wiley, New York, 1995.
2. E. Kapon (Ed.). *Semiconductor Lasers I.* Academic Press, San Diego, 1999.

3. E. Kapon (Ed.). *Semiconductor Lasers II.* Academic Press, San Diego, 1999.
4. R. Diehl (Ed.). *High-Power Diode Lasers: Fundamentals, Technology, Applications.* Springer, Berlin, 2000.
5. G. H. B. Thompson. *Physics of Semiconductor Laser Devices.* Wiley, New York, 1980.
6. A. Yariv. *Quantum Electronics (3rd ed.).* J. Wiley, New York, 1989.
7. M. Kondow, K. Uomi, A. Niwa, T. Kitatani, S. Watahiki, and Y. Yazawa. GaInAsN: A novel material for long wavelength laser diodes with excellent high temperature performance. *Jpn. J. Appl. Phys.*, 35:1273–1275, 1996.
8. S. Sato, Y. Osawa, and T. Saitoh. Room temperature operation of GaInAsN/GaInP double heterostructure laser diodes grown by mocvd. *Jap. J. Appl. Phys.*, 36:2671–2675, 1997.
9. S. Nakatsuka, M. Kondow, T. Kitatani, Y. Yazawa, and M. Okai. Index-guide GaInAsN laser diode for optical communications. *Jap. J. Appl. Phys.*, 37:1380–1383, 1998.
10. K. Nakahara, M. Kondow, T. Kitatani, M.C. Larson, and K. Uomi. 1.3 μm continuous-wave lasing operation in GaInNAs quantum well lasers. *IEEE Photon. Tech. Lett.*, 10:487–488, 1998.
11. D. Bimberg, M. Grundmann, and N. N. Ledentsov. *Quantum Dot Heterostructures.* John Wiley, Chichester, 1998.
12. Y. Arakawa and H. Sakaki. Multidimensional quantum well laser and temperature dependence of its threshold current. *Appl. Phys. Lett.*, 40:939–941, (1982).
13. L. Goldberg, D. Mehuys, M.R. Surette, and D.C. Hall. High-power, near-diffraction-limited large-area traveling-wave semiconductor amplifiers. *IEEE J. Quant. Electr.*, 29:2028–2043, 1993.
14. L. Goldberg and F. Weller. Broad area high power semiconductor optical amplifer. *Appl. Phys. Lett.*, 58:1357–1359, 1991.
15. D. Mehuys, D. F. Welch, and L. Goldberg. 2.0-W cw diffraction-limited tapered amplifier with diode injection. *Electron. Lett.*, 28:1944–1946, 1994.
16. L. Goldberg, D. Mehuys, and D. C. Hall. 3.3 W cw diffraction limited broad area semiconductor amplifier. *Electron. Lett.*, 20:1082–1084, 1992.
17. D. Mehuys, L. Goldberg, and D.F. Welch. 5.25-W cw near-diffraction-limited tapered-stripe semiconductor optical amplifier. *IEEE Photon. Technol. Lett.*, 5:1179–1182, 1993.
18. S. O'Brien, R. Lang, R. Parker, D. F. Welch, and D. Mehuys. 2.2-W continuous-wave diffraction-limited monolithically integrated master oscillator power-amplifier at 854 nm. *IEEE Photon. Technol. Lett.*, 9:440–442, 1997.
19. T. Mukai, Y. Yamamoto, T. Kimura, and W. T. Tsang, editors. *Semiconductors and Semimetals, Vol.22*, chapter Opitcal Amplification by Semiconductor Lasers. Lighwave Communications Tech., Part E, Integrated Optoelectronics, 1985.
20. T. Saitoh and T. Mukai. *Coherence, Amplification and Quantum Effect in Semiconductor Lasers*, chapter Traveling-wave semiconductor laser amplifiers. Wiley, New York, 1991.
21. M. Sagawa, K. Hiramoto, T. Toyonaka, T. Kikiawa, S. Fujisan, and K. Uomi. High-power, highly-reliable operation of InGaAs/InGaAsP 0.98 μm lasers with an exponential flared stripe. *Electron. Lett.*, 32:2277–2279, 1996.

22. G. Bendelli, K. Komori, S. Arai, and Y. Suematsu. A new structure for high-power TW-SLA. *IEEE Photon. Technol. Lett.*, 3:42–44, 1991.
23. R. Parke et al. 2.0 W CW diffraction-limited operation of a monolithically integrated master oscillator power amplifier. *IEEE Photon. Technol. Lett.*, 5:297–300, 1993.
24. J. V. Moloney, R. A. Indik, and C. Z. Ning. Full space-time simulation for high-brightness semiconductor lasers. *IEEE Photon. Technol. Lett.*, 9:731–733, 1997.
25. O. Hess and T. Kuhn. Maxwell-Bloch equations for spatially inhomogeneous semiconductor lasers II: Spatio-temporal dynamics. *Phys. Rev. A*, 54:3360–3368, 1996.
26. R. J. Lang, A. Hardy, R. Parke, D. Mehuys, S. O'Brien, J. Major, and D. Welch. Numerical analysis of flared semiconductor laser amplifiers. *IEEE J. Quant. Electr.*, 29:2044–2051, 1993.
27. J. N. Walpole. Semiconductor amplifiers and lasers with tapered gain regions. *Opt. Quant. Electron.*, 28:623–645, 1996.
28. E. Gehrig, O. Hess, and R. Wallenstein. Modelling of the performance of high-power diode amplifier systems with an opto-thermal microscopic spatio-temporal theory. *IEEE J. Quant. Electr.*, 35:320–331, 1999.
29. M. Pelton, C Santori, J. Vuckovic, B. Zhang, G. S. Solomon, J. Plant, and Y. Yamamoto. Efficient source of single photons: A single quantum dot in a micropost microcavity. *Phys. Rev. Lett.*, 89:233602, 2002.
30. B. Gayral, J.M. Gerard, A. Lemaitre, C. Dupuis, L. Manin, and J. L. Pelouard. look it up. *Appl. Phys. Lett.*, 75:1908, 1999.
31. E. Waks, K. Inoue, C. Santori, D. Fattal, J. Vuckovic, G. S. Solomon, J. Plant, and Y. Yamamoto. Quantum cryptography with a single photon turnstile. *Nature*, 420:762, 2002.
32. J. D. Joannopoulos, R. D. Meade, and J. N. Winn. *Photonic Crystals.* Princeton University Press, Princeton, New Jersey, USA, 1995.
33. K. Sakoda. *Optical Properties of Photonic Crystals.* Springer-Verlag, Berlin, 2001.

2 Semiconductor Laser Theory: The Maxwell–Bloch Equations

In recent years numerous theories have been set up to desribe and simulate the behavior of semiconductor lasers. In principal, these theories can be associated with four levels of description. Listed in order of increasing complexity these are

1. Phenomenological rate equations for the (spatially homogeneous) carrier and photon dynamics
2. Stationary or dynamic wave equations for the optical fields with linear phenomenological parameters for the active semiconductor medium
3. Semiclassical laser theory on the basis of Maxwell–Bloch equations (with and without explicit time dependence) for the spatio-temporal dynamics of the optical fields and the electron–hole plasma
4. Quantum theoretical description of the full light and matter dynamics in the semiconductor laser with consideration of quantum fluctuations.

The first two levels of description are sufficient for the simulation of the behavior of "simple" semiconductor lasers, i.e. single-stripe bulk lasers with small (i.e. a few μm) width of the active area that are driven in a regime where nonlinear effects are still negligible (i.e. low power, continuous-wave operation, moderate pump level). The advantage of the semiclassical Maxwell–Bloch theory (third level of description) is the spatially and temporally resolved description of the coupled light–matter dynamics. This theory provides, in particular, a systematic inclusion of all physical processes relevant in a given material system and of nonlinear effects with their respective characteristic interactions lengths and time scales ranging from femtoseconds up to nanoseconds. The semiclassical laser theory consequently is very appropriate for a realistic description of physical properties (e.g. spatial and spectral beam quality) of spatially extended semiconductor lasers. Quantum theoretical descriptions (level 4) are a well-suited basis for the study of coherence properties of lasers.

It is of fundamental interest to understand the complex interplay of microscopic material with macroscopic waveguide and device properties of semiconductor lasers. Theoretical investigation and numerical modelling consequently have to include both, the macroscopic external constraints imposed by a specific type of laser structure and a spatially resolved microscopic description of the interaction between the optical field and the active semiconductor medium. This chapter gives an overview of semiclassical semiconductor laser theory. The theory developed specifically for spatially inhomogeneous semiconductor lasers such as laser arrays, broad-area lasers or VCSELs [1; 2] includes, in contrast to phenomenological modeling, the full space and momentum dependence of the charge carrier distributions and the polarization on the basis of Maxwell–Bloch equations for spatially inhomogeneous semiconductor lasers [3; 4; 5]. The dynamic interaction between the optical fields, the charge carriers and the interband polarization are described on the basis of microscopic spatially resolved Maxwell–Bloch–Langevin equations for spatially inhomogeneous semiconductor lasers. They consider the relevant dynamic internal laser effects such as diffraction, self-focusing, dynamic local carrier generation, carrier recombination by stimulated emission, scattering and heating. In concert, these effects determine on a fundamental level the individual device properties and the performance of semiconductor lasers.

The equations will be set up in a very general way. Consequently they hold for variable material systems and allow an investigation of a multitude of semiconductor lasers and laser systems such as large-area lasers and amplifiers, laser arrays and lasers with delayed optical feedback (i.e. embeded in an external resonator). Macroscopic device properties such as laser geometry, electronic contacts, epitaxial structure, waveguiding properties, mirror design, resonator shape or operation conditions are fully taken into account. Depending on the specific laser structure different theoretical approaches for spatially extended semiconductor lasers on the basis of a semiclassical description can be applied. The hierarchy ranges from microscopic semiconductor laser Maxwell–Bloch models which take into account the spatio-temporal dynamics of the charge carriers and the interband polarization to a more phenomenological description where the physical properties of the active medium are (once derived from microscopic distributions) considered via a spatially dependent linear representation of the gain and induced refractive index change.

The theory presented in this chapter is the basis for the theoretical descriptions and space-time simulations of various types of lasers and laser systems: in-plane lasers (Chaps. 3, 4 and 6), vertical-cavity surface-emitting lasers (Chap. 5), master-oscillator power-amplifier systems (Chap. 7), quantum dot lasers (Chap. 8), lasers with delayed optical feedback (Chap. 9), and optically pumped lasers (Chap. 10).

In the following we will use a semiclassic approach to describe the physical processes in a semiconductor laser. Thereby the macroscopic optical fields and the polarization are described in a classic manner whereas the dynamics of the carrier distribution and the interband polarization are considered quantum-mechanically. We will derive a set of equations consisting of wave equations for the propagation of light fields in the spatially extended resonator of the semiconductor laser and the semiconductor Bloch equations for the microscopic dynamics of the charge carrier plasma in the valence and conduction bands of the active laser medium.

2.1 Optical Field Dynamics

For the description of the optical field dynamics we derive from Maxwell's equations a wave equation for the optical fields $\boldsymbol{E}$ in a dielectric medium [1]

$$\frac{1}{\varepsilon_0}\nabla\nabla\cdot\boldsymbol{P}+\nabla^2\boldsymbol{E}-\frac{1}{c^2}\frac{\partial^2}{\partial t^2}\boldsymbol{E}=\mu_0\frac{\partial^2}{\partial t^2}\boldsymbol{P}. \tag{2.1}$$

Splitting the polarization $\boldsymbol{P}$ into a linear ($\boldsymbol{P}_\mathrm{l}$) and a nonlinear part ($\boldsymbol{P}_\mathrm{nl}$)

$$\boldsymbol{P}=\boldsymbol{P}_\mathrm{l}+\boldsymbol{P}_\mathrm{nl}=\varepsilon_0\chi_\mathrm{l}\boldsymbol{E}+\boldsymbol{P}_\mathrm{nl} \tag{2.2}$$

leads after substitution in (2.1) to

$$\chi_\mathrm{l}\nabla\nabla\boldsymbol{E}+\nabla^2\boldsymbol{E}-\frac{\varepsilon_\mathrm{l}}{c^2}\frac{\partial^2}{\partial t^2}\boldsymbol{E}=-\frac{1}{\varepsilon_0}\left[\nabla\nabla\cdot\boldsymbol{P}_\mathrm{nl}\right]. \tag{2.3}$$

Since typical geometries of the active area in semiconductor lasers favour the propagation of light fields perpendicular to the facets (in the following the z-coordinate) it is a common method [1] to devide the counterpropagating optical fields $\boldsymbol{E}$, the polarization $\boldsymbol{P}$ and the differential operator ∇ into longitudinal and transverse parts with the ansatz

$$\boldsymbol{E}=\mathrm{e}^{\mathrm{i}\beta z-\mathrm{i}\omega t}\left(\boldsymbol{E}_\mathrm{T}+\boldsymbol{e}_z E_z\right), \tag{2.4}$$

$$\boldsymbol{P}=\mathrm{e}^{\mathrm{i}\beta z-\mathrm{i}\omega t}\left(\boldsymbol{P}_\mathrm{T}+\boldsymbol{e}_z P_z\right), \tag{2.5}$$

$$\nabla=\nabla_\mathrm{T}+\boldsymbol{e}_z\frac{\partial}{\partial z}, \tag{2.6}$$

where $\beta=k_z+\mathrm{i}\frac{\alpha}{2}+\delta\beta$ (k_z denotes the unperturbed propagation wavenumber of the optical fields with $k_z=n_\mathrm{l}k_0$), α is the linear absorption and $\delta\beta=\delta n k_0$ denotes the changes in the propagation wavenumber due to carrier and temperature induced refractive index changes and ω is the frequency. This results in a transverse wave equation

$$\begin{aligned}
&\chi_\mathrm{l}\,\nabla_\mathrm{T}\left(\nabla_\mathrm{T}\cdot\boldsymbol{E}_\mathrm{T}+\left(\mathrm{i}\beta+\frac{\partial}{\partial z}\right)E_z\right)+\left(-\beta^2+2\mathrm{i}\beta\frac{\partial}{\partial z}+\frac{\partial^2}{\partial z^2}+\nabla_\mathrm{T}^2\right)\boldsymbol{E}_\mathrm{T}\\
&+\frac{\varepsilon_\mathrm{l}}{c^2}\left(\omega^2+2\mathrm{i}\omega\frac{\partial}{\partial t}-\frac{\partial^2}{\partial t^2}\right)\boldsymbol{E}_\mathrm{T}\\
&=-\frac{1}{\varepsilon_0}\left[\nabla_\mathrm{T}\left(\nabla_\mathrm{T}\cdot\boldsymbol{P}_\mathrm{T,nl}+\left(\mathrm{i}\beta+\frac{\partial}{\partial z}\right)P_{z,\mathrm{nl}}\right)\right.\\
&\left.+\frac{1}{c^2}\left(\omega^2+2\mathrm{i}\omega\frac{\partial}{\partial t}-\frac{\partial^2}{\partial t^2}\right)\boldsymbol{P}_\mathrm{T,nl}\right]
\end{aligned} \tag{2.7}$$

and a longitudinal wave equation

$$\begin{aligned}
&\chi_\mathrm{l}\left(\mathrm{i}\beta+\frac{\partial}{\partial z}\right)\nabla_\mathrm{T}\cdot\boldsymbol{E}_\mathrm{T}+\varepsilon_\mathrm{l}\left(-\beta^2+2\mathrm{i}\beta\frac{\partial}{\partial z}+\frac{\partial^2}{\partial z^2}\right)E_z+\nabla_\mathrm{T}^2E_z\\
&\quad+\frac{\varepsilon_\mathrm{l}}{c^2}\left(+\omega^2+2\mathrm{i}\omega\frac{\partial}{\partial t}-\frac{\partial^2}{\partial t^2}\right)E_z\\
&=\frac{1}{\varepsilon_0}\left[-\left(\mathrm{i}\beta+\frac{\partial}{\partial z}\right)\nabla_\mathrm{T}\cdot\boldsymbol{P}_{\mathrm{T,nl}}+\left(\beta^2-2\mathrm{i}\beta\frac{\partial}{\partial z}-\frac{\partial^2}{\partial z^2}\right)P_{z,\mathrm{nl}}\right.\\
&\quad\left.-\frac{1}{c^2}\left(\omega^2+2\mathrm{i}\omega\frac{\partial}{\partial t}-\frac{\partial^2}{\partial t^2}\right)P_{z,\mathrm{nl}}\right]. \qquad (2.8)
\end{aligned}$$

In the next step we expand the transverse and the longitudinal wave equations in powers of the dimensionless number $f=\frac{w}{l}=\frac{1}{k_z w}\ll 1$ [1] and, correspondingly, the transverse and longitudinal parts of the optical fields and the nonlinear polarization:

$$\begin{aligned}
\boldsymbol{E}_\mathrm{T}&=\boldsymbol{E}_\mathrm{T}^{(0)}+f^2\boldsymbol{E}_\mathrm{T}^{(2)}+\cdots\\
E_z&=fE_z^{(1)}+f^3E_z^{(3)}+\cdots\\
\boldsymbol{P}_\mathrm{T}&=f^2\boldsymbol{P}_\mathrm{T}^{(2)}+f^4\boldsymbol{P}_\mathrm{T}^{(4)}+\cdots\\
P_z&=f^3P_z^{(3)}+f^5P_z^{(3)}+\cdots. \qquad (2.9)
\end{aligned}$$

Combining these expressions with the wave equations results in longitudinal and transverse equations of the orders $O(f), O(f^3), \ldots$. From the longitudinal equation one obtains in first order the expression

$$\mathrm{i}\nabla_{\tilde{\mathrm{T}}}\boldsymbol{E}_{\tilde{\mathrm{T}}}^{(0)}=E_{\tilde{z}}^{(1)}. \qquad (2.10)$$

This can be inserted into the transverse equation which then reads in first order:

$$\begin{aligned}
&\pm\frac{\partial}{\partial z}\boldsymbol{E}(\boldsymbol{r},t)+\frac{n_\mathrm{l}}{c}\frac{\partial}{\partial t}\boldsymbol{E}(\boldsymbol{r},t)\\
&=\frac{\mathrm{i}}{2k_z}\frac{\partial^2}{\partial x^2}\boldsymbol{E}(\boldsymbol{r},t)-\left(\frac{\alpha}{2}+\mathrm{i}\eta\right)\boldsymbol{E}(\boldsymbol{r},t)+\frac{\mathrm{i}}{\epsilon_0}\frac{k_z}{2n_\mathrm{l}^2}\boldsymbol{P}_\mathrm{nl}(\boldsymbol{r},t). \qquad (2.11)
\end{aligned}$$

In first order Maxwell's wave equation for the optical fields $\boldsymbol{E}(\boldsymbol{r},t)$ and the polarization $\boldsymbol{P}_\mathrm{nl}(\boldsymbol{r},t)$ is purely transverse, but may nevertheless be transversely and longitudinally dependent.

In (2.11) α is the linear absorption and η includes static and dynamic changes in the permittivity affecting both the refractive index and the propagation wavenumber. The static waveguiding structure is a consequence of lateral and vertical confinement of the active area. Its spatial dependence can be obtained from static perturbation theory and then be included in effective parameters such as the background refractive index n_l of (2.11). Dynamic changes arise from the time-dependent carrier and polarization dynamics as well as thermal interactions. Their influence on the permittivity ε and refractive index are deduced from the microscopic carrier dynamics and considered in η whereas the change of the emission frequency is included in the microscopic equations which will be described in Sect. 2.2.

Since the active area of an in-plane laser is given in the form of a Fabry–Perot resonator the description of the light propagation in the active layer requires the simultaneous consideration of optical fields belonging to forward and backward propagation. The optical fields and the polarization are thus composed of two waves, $\boldsymbol{E}^+$ and $\boldsymbol{E}^-$ which counterpropagate in the positive ('+') and negative ('−') z direction in the optical resonator:

$$\begin{aligned}\boldsymbol{E} &= \frac{1}{2}(\boldsymbol{E}^+\mathrm{e}^{\mathrm{i}k_z-\mathrm{i}\omega t} + \boldsymbol{E}^-\mathrm{e}^{-\mathrm{i}k_z-\mathrm{i}\omega t} + \mathrm{c.c.})\\ \boldsymbol{P}_{\mathrm{nl}} &= \frac{1}{2}(\boldsymbol{P}_{\mathrm{nl}}^+\mathrm{e}^{\mathrm{i}k_z-\mathrm{i}\omega t} + \boldsymbol{P}_{\mathrm{nl}}^-\mathrm{e}^{-\mathrm{i}k_z-\mathrm{i}\omega t} + \mathrm{c.c.}).\end{aligned}$$

In (2.11) the polarization is the sum of microscopic dipoles resulting from the interaction of electrons and holes in the valence and conduction bands of the semiconductor medium. In general, these microscopic dipoles have a momentum of $j = \pm 1/2$ leading to right (σ^+) and left (σ^-) polarized light, respectively. Substituting (2.12) in (2.11) we thus finally obtain in dimensional quantities the following paraxial wave equation for the counterpropagating optical fields $\boldsymbol{E}^\pm(\boldsymbol{r},t)$ [4]

$$\begin{aligned}&\pm\frac{\partial}{\partial z}\boldsymbol{E}^\pm(\boldsymbol{r},t) + \frac{n_\mathrm{l}}{c}\frac{\partial}{\partial t}\boldsymbol{E}^\pm(\boldsymbol{r},t)\\ &= \frac{\mathrm{i}}{2k_z}\frac{\partial^2}{\partial x^2}\boldsymbol{E}^\pm(\boldsymbol{r},t) - \left(\frac{\alpha}{2} + \mathrm{i}\eta\right)\boldsymbol{E}^\pm(\boldsymbol{r},t) + \frac{\mathrm{i}}{\epsilon_0}\frac{k_z}{2n_\mathrm{l}^2}\boldsymbol{P}_{\mathrm{nl}}^\pm(\boldsymbol{r},t),\end{aligned} \tag{2.12}$$

where

$$\boldsymbol{P}_{\mathrm{nl}}^\pm(\boldsymbol{r},t) = 2V^{-1}\sum_{\boldsymbol{k}} \boldsymbol{d}_{\mathrm{cv}}(\boldsymbol{k})p_{\mathrm{nl}}^\pm(\boldsymbol{k},\boldsymbol{r},t) \tag{2.13}$$

is the polarization which depends on the microscopic interband polarization $p_{\mathrm{nl}}^\pm(\mathbf{k},\boldsymbol{r},t)$. It is the source of the optical fields and generally describes nonlinear spatio-temporal variations of the gain and refractive index. The amplitudes of the fields in (2.12) are given by

$$\begin{aligned}\boldsymbol{E}^\pm &= (E_{\sigma+}^\pm, E_{\sigma-}^\pm)\\ \boldsymbol{P}_{\mathrm{nl}}^\pm &= (P_{\mathrm{nl},\sigma+}^\pm, P_{\mathrm{nl},\sigma-}^\pm)\\ \boldsymbol{d}_{\mathrm{cv}}(\boldsymbol{k}) &= (\boldsymbol{d}_{\mathrm{cv},\sigma^+}(\boldsymbol{k}), \boldsymbol{d}_{\mathrm{cv},\sigma^-}(\boldsymbol{k})).\end{aligned} \tag{2.14}$$

$\boldsymbol{k}$ is the carrier wavevector, the position vector $\boldsymbol{r} = (x,z)$ denotes the lateral (x) and longitudinal (z) direction, V is the normalization volume of the crystal and $\boldsymbol{d}_{\mathrm{cv}}(\boldsymbol{k})$ is the optical dipole matrix element. In (2.12) k_z denotes the (unperturbed) wavenumber of the propagating fields, n_l is the background refractive index of the active layer, and L the length of the structure. The parameter α considers the linear absorption and η includes lateral and vertical variations of the refractive index due to the waveguide structure as well as dynamic induced changes due to the carrier dynamic and thermal interactions. Γ denotes the optical confinement factor.

In typical semiconductor laser devices the active layer imposes lateral and longitudinal boundary conditions on the optical fields. The light propagating in the longitudinal direction is partially reflected and partially transmitted, depending on the device reflectivities, R_1 and R_2, leading to

$$\begin{aligned} \boldsymbol{E}^{+}(x, z=0, t) &= -\sqrt{R_1}\boldsymbol{E}^{-}(x, z=0, t) \\ \boldsymbol{E}^{-}(x, z=L, t) &= -\sqrt{R_2}\boldsymbol{E}^{+}(x, z=L, t) \end{aligned} \tag{2.15}$$

where L denotes the length of the device. R_1 and R_2 may be the natural reflectivities given by the material-air boundaries or represent effective facet reflectivites defined by e.g. a Bragg mirror (as used in VCSELs) or antireflection coating (as required in laser amplifiers). At the lateral edges, the active area is surrounded by (unpumped) layers of semiconductor media which strongly absorb the fields (absorption coefficient α_{w}). This is represented by the lateral boundary conditions

$$\frac{\partial \boldsymbol{E}^{\pm}}{\partial x} = -\alpha_{\mathrm{w}}\boldsymbol{E}^{\pm} \tag{2.16}$$

at $x = w/2$ (w is the full lateral width of the laser structure) as well as

$$\frac{\partial \boldsymbol{E}^{\pm}}{\partial x} = +\alpha_{\mathrm{w}}\boldsymbol{E}^{\pm} \tag{2.17}$$

at $x = -w/2$.

2.2 Carrier Dynamics

The dynamics of the nonlinear polarization is determined by the microscopic interband polarizations $p_{\mathrm{nl}}^{\pm}(\boldsymbol{k}, \boldsymbol{r}, t)$. The polarizations in turn depend on the dynamics of the distributions of electrons (e) and holes (h) $f^{\mathrm{e,h}}(\boldsymbol{k}, \boldsymbol{r}, t)$. Their interplay and the coupling with the optical field is governed by the semiconductor Bloch equations

$$\begin{aligned} \frac{\partial}{\partial t} f^{\mathrm{e,h}}(\boldsymbol{k}, \boldsymbol{r}, t) = \; & g(\boldsymbol{k}, \boldsymbol{r}, t) \\ & -\tau_{\mathrm{e,h}}^{-1}(\boldsymbol{k}, N)\left[f^{\mathrm{e,h}}(k, \boldsymbol{r}, t) - f_{\mathrm{eq}}^{\mathrm{e,h}}(\boldsymbol{k}, \boldsymbol{r}, t)\right] + \Lambda^{\mathrm{e,h}}(\boldsymbol{k}, \boldsymbol{r}, t) \\ & -\Gamma_{\mathrm{sp}}(\boldsymbol{k}, T_{\mathrm{l}}) f^{\mathrm{e}}(\boldsymbol{k}, \boldsymbol{r}, t) f^{\mathrm{h}}(\boldsymbol{k}, \boldsymbol{r}, t) - \gamma_{\mathrm{nr}} f^{\mathrm{e,h}}(\boldsymbol{k}, \boldsymbol{r}, t) \end{aligned} \tag{2.18}$$

$$\begin{aligned} \frac{\partial}{\partial t} p_{\mathrm{nl}}^{\pm}(\boldsymbol{k}, \boldsymbol{r}, t) = & - \left[\mathrm{i}\bar{\omega}(\boldsymbol{k}, T_{\mathrm{l}}) + \tau_{\mathrm{p}}^{-1}(\boldsymbol{k}, N)\right] p_{\mathrm{nl}}^{\pm}(\boldsymbol{k}, \boldsymbol{r}, t) \\ & + \frac{1}{\mathrm{i}\hbar}\boldsymbol{d}_{\mathrm{cv}}(\boldsymbol{k})\boldsymbol{E}^{\pm}(\boldsymbol{r}, t)\left[f^{\mathrm{e}}(\boldsymbol{k}, \boldsymbol{r}, t) + f^{\mathrm{h}}(\boldsymbol{k}, \boldsymbol{r}, t)\right]. \end{aligned} \tag{2.19}$$

The microscopic generation rate is given by

$$g(\boldsymbol{k},\boldsymbol{r},t) = -\frac{1}{4\mathrm{i}\hbar}\boldsymbol{d}_{\mathrm{cv}}(\boldsymbol{k})\left[\boldsymbol{E}^{+}(\boldsymbol{r},t)p_{\mathrm{nl}}^{+*}(\boldsymbol{k},\boldsymbol{r},t) + \boldsymbol{E}^{-}(\boldsymbol{r},t)p_{\mathrm{nl}}^{-*}(\boldsymbol{k},\boldsymbol{r},t)\right] \\ +\frac{1}{4\mathrm{i}\hbar}\boldsymbol{d}_{\mathrm{cv}}^{*}(\boldsymbol{k})\left[\boldsymbol{E}^{+*}(\boldsymbol{r},t)p_{\mathrm{nl}}^{+}(\boldsymbol{k},\boldsymbol{r},t) + \boldsymbol{E}^{-*}(\boldsymbol{r},t)p_{\mathrm{nl}}^{-}(\boldsymbol{k},\boldsymbol{r},t)\right]. \tag{2.20}$$

$f_{\mathrm{eq}}^{\mathrm{e,h}}(\boldsymbol{k},\boldsymbol{r},t)$ denote the carrier distributions in thermal equilibrium with the lattice, each given by the corresponding Fermi distribution. For the description of a vertical-cavity surface-emitting laser (2.18) has to be calculated for carriers belonging to right and left circularly polarized light, respectively. The respective equations are then coupled via the carrier relaxation towards their equilibrium. In (2.19) $\Gamma_{\mathrm{sp}}(\boldsymbol{k},T_{\mathrm{l}})$ is the spontaneous recombination coefficient and γ_{nr} denotes the rate due to nonradiative recombination. The microscopic pump term includes the pump-blocking effect

$$\Lambda^{\mathrm{e,h}}(\boldsymbol{k},\boldsymbol{r},t) = \Lambda(\boldsymbol{r},t)\frac{f_{\mathrm{eq}}^{\mathrm{e,h}}(\boldsymbol{k},\boldsymbol{r},t)\left(1-f^{\mathrm{e,h}}(\boldsymbol{k},\boldsymbol{r},t)\right)}{V^{-1}\sum_{\boldsymbol{k}} f_{\mathrm{eq}}^{\mathrm{e,h}}(\boldsymbol{k},\boldsymbol{r},t)\left(1-f^{\mathrm{e,h}}(\boldsymbol{k},\boldsymbol{r},t)\right)}, \tag{2.21}$$

where the macroscopic pump term

$$\Lambda(\boldsymbol{r},t) = \frac{\eta_{\mathrm{eff}}}{ed}\mathcal{J}(\boldsymbol{r},t) \tag{2.22}$$

depends on the space and time dependent injection current density $\mathcal{J}(\boldsymbol{r},t)$, with η_{eff} being the quantum efficiency and d the thickness of the active layer. The scattering rates $\tau_{\mathrm{e,h}}^{-1}(\boldsymbol{k},N)$ and $\tau_{\mathrm{p}}^{-1}(\boldsymbol{k},N)$ in (2.18) are microscopically determined [4] and include carrier–carrier scattering mechanisms and the interaction of carriers with optical phonons (LO phonons). Both depend on wavenumber and carrier density and thus vary spatio-temporally within the active semiconductor layer. The detuning between the cavity frequency ω and the transition frequency ω_{T}

$$\hbar\bar{\omega}(\boldsymbol{k},T_{\mathrm{l}}) = \mathcal{E}_{\mathrm{g}}(T_{\mathrm{l}}) + \frac{\hbar^2\boldsymbol{k}^2}{2m_{\mathrm{r}}} + \delta\mathcal{E}(N,T_{\mathrm{l}}) - \hbar\omega \\ = \hbar(\omega_{\mathrm{T}} - \omega) \tag{2.23}$$

contains via $\mathcal{E}_{\mathrm{g}}$ and $\delta\mathcal{E}$ the dependence of the semiconductor bandgap on the density of charge carriers N and on the lattice temperature T_{l}. The variation of the bandgap with carrrier density generally is a complex many-body process [6]. It may be approximated by [7]

$$\delta\mathcal{E}(N,T_{\mathrm{l}}) = \mathcal{E}_0\frac{-a\left(Na_0^3\mathcal{E}_0^2\right)^{1/4}}{\left(Na_0^3\mathcal{E}_0^2 + b^2\left(k_{\mathrm{B}}T_{\mathrm{l}}\right)^2\right)^{1/4}}, \tag{2.24}$$

with the exciton binding energy $\mathcal{E}_0 = m_{\mathrm{r}}e^4/(2\epsilon_0^2\hbar^2)$, the exciton Bohr radius $a_0 = \hbar^2\epsilon/(e^2 m_{\mathrm{r}})$, the lattice temperature T_{l}, Boltzmann's constant k_{B}, the effective mass m_{r}, and the numerical factors $a = 4.64$ and $b = 0.107$. The dependence on the lattice temperature can be expressed as [8]

$$\mathcal{E}_g(T_l) = \mathcal{E}_g(0) - v_1 T_l^2/(T_l + v_2) \tag{2.25}$$

where $\mathcal{E}_g(0)$ (= 1.519 eV for GaAs) is the bandgap at $T_l = 0$ K, and v_1, v_2 are two material parameters given by $v_1 = 5.405 \times 10^{-4}$ eV/K and $v_2 = 204$ K for GaAs.

The Wigner distributions describe the microscopic semiconductor dynamics and are functions of space and momentum. As has been discussed in [4], the spatial transport of charge carriers usually occurs on a much slower time scale in the picosecond up to the nanosecond regime (10 ps,. . .,10 ns) than the $\boldsymbol{k}$-space relaxation of the microscopic variables towards their local quasi-equilibrium values occuring on a femtosecond time scale. This typical separation of time scales between the $\boldsymbol{k}$-space and the $\boldsymbol{r}$-space dynamics allows us to treat both regimes separately, with the influence of spatial gradients on the $\boldsymbol{k}$-space dynamics often being negligible. Transport of carriers thus effectively occurs as ambipolar transport of electrons and holes allowing derivation of a macroscopic transport equation for the carrier density $N(\boldsymbol{r}, t)$ [4] and a corresponding relaxation equation for the energy densities $u^{\mathrm{e,h}}(\boldsymbol{r}, t)$ [9]

$$\begin{aligned} \frac{\partial}{\partial t} N(\boldsymbol{r}, t) = {} & \nabla \cdot (D_f \nabla N(\boldsymbol{r}, t)) + \Lambda(\boldsymbol{r}, t) + G(\boldsymbol{r}, t) \\ & - \gamma_{nr} N(\boldsymbol{r}, t) - W(\boldsymbol{r}, t) \end{aligned} \tag{2.26}$$

$$\begin{aligned} \frac{\partial}{\partial t} u^{\mathrm{e,h}}(\boldsymbol{r}, t) = {} & \Lambda_u^{\mathrm{e,h}}(\boldsymbol{r}, t) + G_u^{\mathrm{e,h}}(\boldsymbol{r}, t) \\ & - \gamma_{nr} u^{\mathrm{e,h}}(\boldsymbol{r}, t) - W_u^{\mathrm{e,h}}(\boldsymbol{r}, t) - R^{\mathrm{e,h}}(\boldsymbol{r}, t) \end{aligned} \tag{2.27}$$

with spatially structured carrier injection (Λ and $\Lambda_u^{\mathrm{e,h}}$), carrier recombination by stimulated emission (G and $G_u^{\mathrm{e,h}}$), nonradiative recombination ($\gamma_{nr} N$ and $\gamma_{nr} u^{\mathrm{e,h}}$), spontaneous emission (W and $W_u^{\mathrm{e,h}}$) and carrier–phonon relaxation ($R^{\mathrm{e,h}}$). The recombination of charge carriers, which occurs at the lateral edges of the structure, is modeled by using characteristic values of the surface recombination velocity coefficient v_{sr} via the boundary conditions

$$\frac{\partial N}{\partial x} = -v_{sr} N \tag{2.28}$$

at $x = \frac{w}{2}$ and

$$\frac{\partial N}{\partial x} = +v_{sr} N \tag{2.29}$$

at $x = -\frac{w}{2}$. In (2.26) D_f is the ambipolar diffusion coefficient given by

$$D_f = \frac{\sigma^{\mathrm{h}} D^{\mathrm{e}} + \sigma^{\mathrm{e}} D^{\mathrm{h}}}{\sigma^{\mathrm{e}} + \sigma^{\mathrm{h}}} , \tag{2.30}$$

with the conductivities of electrons and holes, $\sigma^{\mathrm{e,h}}$ [4]. The macroscopic gain is given by

$$G(\boldsymbol{r}, t) = \chi'' \epsilon_0 / 2\hbar \sum_{\sigma=(\sigma^+,\sigma^-)} \sum_{\nu=(+,-)} \Big(|E_\sigma^\nu(\boldsymbol{r}, t)|^2$$

$$-\frac{1}{4}\hbar\,\mathrm{Im}\left[-E^{\nu}_{\sigma}(\boldsymbol{r},t)P^{\nu *}_{\sigma\,\mathrm{nl}}(\boldsymbol{r},t)\right]\Big), \tag{2.31}$$

where Im[...] denotes the imaginary part and the spontaneous emission reads

$$W(\boldsymbol{r},t) = V^{-1}\sum_{\boldsymbol{k}} \Gamma_{\mathrm{sp}}(\boldsymbol{k},T_l)f^{\mathrm{e}}(\boldsymbol{k},\boldsymbol{r},t)f^{\mathrm{h}}(\boldsymbol{k},\boldsymbol{r},t), \tag{2.32}$$

with a phenomenological rate of spontaneous emission [10]. The corresponding expressions in the heat equations are, with $\epsilon^{\mathrm{e,h}}(\boldsymbol{k}) = \hbar^2\boldsymbol{k}^2/2m_{\mathrm{e,h}}$,

$$\Lambda^{\mathrm{e,h}}_{u}(\boldsymbol{r},t) = V^{-1}\sum_{\boldsymbol{k}} \epsilon^{\mathrm{e,h}}(\boldsymbol{k})\Lambda^{\mathrm{e,h}}(\boldsymbol{k},\boldsymbol{r},t) \tag{2.33}$$

$$\begin{aligned} G^{\mathrm{e,h}}_{u}(\boldsymbol{r},t) = &\sum_{\sigma=(\sigma^+,\sigma^-)}\ \sum_{\nu=(+,-)} -\frac{1}{2\hbar^2 V}\left[|E^{\nu}{}_{\sigma}(\boldsymbol{r},t)|^2\right] \\ &\times V^{-1}\sum_{\boldsymbol{k}} \epsilon^{\mathrm{e,h}}(\boldsymbol{k})\frac{\tau_{\mathrm{p}}^{-1}(\boldsymbol{k},N)}{\bar{\omega}^2(\boldsymbol{k},T_{\mathrm{l}}) + \tau_{\mathrm{p}}^{-2}(\boldsymbol{k},N)}\,|\boldsymbol{d}_{\mathrm{cv}}(\boldsymbol{k})|^2 \\ &\cdot\left[f^{\mathrm{e}}(\boldsymbol{k},\boldsymbol{r},t) + f^{\mathrm{h}}(\boldsymbol{k},\boldsymbol{r},t) - 1\right] \end{aligned} \tag{2.34}$$

$$W^{\mathrm{e,h}}_{u}(\boldsymbol{r},t) = V^{-1}\sum_{\boldsymbol{k}} \epsilon^{\mathrm{e,h}}(\boldsymbol{k})\Gamma_{\mathrm{sp}}(\boldsymbol{k},T_{\mathrm{l}})f^{\mathrm{e}}(\boldsymbol{k},\boldsymbol{r},t)f^{\mathrm{h}}(\boldsymbol{k},\boldsymbol{r},t). \tag{2.35}$$

The relaxation term

$$\begin{aligned} R^{\mathrm{e,h}}(\boldsymbol{r},t) = &\sum_{\mathrm{ph}} \frac{1}{\tau^{\mathrm{e,h}}_{\mathrm{ph}}}\left(u^{\mathrm{e,h}}(\boldsymbol{r},t) - u^{\mathrm{e,h}}_{\mathrm{eq}}(\boldsymbol{r},t)\right) \\ &+ V^{-1}\sum_{\boldsymbol{k}} \epsilon^{\mathrm{e,h}}(\boldsymbol{k})\frac{1}{\tau^{\mathrm{e,h}}_{\mathrm{po}}(\boldsymbol{k},N)}\left(f^{\mathrm{e,h}}(\boldsymbol{k},\boldsymbol{r},t) - f^{\mathrm{e,h}}_{\mathrm{eq}}(\boldsymbol{k},\boldsymbol{r},t)\right) \end{aligned} \tag{2.36}$$

includes carrier–phonon interactions, where, in particular, the polar optical phonons ($\tau^{\mathrm{e,h}}_{\mathrm{po}}$) are considered in dependence on density and wavenumber.

2.3 Temperature Dynamics

The existence and interplay of thermal effects are known to play an important role in the performance of semiconductor lasers. Among all laser configurations it is, in particular, the high-power semiconductor amplifier system and the VCSEL that are strongly affected by temperature dynamics. In these systems the nonlinear coupling of the optical field to the spatio-temporally varying temperature distribution has a significant influence on the over-all laser properties [9]. In the case of the laser amplifier the spatially varying power distribution and the high light-induced changes in carrier density and refractive index lead to the formation of characteristic spatio-temporally varying temperature profiles that strongly determine the spatio-temporal dynamics of the propagating light fields and consequently affect the properties of the emitted amplified radiation. In paticular in a VCSEL it is the transverse carrier–light field dynamics microscopically coupled to the interband dipole dynamics and to a temperature-dependent bandgap that are responsible for thermal effects such as shift of the emission frequency or thermal switch-off.

On the basis of a grand canonical ensemble interpretation having both energy and particle contact and exchange with the environment, the carrier density N and the carrier energies $u^{\mathrm{e,h}}$ can be expressed as functions of the independent variables of the chemical potentials $\mu^{\mathrm{e,h}}$ and the plasma temperatures $T_{\mathrm{pl}}^{\mathrm{e,h}}$. Solving the resulting set of equations for the plasma temperatures $T_{\mathrm{pl}}^{\mathrm{e,h}}$ leads to [9]

$$\dot{T}_{\mathrm{pl}}^{\mathrm{e,h}}(\boldsymbol{r},t) = J_u^{\mathrm{e,h}}(\boldsymbol{r},t)\dot{u}^{\mathrm{e,h}}(\boldsymbol{r},t) - J_N^{\mathrm{e,h}}(\boldsymbol{r},t)\dot{N}(\boldsymbol{r},t), \tag{2.37}$$

where $\dot{[\ldots]}$ denotes the temporal derivative and the Jacobian derivates $J_u^{\mathrm{e,h}}$ and $J_N^{\mathrm{e,h}}$ are given by

$$J_u^{\mathrm{e,h}}(\boldsymbol{r},t) = \frac{\partial N^{\mathrm{e,h}}}{\partial \mu^{\mathrm{e,h}}}\left(\frac{\partial u^{\mathrm{e,h}}(\boldsymbol{r},t)}{\partial T_{\mathrm{pl}}^{\mathrm{e,h}}(\boldsymbol{r},t)}\frac{\partial N^{\mathrm{e,h}}(\boldsymbol{r},t)}{\partial \mu^{\mathrm{e,h}}(\boldsymbol{r},t)} - \frac{\partial u^{\mathrm{e,h}}(\boldsymbol{r},t)}{\partial \mu^{\mathrm{e,h}}(\boldsymbol{r},t)}\frac{\partial N^{\mathrm{e,h}}(\boldsymbol{r},t)}{\partial T_{\mathrm{pl}}^{\mathrm{e,h}}(\boldsymbol{r},t)}\right)^{-1}, \tag{2.38}$$

$$J_N^{\mathrm{e,h}}(\boldsymbol{r},t) = \frac{\partial u^{\mathrm{e,h}}}{\partial \mu^{\mathrm{e,h}}}\left(\frac{\partial u^{\mathrm{e,h}}(\boldsymbol{r},t)}{\partial T_{\mathrm{pl}}^{\mathrm{e,h}}(\boldsymbol{r},t)}\frac{\partial N^{\mathrm{e,h}}(\boldsymbol{r},t)}{\partial \mu^{\mathrm{e,h}}(\boldsymbol{r},t)} - \frac{\partial u^{\mathrm{e,h}}(\boldsymbol{r},t)}{\partial \mu^{\mathrm{e,h}}(\boldsymbol{r},t)}\frac{\partial N^{\mathrm{e,h}}(\boldsymbol{r},t)}{\partial T_{\mathrm{pl}}^{\mathrm{e,h}}(\boldsymbol{r},t)}\right)^{-1}. \tag{2.39}$$

The dynamics of the lattice temperature is given by

$$\begin{aligned}
\dot{T}_{\mathrm{l}}(\boldsymbol{r},t) = {} & -\gamma_{\mathrm{a}}\left[T_{\mathrm{l}}(\boldsymbol{r},t) - T_{\mathrm{a}}\right] + \sum_{\mathrm{ph}} \frac{1}{\tau_{\mathrm{ph}}^{\mathrm{e}}}\left[T_{\mathrm{pl}}^{\mathrm{e}}(\boldsymbol{r},t) - T_{\mathrm{l}}(\boldsymbol{r},t)\right] \\
& + J_u^{\mathrm{e}}(\boldsymbol{r},t)\pi^{-2}V^{-1}\sum_{\boldsymbol{k}} \epsilon^{\mathrm{e}}(\boldsymbol{k})\frac{1}{\tau_{\mathrm{po}}^{\mathrm{e}}}\left(f^{\mathrm{e}}(\boldsymbol{k},\boldsymbol{r},t) - f_{\mathrm{eq}}^{\mathrm{e}}(\boldsymbol{k},\boldsymbol{r},t)\right) \\
& + \sum_{\mathrm{ph}} \frac{1}{\tau_{\mathrm{ph}}^{\mathrm{h}}}\left[T_{\mathrm{pl}}^{\mathrm{h}}(\boldsymbol{r},t) - T_{\mathrm{l}}(\boldsymbol{r},t)\right] \\
& + J_u^{\mathrm{h}}(\boldsymbol{r},t)\pi^{-2}V^{-1}\sum_{\boldsymbol{k}} \epsilon^{\mathrm{h}}(\boldsymbol{k})\frac{1}{\tau_{\mathrm{po}}^{\mathrm{h}}}\left(f^{\mathrm{h}}(\boldsymbol{k},\boldsymbol{r},t) - f_{\mathrm{eq}}^{\mathrm{h}}(\boldsymbol{k},\boldsymbol{r},t)\right) \\
& + \hbar\omega\gamma_{\mathrm{nr}}N(\boldsymbol{r},t) + \frac{\mathcal{J}^2 A^2 R}{c_{\mathrm{q}} V_{\mathrm{BAL}}}
\end{aligned} \tag{2.40}$$

where $-\gamma_{\mathrm{a}}$ describes the relaxation to the ambient temperature T_{a}, c_{q} is the specific heat, R denotes the total resistance, A is the cross section, and V_{BAL} is the volume of the active zone.

The approach taken here incorporates the full microscopic spatio-temporal dynamics together with the relevant macroscopic properties of typical semiconductor lasers. It generally applies to semiconductor laser systems with a

Table 2.1. Fundamental material and device parameters of GaAs-based semiconductor lasers

parameter	physical quantity	value
n_c	refractive index of the cladding layers (GaAlAs)	3.35
n_l	refractive index of active layer (GaAs)	3.59
λ	laser wavelength	815 nm – 930 nm
R_1	front facet mirror reflectivity	$10^{-4} \ldots 1$
R_2	back facet mirror reflectivity	$10^{-4} \ldots 1$
τ_nr	nonradiative recombination time	5 ns
a_0	exciton Bohr radius	1.243×10^{-6}cm
m_0	mass of the electron	$9.1093879 \times 10^{-31}$ kg
m_e	effective electron mass	$0.067\ m_0$
m_h	effective hole mass	$0.246\ m_0$
$\mathcal{E}_\mathrm{g}(0)$	semiconductor energy gap at $T = 0$ K	1.519 eV
η_i	injection efficiency	0.5
Γ	confinement factor	0.55/0.54
α_w	absorption	30 cm^{-1}
v_sr	surface recombination velocity	10^6 m/s

large variety of geometries and active material systems. To be specific, in this book we will focus on typical III-V semiconductor material systems and use the relevant parameters for the GaAs-based system [11]. The relevant material and structural parameters used in the simulations are given in Table 2.1.

The system of (2.12) and (2.18) describing the microscopic and macroscopic carrier–light field dynamics in the active area of semiconductor lasers is the basis for the numerical simulations presented in the following chapters. Suitable modifications of the equations for realistic consideration of the individual system properties and device configurations allow both, analysis of the laser-internal physical interactions and numerical modeling of the performance of semiconductor lasers based on microscopic principles.

References

1. O. Hess. *Spatio-temporal dynamics of semiconductor lasers.* Wiss. - und Technik-Verlag, Berlin, 1993.
2. O. Hess and T. Kuhn. Spatio-temporal dynamics of semiconductor lasers: Theory, modeling and analysis. *Prog. Quant. Electr.*, 20:85–179, 1996.
3. O. Hess, S.W. Koch, and J.V. Moloney. Filamentation and beam propagation in broad-area semiconductor lasers. *IEEE J. Quant. Electr.*, 31:35–43, 1995.
4. O. Hess and T. Kuhn. Maxwell-Bloch equations for spatially inhomogeneous semiconductor lasers I: Theoretical description. *Phys. Rev. A*, 54:3347–3359, 1996.

5. O. Hess and T. Kuhn. Maxwell-Bloch equations for spatially inhomogeneous semiconductor lasers II: Spatio-temporal dynamics. *Phys. Rev. A*, 54:3360–3368, 1996.
6. F. Jahnke, M. Kira, and S. W. Koch. Linear and nonlinear optical properties of excitons in semiconductor quantum wells and microcavities. *Z. Phys. B*, 104:559–572, 1997.
7. R. Zimmermann. Nonlinear optics and the mott trasition in semiconductors. *Phys. Stat. Sol. B*, 146:371–381, 1988.
8. Y. P. Varshni. Temperature dependence of the energy gap in semiconductors. *Physica*, 34:149–154, 1967.
9. E. Gehrig and O. Hess. Nonequilibrium spatio-temporal dynamics of the Wigner-distributions in broad-area semiconductor lasers. *Phys. Rev. A*, 57:2150–2163, 1998.
10. G. H. B. Thompson. *Physics of Semiconductor Laser Devices.* Wiley, New York, 1980.
11. S. Adachi. GaAs, AlAs and $Al_xGa_{1-x}As$: Material parameters for use in research and device applications. *J. Appl. Phys*, 59:R1–R29, 1985.

3 The Active Carrier Plasma: Ultrafast Microscopic Dynamics and Spatio-Spectral Gratings

The physical properties of semiconductor lasers are determined by a multitude of coupled physical processes in the active medium. Microscopic scattering, carrier heating and spatio-spectral hole burning lead to continuous dynamic changes in the charge carrier plasma. In combination with the external constraints given by the laser geometry and the physical properties of the active medium they are responsible for a dynamic interplay of light and matter. The nonlinear interaction of spatial with temporal degrees of freedom leads to non-equilibrium carrier distributions and to complex or even chaotic spatio-temporal light field dynamics.

It is highly desirable to understand the nature of these complex temporal and spatio-temporal processes and, building on this knowledge, to develop schemes to control and suppress them (see Chaps. 7 and 9).

In this chapter we will analyze the dynamic physical processes in the active charge carrier plasma of a semiconductor laser. Calculation and interpretation of the so-called Wigner distributions – representing spatio-spectral gain and induced refractive index – visualize spatio-spectral hole burning, carrier heating and relaxation. The basis of our analysis will be the Maxwell–Bloch equations (see Chap. 2) that take into account the various spatio-temporal optical, kinetic and thermal interactions. Extending this approach, we introduce the Maxwell–Bloch–Langevin equations that consider, in particular, the interplay of spontaneous and stimulated emission as well as light and dipole fluctuations relevant for e.g. high-power semiconductor lasers and ultrashort-time dynamics.

3.1 Semiconductor Laser Maxwell–Bloch–Langevin Equations: A Dual-Field Approach

In order to gain insight into the interplay of microscopic excitation and relaxation dynamics in the charge carrier plasma of a semiconductor laser one has to analyze the propagation of a (ultra-) short light pulse. Via dynamic coupling to the carrier system the pulse burns a spatio-spectral hole and thereby creates a highly non-equilibrium carrier distribution. The heated charge carrier plasma then relaxes via a multitude of microscopic carrier–carrier and carrier–phonon scattering processes to the quasi-equilibrium given by the respective Fermi–Dirac distributions. For a self-consistent inclusion of the spatio-temporal dynamics and interplay of both, an optically injected propagating light signal and light from spontaneous emission processes we use a Dual-Field wave description. The spatio-temporal dynamics of the counter-propagating optical fields $\boldsymbol{E}^{\pm}(\mathbf{r},t)$ of the injected light signal (sig) and the amplified spontaneous emission (se) are then given by [1] the wave equation:

$$\begin{aligned}\frac{\partial}{\partial z}\boldsymbol{E}^{\pm}_{\mathrm{sig,se}}(\mathbf{r},t)+\frac{n_{\mathrm{l}}}{c}\frac{\partial}{\partial t}\boldsymbol{E}^{\pm}_{\mathrm{sig,se}}(\mathbf{r},t)&=\frac{\mathrm{i}}{2}\frac{1}{K_z}\frac{\partial^2}{\partial x^2}\boldsymbol{E}^{\pm}_{\mathrm{sig,se}}(\mathbf{r},t)\\&\quad-\eta\boldsymbol{E}^{\pm}_{\mathrm{sig,se}}(\mathbf{r},t)+\frac{\mathrm{i}}{\epsilon_0}\frac{K_z}{2n_{\mathrm{l}}^2}\Gamma\boldsymbol{P}^{\pm}_{\mathrm{sig,se}}(\mathbf{r},t)\\&\quad+F_{\mathrm{E}}q^{\mathrm{E}}(\mathbf{r},t).\end{aligned}\tag{3.1}$$

$\boldsymbol{P}^{\pm}(\mathbf{r},t)$ represents the polarization that contains all nonlinear spatio-temporal gain and refractive index variations. $F_{\mathrm{E}}q^{\mathrm{E}}(\mathbf{r},t)$ is a Langevin noise term taking light field fluctuations into account [2].

The injection of optical pulses induces a continuous interplay between the propagating light fields and the active medium. It is taken into account via the boundary conditions. The laterally dependent injected light field at the laser facets $z=0$ or $z=L$ may conveniently be expressed in terms of modified boundary conditions

$$\begin{aligned}\boldsymbol{E}^{+}(x,z=0,t)&=-R_1\boldsymbol{E}^{-}(x,z=0,t)+\boldsymbol{E}_{\mathrm{inj}}(x,z=0,t),\\\boldsymbol{E}^{-}(x,z=L,t)&=-R_2\boldsymbol{E}^{+}(x,z=L,t)+\boldsymbol{E}_{\mathrm{inj}}(x,z=L,t)\end{aligned}\tag{3.2}$$

where

$$\boldsymbol{E}_{\mathrm{inj}}(x,z=0,L,t)=\boldsymbol{E}^{0}_{\mathrm{inj}}(x,z=0,L,t)\times\exp\left[\mathrm{i}K_z\frac{(x-x_0)^2}{2R(z)}\right]\tag{3.3}$$

with

$$\begin{aligned}\boldsymbol{E}^{0}_{\mathrm{inj}}(x,z=0,L,t)&=\mathcal{E}^{0}_{\mathrm{inj}}(x,z=0,L,t)\\&\quad\times T_{1,2}\exp\left[-\mathrm{i}\omega_{\mathrm{inj}}(z=0,L)t\right]\exp\left[-\frac{(\boldsymbol{x}-\boldsymbol{x}_0)^2}{w_0^2}\right]\\&\quad\times\exp\left[\frac{\mathrm{i}2\pi}{\lambda}(x-x_0)\sin\left(\alpha_{\mathrm{inj}}\left(z=0,L\right)\right)\right]\\R(z)&=z+\frac{z_0^2}{z}.\end{aligned}\tag{3.4}$$

In (3.3) we have assumed a Gaussian shaped monofrequent optical injection. $T_{1,2}=1-R_{1,2}$ is the transmission of the front and back facet and $x-x_0$ is the transverse (i.e. parallel to the facets) distance from the lateral injection position x_0. $\omega_{\mathrm{inj}}(z=0,L)=2\pi c/\left(\lambda_{\mathrm{inj}}(z=0,L)\right)$ and $\alpha_{\mathrm{inj}}(z=0,L)$ are the frequency and the injection angle of the injected light field, respectively, $z_0=\pi w_0^2/\lambda_{\mathrm{inj}}$ is the Rayleigh range, $R(z)$ is the radius of curvature of areas of constant phase and $2w_0$ the beam waist of the injected beam. (3.4) describes the transversely varying amplitude of the injected optical field after transmission in the active layer, including the phase change between different transverse positions due to the angle of incidence. Note that (3.3) very generally represents the dynamic spatio-temporally resolved injection of an optical beam or pulse. In combination with the wave equation (3.1) this allows us to analyze the physical effects during the interaction of one or several light fields with the carrier dynamics such as in the case of dynamic wave mixing when using injection of various different light beams from one side or opposite sides.

Similar as in (2.18), the Bloch equations

$$\begin{aligned}
\frac{\partial}{\partial t}p^{\pm}(\boldsymbol{k},\boldsymbol{r},t) &= -\left[\mathrm{i}\bar{\omega}(\boldsymbol{k},T_{\mathrm{l}}) + \tau_{\mathrm{p}}^{-1}(\boldsymbol{k},N)\right]p^{\pm}(\boldsymbol{k},\boldsymbol{r},t) \\
&\quad + \frac{1}{\mathrm{i}\hbar}\mathcal{U}^{\pm}(\boldsymbol{k},\boldsymbol{r},t)\left[f^{\mathrm{e}}(\boldsymbol{k},\boldsymbol{r},t) + f^{\mathrm{h}}(\boldsymbol{k},\boldsymbol{r},t) - 1\right] \\
&\quad - \frac{1}{\mathrm{i}\hbar}\delta\mathcal{U}^{\pm}(\boldsymbol{r},t) + F_{\mathrm{p}}q^{\mathrm{p}}(\boldsymbol{k},\boldsymbol{r},t) \\
\frac{\partial}{\partial t}f^{\mathrm{e,h}}(\boldsymbol{k},\boldsymbol{r},t) &= g(\pm\boldsymbol{k},\boldsymbol{r},t) + \Lambda^{\mathrm{e,h}}(\boldsymbol{k},\boldsymbol{r},t) - \gamma_{\mathrm{nr}}f^{\mathrm{e,h}}(\boldsymbol{k},\boldsymbol{r},t) \\
&\quad - \Gamma_{\mathrm{sp}}f^{\mathrm{e}}(\pm\boldsymbol{k},\boldsymbol{r},t)f^{\mathrm{h}}(\mp\boldsymbol{k},\boldsymbol{r},t) \\
&\quad - \sum_{\mathrm{ph}}\tau_{\mathrm{e,h}}^{\mathrm{ph}\,-1}\left[f^{\mathrm{e,h}}(\boldsymbol{k},\boldsymbol{r},t) - f_{\mathrm{eq}}^{\mathrm{e,h}}(\boldsymbol{k},\boldsymbol{r},t,T_{\mathrm{l}})\right] \\
&\quad - \tau_{\mathrm{e,h}}^{-1}\left[f^{\mathrm{e,h}}(\boldsymbol{k},\boldsymbol{r},t) - f_{\mathrm{eq}}^{\mathrm{e,h}}(\boldsymbol{k},\boldsymbol{r},t,T_{\mathrm{pl}}^{\mathrm{e,h}})\right].
\end{aligned} \tag{3.5}$$

describe the spatio-temporal dynamics of the microscopic interband polarizations $p_k^{\pm} = p^{\pm}(\mathbf{k},\mathbf{r},t)$ and the Wigner distributions of the electrons (e) and holes (h) $f_k^{\mathrm{e,h}} = f^{\mathrm{e,h}}(\mathbf{k},\mathbf{r},t)$.

The fields $\boldsymbol{E}_{\mathrm{sig}}^{q\;\pm}$ and $\boldsymbol{E}_{\mathrm{se}}^{q\;\pm}$ determine the local fields $\boldsymbol{U}^{\pm}(\boldsymbol{k},\boldsymbol{r},t) = \int_{\boldsymbol{q}}\left[\boldsymbol{E}_{\mathrm{sig}}^{q\;\pm}(t)\;\mathrm{e}^{\mathrm{i}\boldsymbol{q}\boldsymbol{r}} + \boldsymbol{E}_{\mathrm{se}}^{q\;\pm}(t)\;\mathrm{e}^{\mathrm{i}\boldsymbol{q}\boldsymbol{r}}\right] + \delta\boldsymbol{U}^{\pm}(\boldsymbol{k},\boldsymbol{r},t)$ (where $\boldsymbol{q}$ is the relative wavevector between conduction and valence bands. The Coulomb-induced field contributions $\delta\mathcal{U}^{\pm}$ are treated on the level of the Hartree–Fock approximation [3].

In (3.5), $g = \frac{1}{4\hbar}\mathrm{Im}\left[\mathcal{U}^{+}p^{+*} + \mathcal{U}^{-}p^{-*}\right]$ is the generation rate, $\boldsymbol{k}$ denotes the carrier-momentum wavenumber and $f_{\mathrm{eq}}^{\mathrm{e,h}}$ is the carrier distribution in thermal equilibrium with the carrier plasma ($T_{\mathrm{pl}}^{\mathrm{e,h}}$) and with the lattice ($T_{\mathrm{l}}$) given by the corresponding Fermi functions. The spatio-temporal temperature dynamics thereby may be calculated on the basis of hydrodynamic equations derived from the Boltzmann equation (see Sect. 2.2). $F_{\mathrm{p}}q^{\mathrm{p}}(\boldsymbol{k},\boldsymbol{r},t)$ simulates dipole fluctuations [2]. The Maxwell–Bloch–Langevin equations consisting of the Dual-Field wave equation (3.1) and the Bloch equation (3.5) are the basis of the fundamental analysis of the carrier and light field dynamics in the picosecond and femtosecond regimes discussed in the following Sects. 3.2 and 3.3, respectively.

3.2 Nonlinear Gain and Index-Dynamics: Picosecond Pulse Propagation

The Maxwell–Bloch–Langevin equations provide a realistic description of the spatio-temporal behavior of semiconductor lasers. In particular, they allow a microscopic analysis of the intraband and interband dynamics in the charge carrier plasma. The resulting dynamic wave mixing processes and gain competition have a strong influence on the emission properties of spatially extended (edge-emitting and vertical-emitting) lasers and laser amplifier configurations [4]. Specifically, we will discuss the spatio-temporal dynamics of an InGaAs broad-area semiconductor laser amplifier with optical injection of a picosecond light pulse.

3.2.1 Time Scales

The relevant time scales of a semiconductor laser that are associated with its gain medium are, in particular,

(1) the relaxation time of the carriers ($\tau^{e,h}$) and the interband polarization (τ_p)
(2) the pump rate ($\Lambda^{e,h}$),
(3) the nonradiative recombination (τ_{nr}) and
(4) the spontaneous emission (Γ_{sp}^{-1}).

$\tau^{e,h}$ and τ_p represent the fast response of the medium (50 ... 200 fs) while τ_{nr} and Γ_{sp}^{-1} determine the slow dynamics of the spatial density (ps ... ns).

The injection of a light pulse enables a probing of the ultrafast carrier dynamics. If the pulse duration τ_{pulse} is significantly larger than τ_p, $\tau_{pulse} \gg \tau_p$, the dynamic changes in the carrier distribution $f^{e,h}$ are predominantly determined by the generation rate g, i.e. the induced recombination. A pulse that has been injected into the active layer modifies the carrier distribution functions $f^{e,h}$ and thereby reduces the spectral inversion $[f^e + f^h - 1]$ (see (3.5)). Carrier injection via the pump current and relaxation via carrier–carrier and carrier–phonon scattering on the other hand lead to a partial refilling of the spatio-spectral hole created by the injected light pulse. As a consequence, a spatially and temporally dependent quasi-equilibrium is established which determines – via the nonlinear interband polarization – the spectral gain and the induced refractive index changes. These spatially and temporally varying quantities imply dynamic changes in amplitude and phase of the pulse resulting in temporally dependent amplification as well as dynamic phase and frequency changes within the pulse (the so-called chirp).

For pulses with a duration of several picoseconds the carrier injection via the pump rate cannot re-establish the inversion as long as the pulse is still present (typical times for the re-establishment of the inversion are 0.1 to 0.5 ns) so that the influence of the terms in (2.18) describing the build-up and decay of the carrier density in the active layer ($\Lambda^{e,h}$, $\gamma_{nr}^{e,h}$, $\Gamma_{sp} f^e f^h$) is small. In the following discussion we will focus on the spatio-temporal dynamics of a light pulse in the picosecond regime. From its propagation characteristics we will see that a combination of spatial and spectral saturation effects play an important role.

3.2.2 Non-Equilibrium Wigner Distributions

The dynamic changes in amplitude and phase a pulse experiences during propagation in a semiconductor laser are determined by both the *spatial* distributions of gain and refractive index and the microscopic *spectral* distributions of the spatio-spectral gain and refractive index. The latter are determined by microscopic changes in the carrier distributions of electrons and holes in the valence and conduction bands and the nonlinear interband polarization. Visualization of the spectrally resolved distributions may thus yield important interpretations of physical properties that determine the properties of the emitted radiation. Thereby the variations in propagation direction (z) as well as the dependence on frequency (k) and time (t) are important for the over-all behavior of the device.

The dynamics of the spatio-spectral gain and refractive index distributions originate from microscopic changes in the electron–hole plasma [5]. In the following we will analyze both the small-signal and gain saturation regimes on the basis of microscopic distribution functions of the spectral gain and the induced refractive index dispersion. We consider a light pulse with duration of 7.5 ps (full width at half maximum, "FWHM") and a total input power of $P_{in} = 0.02\ P_s$ (where P_s denotes the power required for complete saturation of the gain) that propagates in a semiconductor laser amplifier (length 2 mm, width 100 μm). The injection current density is chosen such that the carrier density in the active area of the amplifier is significantly above the transparency value N_t ($N \approx 3.0\ N_t$).

In order to analyze the microscopic carrier dynamics it is convenient to define the Wigner distributions [5] of electrons (e) and holes (h), $\delta f^{e,h} = f^{e,h} - f_{eq}^{e,h}$. Calculation of these distributions for a specific situation (e.g. the propagation of a light pulse) directly reflect the deviation of the carrier distributions in the bands from their quasi-equilibrium. Figure 3.1 displays typical snapshots of the spatially and spectrally dependent Wigner distributions $\delta f^{e,h}(k,z,t_0) = f^{e,h}(k,z,t_0) - f_{eq}^{e,h}(k,z,t_0)$ of electrons (Fig 3.1(a–c)) and holes (Fig 3.1(d–f)) during the propagation of the pulse in a broad-area amplifier at time intervalls of 8 ps. The frequency of the injected light field of the pulse is chosen to be in the center of the gain maximum of the semiconductor laser. The figures show the mutual dependence of the distributions

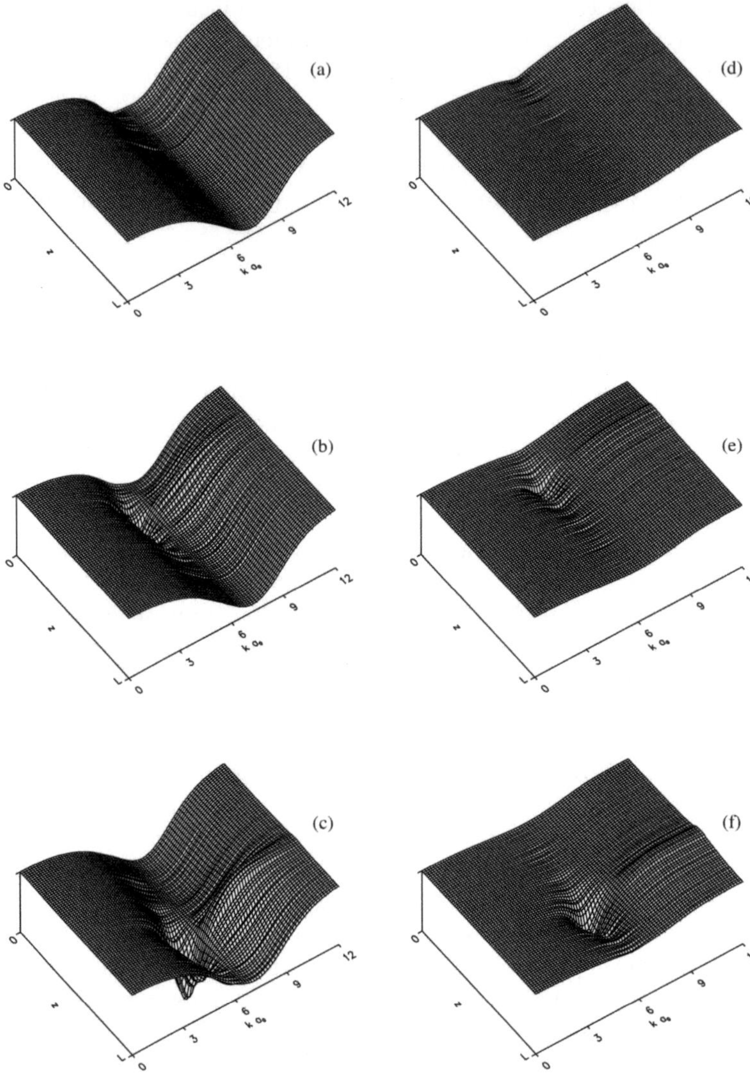

Fig. 3.1. Pulse propagation in the linear gain regime: temporal snapshots of the nonequilibrium Wigner distributions of electrons $\delta f^{\mathrm{e}}(k,z,t_i)$ (*left*) and holes $\delta f^{\mathrm{h}}(k,z,t_i)$ (*right*). Shown are the distributions at (**a**, **d**) $t_1 = 2$ ps, (**b**, **e**) $t_2 = 10$ ps und (**c**, **f**) $t_3 = 18$ ps after the injection of a pulse with duration of 7.5 ps and input power 0.02 P_{s}

on wavenumber k (to the right) and on the longitudinal propagation direction z in the lateral center ($x = 0$) of the broad-area laser amplifier. Due to the different effective masses of electrons and holes and due to the different scattering rates the distributions of electrons and holes are in general not identical. Indeed, Fig. 3.1 clearly sheds light on the different spectral contributions of the light pulse and the ASE. In concert they induce a characteristic excitation of the carriers from their quasi-equilibrium distributions $f_{\mathrm{eq}}^{\mathrm{e,h}}$ (given by the respective Fermi–Dirac statistics). According to its spectral bandwidth (in the example: 0.17 nm) the propagating light pulse leads to a selective reduction of the spectral inversion in the valence and conduction bands causing spectral hole burning (at $k \approx 4$, see e.g. Fig. 3.1(c)). The amplification of spontaneous emission, on the other hand, leads to a reduction of the inversion in a large spectral regime up to $\approx 12\ ka_0$. It is this contribution which – after propagating in the laser – is the origin of a spectrally broad-band "ASE"-level that is part of the emitted radiation.

The relaxation of the carrier distributions in the bands towards their quasi-equilibrium distribution then leads to a partial refilling of the spectral hole. As a consequence new carriers are available for light amplification in the frequency regime of the propagating pulse. Compared to the dynamics of the spatial density (that is given by the spectrally integrated microscopic values in the frequency domain) the carrier distributions in the k-space are mostly determined by fast microscopic scattering processes in the bands (fs to ps). Figure 3.2 shows the Wigner distributions in the gain saturation regime. The propagating pulse induces a deep spectral hole and carrier heating resulting in complex non-equilibrium distributions of the carriers. Since the carrier relaxation is proportional to the deviation of the carrier distribution from quasi-equilibrium (i.e. $\propto f^{\mathrm{e,h}}(k,z,t) - f_{\mathrm{eq}}^{\mathrm{e,h}}(k,z,t)$), an increased relaxation of the carriers towards their quasi-equilibrium occurs. As a consequence, carriers that initially are in frequency regimes outside the spectral bandwidth of the injected light field can reach the spectral hole previously burnt by the pulse via scattering processes. They thus contribute to the induced emission of the pulse light field (homogeneous broadening). This process reduces the spectral inversion available for amplification of the broad-band spontaneous emission. It is this gain competion originating from microscopic interaction processes that is responsible for the spectral properties of the emitted radiation that can then be varified and confirmed by measurements of optical spectra.

The strong spectral hole burning created by the injected light field not only leads to a "spectral condensation" of the gain but also to spectral gain saturation. The carrier injection (via the pump current) and the finite carrier relaxation rates cannot re-establish the spectral inversion on such short time scales. This impedes immediate refilling of the spatio-spectral trench burnt by the pulse. As a consequence, the spatial (longitudinal) extension of the trench in the Wigner distributions may be significantly larger than the respective extension of the light field. For high carrier density and large propagation length this effect, that may be interpreted as spectral gain saturation, can cause an increase of the pulse duration. Furthermore, via the light

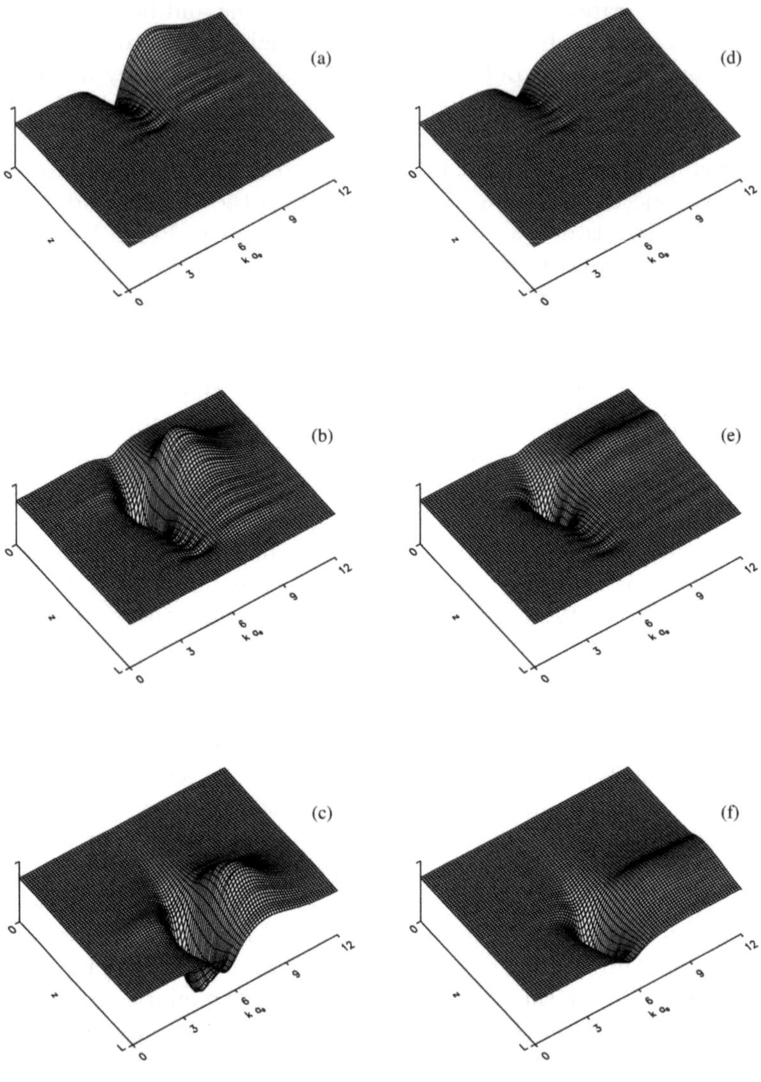

Fig. 3.2. Pulse propagation in the gain saturation regime: temporal snapshots of the nonequilibrium Wigner distributions of electrons $\delta f^{\mathrm{e}}(k, z, t_i)$ (*left*) and holes $\delta f^{\mathrm{h}}(k, z, t_i)$ (*right*). Displayed are the distributions at (**a**, **d**) $t_1 = 2$ ps, (**b**, **e**) $t_2 = 10$ ps und (**c**, **f**) $t_3 = 18$ ps after the injection of a pulse with duration of 7.5 ps and input power 2.0 P_{s}

field propagation the carrier relaxation rates are transferred to corresponding distances leading to the formation of characteristic longitudinal structures.

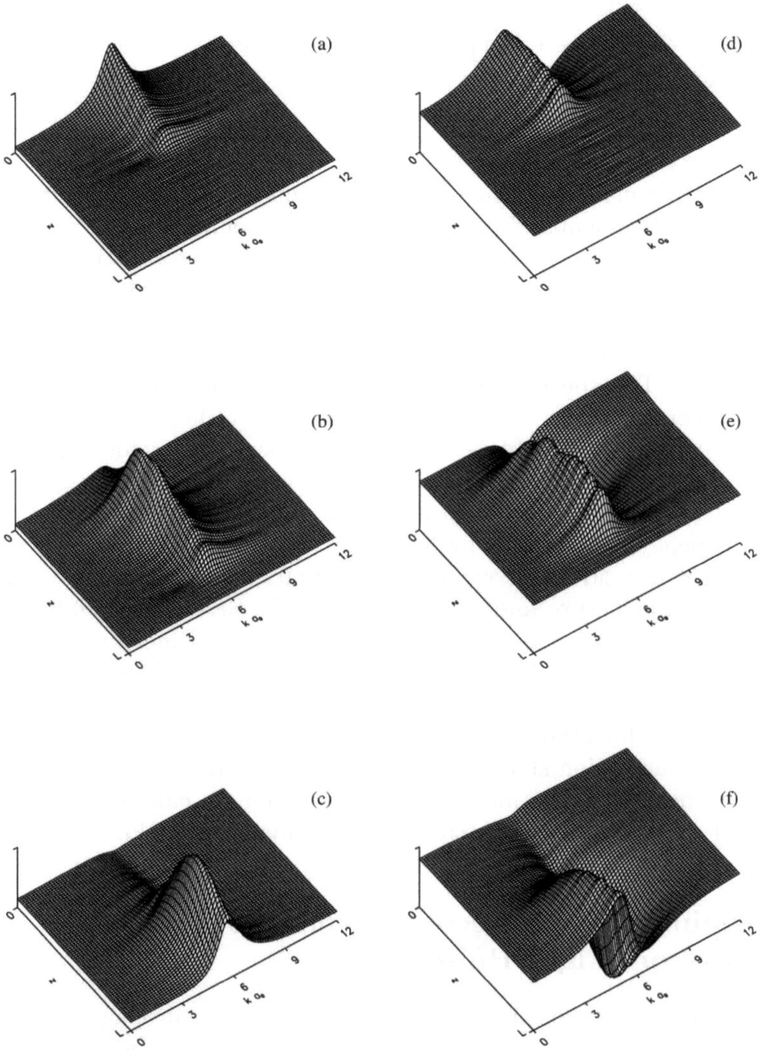

Fig. 3.3. Pulse propagation in the gain saturation regime: temporal snapshots of the spectral gain $g(k, z, t_i)$ (*left*) and the induced index dispersion $\delta n(k, z, t_i)$ (*right*). Shown are the distributions at (**a**, **d**) $t_1 = 2$ ps, (**b**, **e**) $t_2 = 10$ ps und (**c**, **f**) $t_3 = 18$ ps after the injection of a pulse with duration 7.5 ps and input power 2.0 P_s

The influence of the carrier dynamics in the frequency domain on the spectral properties of the pulse can be analyzed on the basis of the microscopic gain and the refractive index dispersion shown in Fig. 3.3. The nonequilibrium distributions of gain and index are via

$$g \propto \frac{\gamma_k^{\mathrm{p}}}{\bar{\omega}_k^2 + \gamma_k^{\mathrm{p}2}} \left[\left| \boldsymbol{E}^+ \right|^2 + \left| \boldsymbol{E}^- \right|^2 \right] \cdot \left[f_k^{\mathrm{e}} + f_k^{\mathrm{h}} - 1 \right] , \tag{3.6}$$

$$\delta n \propto n_1 \frac{\bar{\omega}_k}{\bar{\omega}_k^2 + \gamma_k^{\mathrm{p}2}} \cdot \left[f_k^{\mathrm{e}} + f_k^{\mathrm{h}} - 1 \right] \tag{3.7}$$

related to the microscopic carrier distributions of electrons and holes that are dynamically calculated on the basis of the semiconductor Bloch equations. The saturation behavior is also reflected in these distributions. The hole burning leads in the spectral gain to a slight spectral broadening and to a spectral shift towards smaller wavenumbers (see e.g. Fig. 3.3(b)). The spectral width is thereby given by the relaxation rates of the carrier scattering processes. The spectral shift originates from the saturation of the gain in the frequency domain. The depth of the spectral hole located at the frequency of the injected light field reaches its maximum value (given by the injection current and the carrier relaxation) during the propagation. As a result, the light field escapes towards the low-frequency regime. The direction of the frequency drift is thereby based on two effects. First, a preference of the low-energetic side is given by the occupation of the bands determined by both band structure and the Pauli-principle. Second, the induced spectrally dependent refractive index arising at the same time (see in particular Fig. 3.3(e)) draws the light towards the low-energy side. Figure 3.3(d) shows the increase of the refractive index due to carrier recombination during the first time step, which then via spectral hole burning and carrier heating transforms into a dispersion shape (Fig. 3.3(e)). The refractive index lowering at high frequencies and the increase near the bandgap can be interpreted as dynamic focussing and defocussing in the frequency domain.

3.3 Nonlinear Femtosecond Dynamics of Ultrashort Light Pulses

The theoretical and experimental investigation of the propagation of ultrashort light pulses in active semiconductor media is of fundamental importance for an analysis of microscopic carrier dynamics and nonlinear effects that directly determine the modulation response and mode stability [6] of semiconductor lasers. In particular, they are spatio-spectral hole burning, carrier heating and dynamic phase, affecting both the spectrum and the (temporal) shape of a light pulse during its propagation. Indeed, with recent progress in both microscopic simulations and experimental femtosecond spectroscopy techniques it is possible to investigate ultrafast carrier dynamics in semiconductor lasers and to identify and analyze the fundamental physical effects. A theoretical study and numerical analysis of ultrafast laser dynamics consequently is of fundamental interest for the design of nonlinear elements such as modulators and switches or for the observation of nonlinear effects such as the formation of solitons [7]. In this section we discuss ultrashort carrier dynamics on the basis of nonlinear amplification and propagation of ultrashort optical pulses ($\approx$ 250 fs) in semiconductor lasers.

3.3.1 Ultrafast Gain Dynamics

An ultrashort light pulse propagating in a semiconductor laser directly couples to the carrier populations in the valence and conduction bands. Thereby it induces highly nonequilibrium carrier distributions. As a consequence, dynamic changes in both gain and refractive index arise that relax on time scales of a few picoseconds towards new equilibrium distributions. On such short time scales the characteristic dynamics of the charge carriers leads to strong distortions of the temporal and spectral profiles of the pulse. Figure 3.4 shows the Wigner distributions of the electrons, the microscopic gain and the induced refractive index at the output facet of a semiconductor laser amplifier that has been optically excited by an ultrashort light pulse (duration 250 fs). The injection current density has been set to $2.5 j_{\mathrm{thr}}$ The central pulse frequency is located within the amplifier gain bandwidth so that the pulse is amplified during its propagation. Shown are the dependence of the gain, refractive index and carrier distributions $\delta f^{\mathrm{e}} = f^{\mathrm{e}} - f^{\mathrm{e}}_{\mathrm{eq}}$ on wavenumber (in units of the Bohr radius a_0, $k = 0$ denotes the bandgap) and on time ($t = 0$ denotes the passage of the pulse maximum through the output facet). The calculated results for δf^{e} displayed in Fig. 3.4(a) show the physical processes that affect the spectral and temporal properties of a light pulse during its propagation. The optical excitation induced by the light pulse leads to spatio-spectral hole burning (negative values of δf^{e}) and carrier heating (positive values). The heated nonequilibrium distributions then relax via carrier–carrier and carrier–phonon scattering towards the quasi-equilibrium given by the respective Fermi–Dirac distributions. The partial refilling of the hole via carrier relaxation and carrier injection typically occurs on time scales of a few hundred fs up to a few ps. These processes consequently determine the temporal and spectral shape of the spatio-spectral trench burnt by the pulse. For high input power levels and high inversion the spectral and temporal extension of the hole burning may thus significantly exceed the duration and spectral bandwidth of the pulse. The highly non-equilibrium carrrier dynamics is also reflected in the microscopic gain and induced index displayed in Fig. 3.4(b) and (c), respectively. The spectral hole burning and carrier heating induced by the leading part of the pulse shapes the (spectrally and temporally varying) gain (Fig. 3.4(b)) available for the trailing part of the pulse. At the same time the spatio-spectral reduction of the inversion in the bands induces a dynamic spectrally dependent refractive index distribution (Fig. 3.4(c)). In particular, the reduction of the spectral inversion and the accumulation of heated carrier states leads to a spectral broadening (reflected in the spectrally dispersive shape of the induced index) and to a temporal

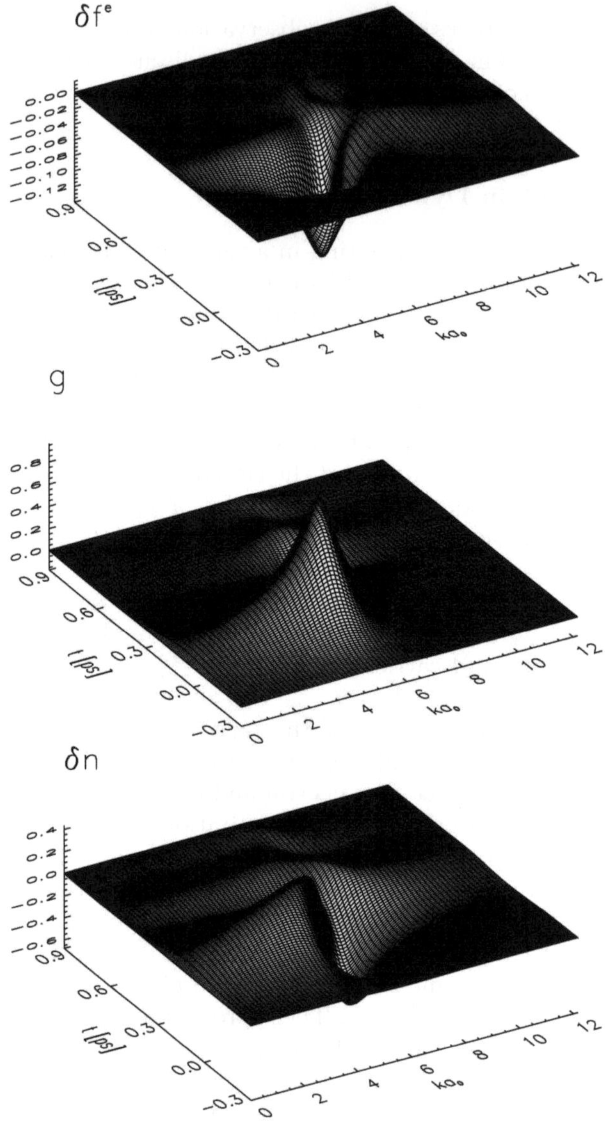

Fig. 3.4. Dynamic Wigner distributions of the electron distribution, the microscopic gain and the induced refractive index

frequency shift (chirp). Furthermore, the dynamic carrier relaxation of the excited charge carrier plasma leads to modulations in the trailing parts of the microscopic gain and the induced refractive index that show a characteristic drift with increasing time. The amplitude and the temporal shape of this drift are given by the carrier injection rate (i.e. absorbing or amplification regime) and by the frequency detuning between the injected light field and the frequency of the free-running laser.

3.3.2 Ultrafast Nonlinear Propagation Effects

During the propagation of an ultrashort light pulse in a waveguide structure the nonlinear material properties and the coupling to the transverse and longitudinal degrees of freedom of the waveguide structure determine the spatio-temporal light field dynamics. In particular, in addition to the geometrical influences it is the nonlinear gain saturation and index changes that lead to a reshaping and a temporal broadening of the propagating light pulse. The saturation behavior thereby depends on a variety of physical interactions whose traces can be found in both the space and frequency domains. The combination and mutual influence of the underlying microscopic and macroscopic changes determine important output properties such as power, shape, duration and spectrum of the amplified pulse. The microscopic interactions occuring on very short time scales (a few 100 fs) in the charge carrier plasma are spatio-spectral hole burning, carrier heating as well as the dynamic spatio-spectral wave mixing between the propagating light pulse and the ASE. They are responsible for a time-dependent phase shift between the microscopic gain and the induced index that lead to fast self phase modulation. The slow macroscopic changes are the long-lived reduction of the carrier density in the active area leading to spatial gain and index profiles. In combination the slow and fast phase changes lead to distortions of the pulse shape, spectral broadening and interference effects in the optical spectrum.

In order to visualize the spatio-temporally varying light propagation we focus on the laser-internal intensity distribution during the propagation of an ultrashort light pulse in an inverted $(j = 2.5j_{\mathrm{thr}})$ semiconductor waveguide (length 650 μm). Figure 3.5 shows temporal snapshots of the intensity taken in the front (a), middle (b) and back part (c) of the active area during the propagation of a light pulse with energy 30 pJ and duration 250 fs. Note that only the spatial region around the respective pulse position and not the entire active area is plotted. The snapshots show the influence of the microscopic amplification, saturation and induced index dynamics on the signal amplification. During its propagation in the waveguide structure the pulse is significantly reshaped via its nonlinear continuous interaction with the microscopic charge carrier plasma. The leading part of the pulse experiences the initial gain whereas the trailing part experiences a gain that has been reduced by the leading part. As a consequence the leading part of the pulse is amplified to a larger degree than the trailing part, leading to a steep rise in the front part and a significant broadening of the pulse in the propagation direction. Additionally, the longitudinally varying amplification leads via carrier scattering, diffusion and light diffraction to transverse modulations in the pulse shape (see Fig. 3.5(c)). The carrier thermalization following the pulse-induced excitation of the medium causes a partial re-establishment of

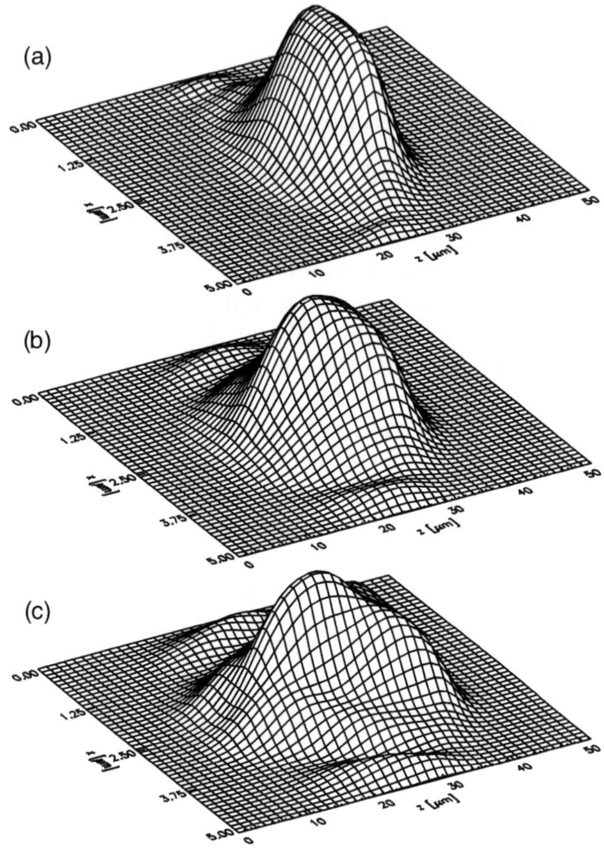

Fig. 3.5. Snapshots of the intensity taken in the front (**a**), middle (**b**) and back part (**c**) of the active area during the propagation of a light pulse (with energy 30 pJ and duration 250 fs)

the spectral gain. These time-dependent gain changes may result in characteristic modulations that can be seen in Fig. 3.5(c).

3.3.3 Dynamic Spectral Properties

Direct experimental access to ultrafast nonlinear effects in semiconductor lasers can be obtained via a pump-probe set-up. This allows both the measurement of the temporal properties of a propagating pulse (i.e. its pulse shape and duration) and of the spectral properties via a frequency resolved optical gating technique (FROG) [8]. By numerically simulating the spatio-temporal propagation of ultrashort pulses we can directly obtain information on the ultrafast gain and index dynamics on the basis of microscopically calculated Wigner functions. Figure 3.6 displays a calculated FROG-plot. The

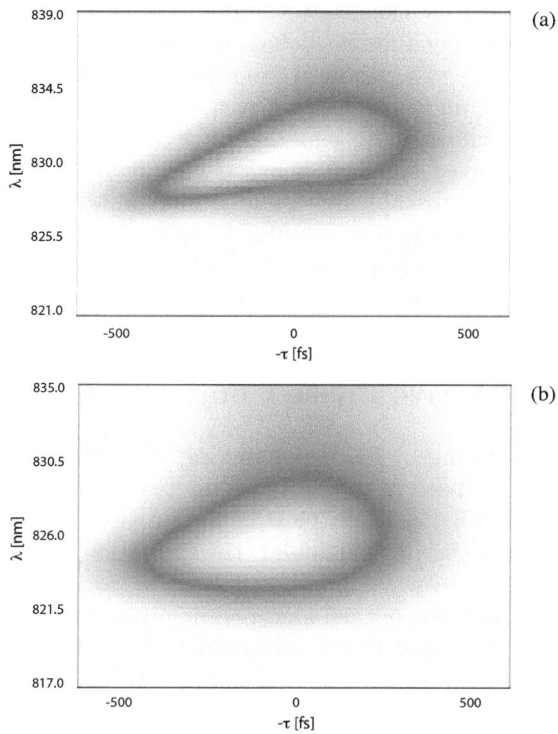

Fig. 3.6. FROG-plots with 30 pJ excitation energy. The injection current is 50 mA and the wavelengths are 830 nm (**a**) and 825 nm (**b**)

semiconductor laser is electrically pumped at its threshold injection current, the energy of the injected (pump) pulse is 30 pJ. In Fig. 3.6 (a) the wavelength of the pump pulse has been set to correspond to the gain maximum of the laser (830 nm). One can see that the gain saturation generated by the resonant pulse leads to a steep rise and via the induced index changes to generation of new frequencies and consequently to a broadening of the spectrum in the leading part. Due to the carrier induced increase of the instantaneous refractive index the spectrum is much broader on the red side of the maximum than on the blue side. This leads to an overall red-shift of the pulse center. The drift of approximately 0.7 nm of the spectral maximum that can be observed in the time interval between $-\tau = 300$ fs and $-\tau = 0$ fs can be attributed to the instantaneous negative refractive index changes in the trailing part of the pulse. In particular, Fig. 3.6(a) clearly shows a blue shift of approximately 1 nm of the central wavelength in the trailing part of the pulse between -200 fs and -400 fs. This effect demonstrates the influence of carrier thermalization on the refractive index.

In Fig. 3.6(b) the wavelength of the pump pulse has been set to 825 nm. This leads to completely different behavior. The blue shift in the trailing part of the pulse is much weaker than in the situation displayed in Fig. 3.6(a). In addition, the spectrum in the leading part of the pulse is less broadened. A comparison of the theory with experiment [9] demonstrates that a spatially and spectrally resolved simulation based on the Maxwell–Bloch–Langevin approach allows a microscopically founded interpretation of the experiments and opens up a visualization of complex laser internal processes such as nonlinear spatially varying light propagation, inhomogeneous gain saturation and phase dynamics, spatio-spectral hole burning, carrier heating and relaxation.

3.4 Ultrafast Active Phase Conjugation

Ultrafast pulse tailoring by four-wave mixing of short optical pulses in semiconductor lasers has been in the center of interest in recent experimental and theoretical works [10; 11; 12] in view of applications for high-speed optical information technologies. In particular, the formation of a phase conjugating mirror (PCM) by multi-wave mixing has been demonstrated [13; 14; 15].

The dynamic ultrafast optical phase-conjugation and interference of counterpropagating light pulses in the active area of a broad-area semiconductor laser can be modelled on the basis of the Maxwell–Bloch–Langevin equations (3.1). Being based on a Dual-Field wave equation they include, in particular, the dynamic interactions of the counterpropagating light fields of the propagating light signal and the amplified spontaneous emission self-consistently with the dynamics of the active nonlinear charge carrier plasma within the laser amplifier. The microscopic internal processes that are responsible for the grating formation are the microscopic spatio-spectral distributions of the nonlinear gain and the induced refractive index variations. In the following we will present results of microscopic simulations on the basis of the Maxwell–Bloch–Langevin equations. Snapshots of the Wigner distributions and the characteristic spatial intensity and index distribution will demonstrate the spatio-spectral dynamics of fields and carrier plasma. They will reveal the underlying physical processes of the ultrafast refractive index grating formation: dynamic spatio-spectral hole burning, carrier heating as well as coupled diffraction and interference of the light beams via the nonlinear induced changes in the spatio-spectral carrier and refractive index distributions.

3.4.1 Spatio-Temporal Wave-Mixing

To elucidate the striking phase-conjugation processes in the following we will specifically analyze a model set-up with the spatio-temporal mixing of two light pulses (spatially Gaussian shaped with a full width at half maximum of 25 μm, pulse duration 4 ps) which are injected from both sides into the (antireflection-coated) model broad-area semiconductor laser amplifier

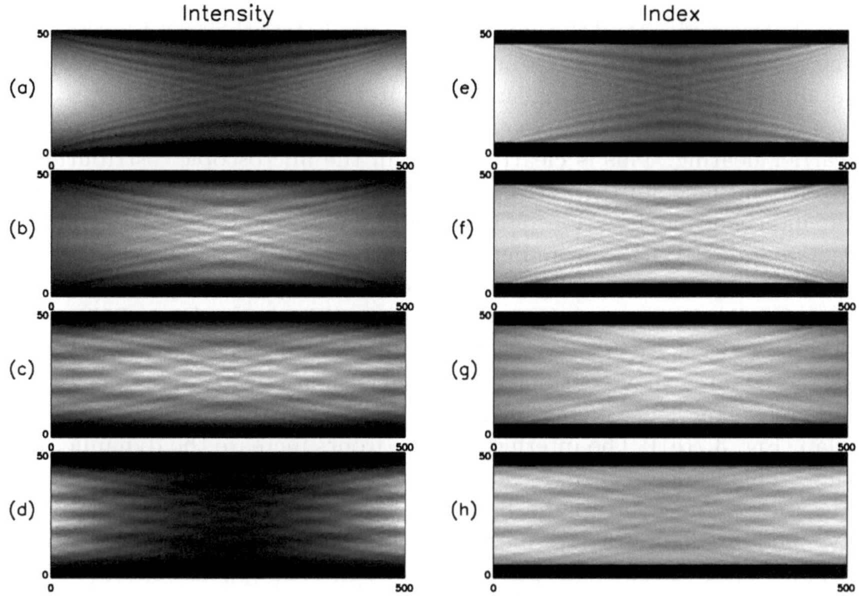

Fig. 3.7. Snapshots of the intensity (*left column*) and induced index distribution (*right column*) in the active area of a broad-area amplifier after optical injection of two light pulses with a duration of 4 ps and an input power P_{in} which is low compared to the saturation power P_s of the active medium ($P_{in} = 0.1P_s$). The time distance between successive plots is 3 ps. In the snapshots light shading corresponds to high local intensity and refractive index

(length $L = 500$ μm, width $w = 50$ μm). For the configuration of Kuerz and Mukai [13], Fig. 3.7 displays snapshots of the simulated intensity (left column) and refractive index distribution (right column) in the inverted medium at time steps of 3 ps. Figure 3.7(a) and (e) show the spatial distributions in the first time step, $t_0 = 0$ ps (t_0 refers to the temporal center of the pulses) where the two counterpropagating pulses with a duration of 4 ps are injected at the facets. The input power P_{in} of the pulses thereby was low compared to the saturation power P_s which is defined as the power required to reduce the density (during the propagation) to its transparency value. Their interaction with the charge carrier plasma leads instantly to the formation of a lateral and longitudinal index grating. In the lateral direction, the diffraction of the light fields in combination with the carrier diffusion in the active area with characteristic lengths in the μm-regime lead via self-focusing and modulation instabilities to transverse modulations in both intensity and refractive index. In the longitudinal (propagation) direction the dynamic interaction of the two counterpropagating pulses with the active medium creates a corrugated carrier density distribution which further enhances the counterpropagation. Consequently the pulses influence each other via the nonlinear coupling between charge carriers and the optical fields even before they meet in the center of the resonator. In the next time step (Fig. 3.7(b) and (f)), the two pulses have almost reached the longitudinal center. Now the optical fields of the pulses induce in addition to spatial also spectral hole burning accompanied by carrier heating. This is clearly visible in the microscopic distributions of electrons and holes in the active semiconductor medium which via the nonlinear interband polarization [5] determine the gain and the carrier-induced refractive index. With continuing propagation the increasing spatial overlap and interference of the light pulses thus creates a dynamic optical grating in the spatial distributions of the total intensity (Fig. 3.7(a–d)) and the refractive index (Fig. 3.7(e–h)). After the two light fields have passed the center of the active area they continue their propagation toward the opposite facet and thereby experience increasing mutual interaction due to the strong spatial overlap (Fig. 3.7(c)). The interfering wave fronts increase the dynamic phase grating. This leads, in turn, to additional lateral and longitudinal structures in the μm-regime. Due to the finite relaxation times of the carriers in the active medium the refractive index distribution remains rather distinct and clear, even after the two light pulses have reached the opposite facet ($t = 6$ ps, Fig. 3.7(d) and (h)). The PCM induced by the dynamic interference of two counterpropagating light fields thus "saves" the light-induced modifications within the carrier system of the active semiconductor.

It is the grating formations displayed in Fig. 3.7 that are responsible for the reflection and deflection in four-wave mixing configurations. The characteristic spatio-temporal dynamics of this PCM orginates from the microscopic coupling of the light fields to the charge carrier plasma and the nonlinear response of the active semiconductor medium. This interplay is described within the frame of the semiconductor Bloch equations.

3.4.2 Nonequilibrium Gain and Refractive Index Distributions

In the following we will analyze the microscopic dynamics of the charge carrier plasma in the active semiconductor. They are governed by a multitude of (partially coherent and incoherent) dynamic intraband and interband processes in the charge carrier system with time scales ranging from the femto to the nanosecond regime. We will focus on the spatio-spectral distributions of the microscopic refractive index δn and gain g. These nonequilibrium distributions are via $g_k \propto \frac{\gamma_k^{\mathrm{p}}}{\bar{\omega}_k^2+\gamma_k^{\mathrm{p}2}} \left[\left|\boldsymbol{E}^+\right|^2 + \left|\boldsymbol{E}^-\right|^2\right] \cdot \left[f_k^{\mathrm{e}} + f_k^{\mathrm{h}} - 1\right]$, $\delta n \propto n_1 \frac{\bar{\omega}_k}{\bar{\omega}_k^2+\gamma_k^{\mathrm{p}2}} \cdot \left[f_k^{\mathrm{e}} + f_k^{\mathrm{h}} - 1\right]$ related to the microscopic carrier distributions of electrons and holes that are dynamically calculated within the frame of the semiconductor Bloch equations.

While the intensity distributions of Fig. 3.7 show the dynamic optical grating formation, Fig. 3.8 and Fig. 3.9 reveal the underlying microscopic processes that are responsible for that characteristic phenomenon: the nonequilibrium distributions of the microscopically induced refractive index

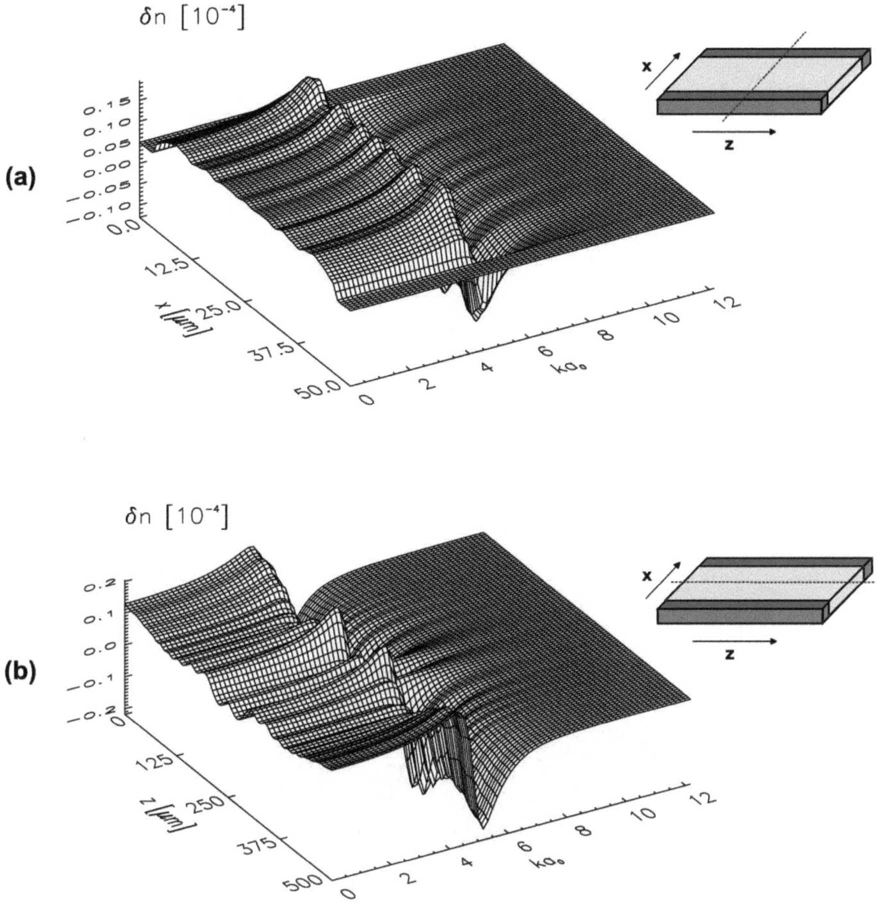

Fig. 3.8. Transverse (**a**) and longitudinal (**b**) dependence of the microscopic refractive index distribution during the molding of the optical pulses. The cuts were taken in the longitudinal and transverse center of the semiconductor laser, respectively (see *insert*)

(Fig. 3.8) and gain (Fig. 3.9) show simultaneous spectral and spatial dependencies, respectively. The lateral (Fig. 3.8(a) and Fig. 3.9(a)) and longitudinal (Fig. 3.8(b) and Fig. 3.9(b)) cuts were taken at $z = L/2$ and $x = w/2$, respectively, i.e. at the longitudinal and lateral centers of the resonator. The snapshots of the microscopic refractive index and gain were taken at the time step of Fig. 3.7(c), i.e. when via the nonlinear active medium the two light pulses strongly interact with each other in the center of the resonator. Positive (negative) values indicate an increase (reduction) of the refractive index and gain. The distributions in Figs. 3.8 and 3.9 show the multitude of coupled

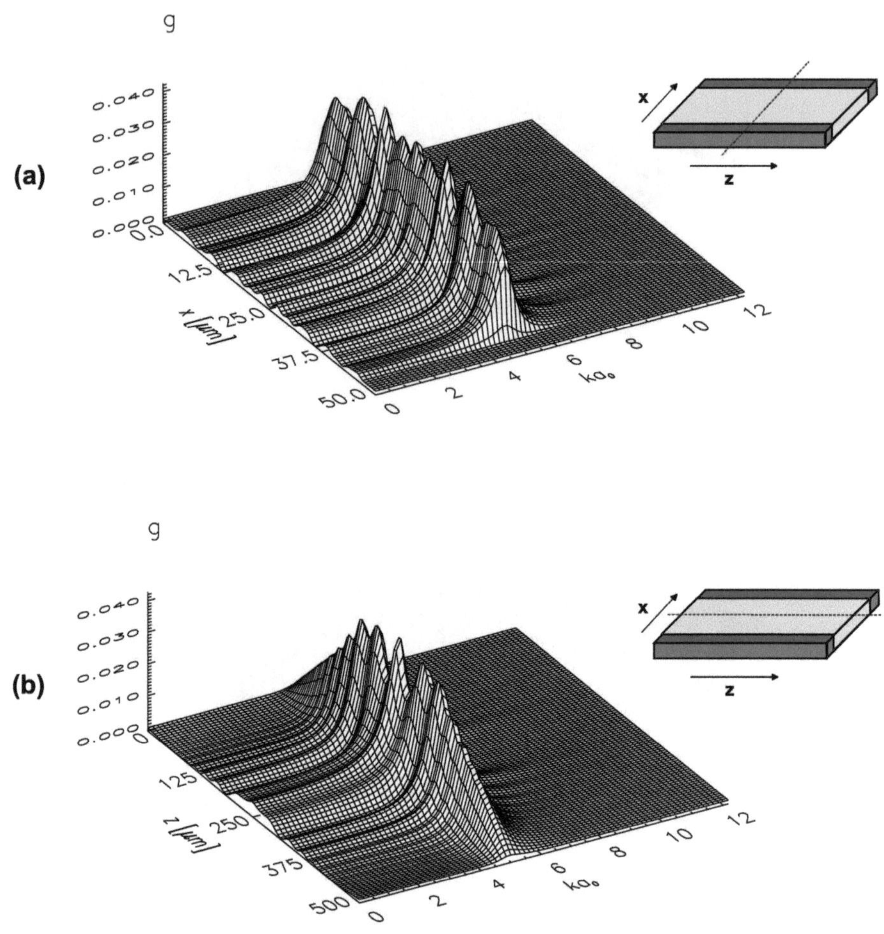

Fig. 3.9. Transverse (**a**) and longitudinal (**b**) dependencies of the microscopic gain distribution during the molding of the optical pulses. The cuts were taken in the longitudinal and transverse centers of the semiconductor laser, respectively (see *insert*)

physical effects that are associated with the simultaneous mixing and propagation of the optical light fields within the amplifier. The propagation of the light pulses in combination with their mutual influence leads to spectral hole burning and spectral broadening. Via the microscopic coupling to the charge carrier plasma the pulses directly modify the spatio-spectral distributions of the refractive index (δn) and gain (g). The longitudinal and the transverse distributions of δn and g show characteristic modulations in both the space and frequency domains originating from diffraction, carrier diffusion and spatio-spectral mixing of the counterpropagating pulses. The counterpropagating pulses leave a particulary distinct spatio-spectral trace in the longitudinal cut of the distribution functions. On the other hand the transverse dependence of δn (Fig. 3.8(a)) shows the induced grating structure. Its periodicity is determined by the diffraction length of the injected light and carrier diffusion. While in the $\mathbf{k}$ (i.e. frequency) domain the combination of carrier–carrier and carrier–phonon scattering processes (with characteristic time scales ranging from $\approx$ 100 fs to ps) causes a thermalization on a picosecond time scale, the real-space ($\mathbf{r}$) transport leads to typical diffusion lengths and laser-internal coherence lengths in the µm-regime. It is the spectrally confined gain in combination with the typical dispersion of δn that spatially and spectrally stabilizes a phase pattern that is required for a phase conjugating mirror. The microscopic spatio-spectral dynamics thus allows grating formation as well as the unique ultrafast response of the phase-conjugation processes.

3.4.3 Spatio-Temporal Dynamics of the Phase Conjugation

The spatio-spectral distributions discussed so far show that the dynamics of the active phase conjugating mirror is determined by both the light field dynamics of the counterpropagating optical pulses and the microscopic processes in the charge carrier plasma. The spatio-temporal dynamics of the light fields is given by the resonator round trip time and the spatial refractive index and gain distribution. They are a result of the reduction of the inversion by induced recombination in the active layer and the injection of carriers via the pump current. These spatial variations occur on relatively slow time scales of typically 1 ps up to a few hundred ps, depending on the current and power of the injected pulses. The microscopic dynamics of the electrons and holes, on the other hand, is given by the fast (100 fs to 2 ps) carrier–carrier and carrier–phonon scattering processes. It is thus the injection current and the input power of the pulses, in particular, which determine the relative influence of slow and fast physical processes on the build-up and decay of the phase conjugating mirror. In the case of Fig. 3.7 the spatial refractive index and gain variations induced by the light pulses are small compared to the fast changes in the microscopic charge carrier system. Thus the optical pattern created by the spatio-spectral interference of the pulses forms an ultrafast PCM whose response time is only limited by the finite relaxation time (i.e. a few hundred fs) of the microscopic charge carrier plasma. The situation changes significantly if we increase the gain (by increasing the injection current) or the input power. As an example Fig. 3.10 shows temporal snapshots (taken at time intervals of 3 ps) of the spatial intensity (left) and refractive index distribution (right) for a semiconductor medium driven above threshold where the injection of optical pulses with high input peak power, $P_{\text{in}} = 0.8P_{\text{s}}$, leads with increasing propagation to a significant reduction of the spatial gain. The spatio-temporal dynamics of the optical pattern is now dominated by slow physical processes such as spatial gain saturation and

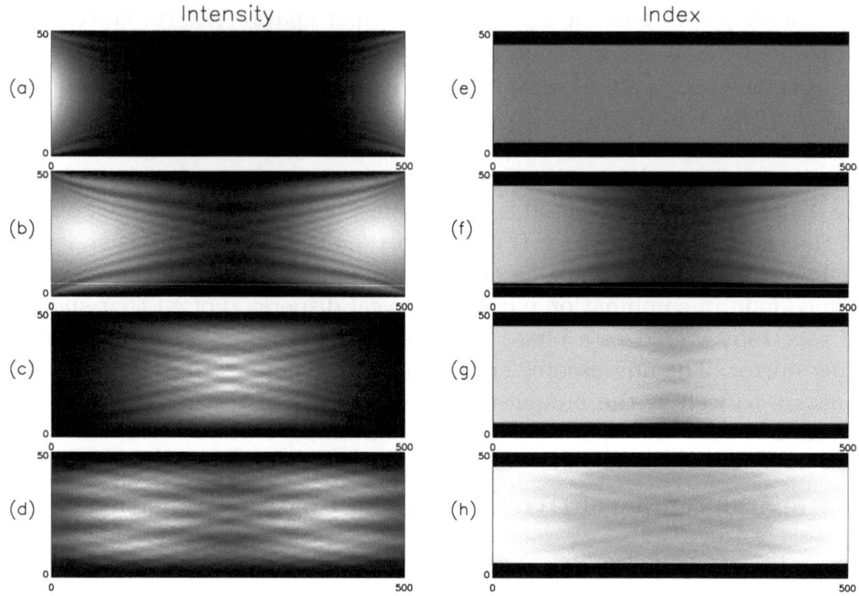

Fig. 3.10. Snapshots of the intensity (*left column*) and induced index distribution (*right column*) in the active area of a broad-area amplifier after optical injection of two light pulses with a duration of 4 ps and high input power ($P_{\text{in}} = 0.8P_{\text{s}}$). The time between successive plots is 3 ps. In the snapshots light shading corresponds to high local intensity and refractive index

carrier injection via the pump current (with typical time scales of a few hundred ps). Thus in contrast to the situation shown in Fig. 3.7 the spatial hole burning induced by the strong optical fields levels out the fast microscopic phase pattern present in Figs. 3.8 and 3.9 thereby significantly reducing the response time of the phase-conjugation processes.

3.4.4 Spatio-Temporal Response of the Phase Conjugating Mirror

The results of our microscopic simulations demonstrate that the nonlinear mixing of two counterpropagating light fields can indeed lead to the fast build-up of a dynamic index and phase grating which are at the heart of a fast phase conjugating mirror. As a final step we now discuss the results of the modelling of a typical pump-probe situation where a probe pulse with duration of 2 ps is injected into this phase conjugating mirror. The mixing of the probe light field with the counterpropagating optical pulses then creates the phase conjugate signal. As an example Fig. 3.11 shows intensity snapshots of the injected probe pulse with a duration 2 ps (left) and the respective phase conjugate signal (right) at time steps of 1.5 ps. The time of the injection was adjusted to the probe pulses, i.e. the three pulses reach

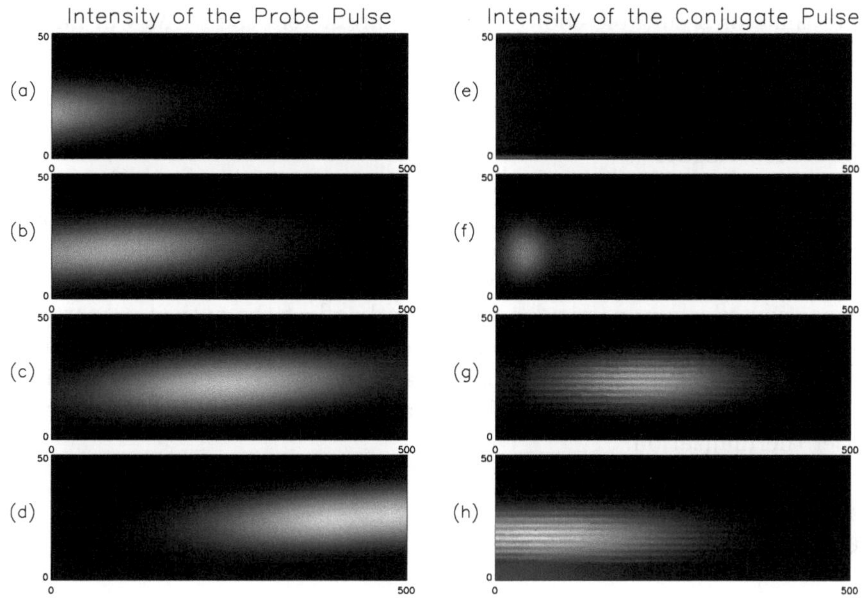

Fig. 3.11. Intensity snapshots of the probe pulse (*left*) and the respective phase conjugate signal (*right*) for a typical pump-probe situation (duration of the counter-propagating pump pulses 4 ps, duration of the probe pulse 2 ps). The time between successive plots is 1.5 ps

the center of the resonator at the same time. The probe pulse with an input peak power of one percent of the pump pulses is injected at an oblique angle of incidence (Fig. 3.11(a)). During the propagation toward the center of the active semiconductor medium the spatio-spectral overlap between the probe pulse and the counterpropagating pump pulses increases. Via the microscopic coupling described in the previous sections this leads to a strong interaction and wave mixing between the respective light fields (Fig. 3.11(b)). As a consequence, the first trace of the formation of a phase conjugate signal can be observed (Fig. 3.11(e)). The spatially and temporally varying energy transfer during the crossing and interference of the pulses finally creates the full phase conjugated light pulse (Fig. 3.11(f)) which propagates anti-parallel to the geometrical light path of the injected probe pulse toward the left facet (Fig. 3.11(g)). As can be seen from Fig. 3.11, the fast response of the PCM built up in the semiconductor laser leads to a conjugate light pulse of approximately the same duration and shape as the original probe pulse.

The spatio-temporal dynamics of the grating formation and the microscopic spatio-spectral distributions of the nonlinear gain and the induced refractive index provide a systematic image of the ultrafast phase conjugating mirror built by the spatio-spectral interference of the light beams and the nonlinear response of the semiconductor material. The simulations demonstrate that for a suitable choice of input parameters the PCM will have an ultrafast time response and very high efficiency.

3.5 Conclusion

We have discussed the spatio-temporal light field dynamics and the ultrafast carrier dynamics in active semiconductor laser waveguides. The results of the microscopic calculations show that the ultrafast carrier dynamics and propagation effects in particular, significantly affect the optical properties of semiconductor lasers and amplifiers. They lead to a hierarchy of nonlinear effects such as inhomogeneous light propagation, space-dependent amplitude and phase changes, spatio-spectral gain saturation, spectral hole burning and carrier thermalization.

References

1. E. Gehrig and O. Hess. Spatio-temporal dynamics of light amplification and amplified spontaneous emission in high-power tapered semiconductor laser amplifiers. *IEEE J. Quant. Electr.*, 37:1345–1355, 2001.
2. H. F. Hofmann and O. Hess. Quantum Maxwell-Bloch equations for spatially inhomogenous semiconductor lasers. *Phs. Rev. A*, 59:2342–2358, 1999.
3. W. W. Chow, S. W. Koch, and M. Sargent III. *Semiconductor-Laser Physics.* Springer-Verlag, Berlin, 1994.
4. E. Gehrig, D. Woll, M. Tremont, A. Robertson, R. Wallenstein, and O. Hess. Saturation behavior and self phase modulation of picosecond pulses in single stripe- and tapered semiconductor laser amplifiers. *J. Opt. Soc. Am. B*, 17:1452–1456, 2000.
5. E. Gehrig and O. Hess. Nonequilibrium spatio-temporal dynamics of the Wigner-distributions in broad-area semiconductor lasers. *Phys. Rev. A*, 57:2150–2163, 1998.
6. G. P. Agrawal and N. K. Dutta. *Long Wavelength Semiconductor Lasers.* Van Nostrand, New York, 1986.
7. E. Gehrig and O. Hess. Propagating spatial optical solitons in semiconductor lasers. *OPN Optics in 1999*, 10:26–27, 1999.
8. Greg Taft, Andy Rundquist, Margaret M. Murnane, Henry C. Kapteyn, Kenneth W. DeLong, Rick Trebino, and Ivan P. Christov. Ultrashort optical waveform measurements using frequency-resolved optical gating. *Optics Lett.*, 20:743–745, 1995.
9. A. Volland, G. Jennemann, I. Fischer, W. Elsäßer, E. Gehrig, and O. Hess. Femtosecond dynamics of active semiconductor waveguides - microscopic analysis and experimental investigations. In *CLEO*, Nice, France, 2000. IEEE Lasers and Electro-Optics Society.
10. S. Diez, A. Mecozzi, and J. Mørk. Bit rate and pulse width dependence of four-wave mixing of short optical pulses in semiconductor optical amplifiers. *Opt. Lett.*, 24:1675–1677, 1999.

11. H. Shi, I. Nitta, A. Schober, P. J. Delfyett, G. Alphonse, and J. Connolly. Demonstration of phase correlation in multiwavelength mode-locked semiconductor diode lasers. *Opt. Lett.*, 24:238–240, 1999.
12. J. M. Tang and K. A. Shore. Characteristics of optical phase conjugation of picosecond pulses in semiconductor optical amplifiers. *IEEE J. Quant. Electr.*, 35:1032–1040, 1999.
13. P. Kürz, R. Nager, and T. Mukai. Highly efficient phase conjugation using spatially nondegenerate four-wave mixing in a broad-area laser diode. *Appl. Phys. Lett.*, 68:1180–1182, 1996.
14. P. Kürz and T. Mukai. Frequency stabilization of a semiconductor laser by external phase-conjugate feedback. *Opt. Lett.*, 21:1369–1371, 1996.
15. P. P. Vasil'ev and I. H. White. Phase-conjugation broad area twin-contact semiconductor laser. *Appl. Phys. Lett.*, 71:40–42, 1997.

4 High-Speed Dynamics of Semiconductor Lasers

4.1 Introduction: Transverse and Longitudinal Mode Dynamics

The dynamics of semiconductor lasers may be very diverse and complex. In the always-present interplay of spatial with temporal degrees of freedom it is, in particular, the effective relevance of the spatial degrees of freedom that is of importance for the spatio-temporal dynamics of semiconductor lasers. The most relevant physical parameter that determines this relevance is the extension of the laser in comparison to characteristic internal length scales. Spatio-temporal effects are generally important for both types of semiconductor lasers, i.e. edge- and vertical-cavity surface-emitting laser (VCSEL) structures. In this chapter we will concentrate on edge-emitting lasers while the following Chap. 5 will be devoted to the spatio-temporal light field and polarization dynamics of VCSELs.

The transverse extension of the active area affects to a high degree the *transverse mode dynamics* that appears in the formation and migration of dynamic filaments and the (spatial and spectral) beam characteristics (nearfield, farfield, optical spectra).

Figure 4.1 displays the temporal behavior of the light field distribution at the output facet of a single-stripe laser (a), a (five stripe) laser array (b) and a broad-area laser (c). Dark shading represents high intensity values. In the single-stripe laser the typical transverse dimension of the active area is approximately 3–5 μm. As consequence the laser stripe supports only one transverse mode leading to a Gaussian shaped beam-profile. However, in the temporal direction the switching of the injection current from zero to a multiple of the threshold value initiates characteristic relaxation oscillations in the emitted intensity (Fig. 4.1(a)).

The situation is significantly changed if we consider a multi-stripe laser where two or more laser stripes, each by itself representing a nonlinear oscillator, are combined to one single device. Increasing the number of transversely coupled laser stripes increases the output power of the laser but at the same time increases this spatio-temporal complexity. The narrow stripe width prevents filamentation within each laser. As an example Fig. 4.1(b)) shows the temporal behavior of the intensity at the output facet of a five-

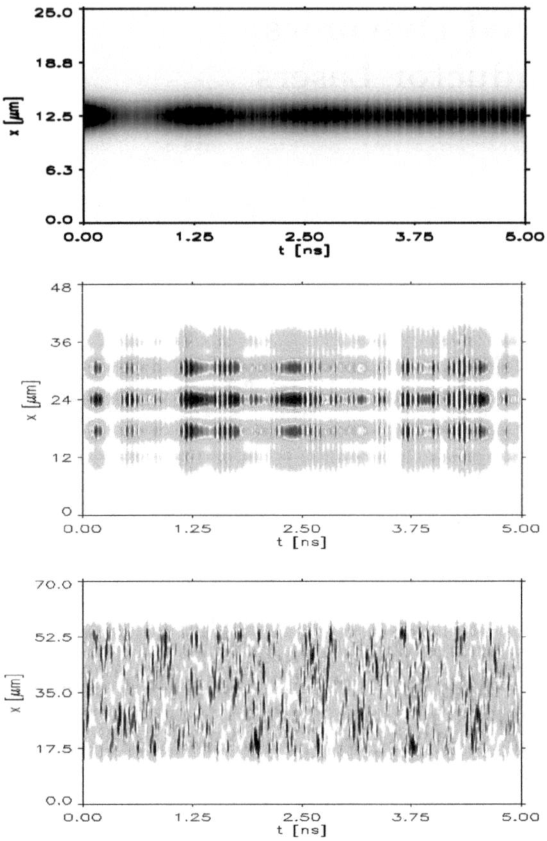

Fig. 4.1. From the single-stripe to the broad-area laser: influence of the transverse degree of freedom on spatio-temporal dynamics. In the calculated temporal behavior of the intensity at the output facet dark shading indicates high intensity values

stripe laser. In spite of the identical pump current the individual stripes each with a transverse width of 5 μm show different dynamics when compared to the single-stripe laser. This behavior originates from the transverse degree of freedom which now plays an important role: the spatio-temporal overlap of the evancescent waves from neighboring laser stripes induces a transverse coupling of the stripes. In combination with carrier diffusion and light diffraction a complex transverse light field dynamics results. As a consequence higher order transverse modes arising in one of the stripes may become unstable via the transverse coupling to the neighboring stripes. Depending on the wave-guiding properties of the structure, the separation of the stripes (typically 4–12 μm) and the carrier injection, the oscillators are strongly, moderately or weakly coupled. The wave-guiding properties are determined by the lateral profile of the refractive index. One distinguishes between anti-guided and positive index-guided arrays. In an anti-guided array the region between the stripes has a higher refractive index than the stripes leading to anti-guiding and to strong interelement coupling. In a positive-index-guided array the stripes have a higher index. As a consequence a strong optical confinement of the individual modes exists. Due to their large active area (approximately 50–150 μm) broad-area lasers allow the generation of high power. However, their lateral width significantly exceeds the laser-internal interaction lengths (μm-regime) given by light diffraction and carrier diffusion. As a consequence the spatio-temporal behavior of the nearfield shows a complex transverse light field dynamics (Fig. 4.1(c)). The dynamic light–matter coupling and the mutual interplay of self-focusing, light diffraction and carrier diffusion lead to the formation of spatially inhomogeneous gain and index profiles. This leads to the formation of light filaments characterized by low gain and high induced refractive index. Via carrier depletion and self-focusing the filament provides itself with a self-induced waveguide. At the same time the carrier injection leads to the formation of high-gain regions surrounding the filament. The neighboring regions of higher gain and low index then give rise to new filaments which after some time destabilize the previous filament via the nonlinear transverse coupling. As a consequence a continuing transverse migration of filaments in space and time can be observed.

The length of the laser device, on the other hand, determines the number and interplay of *longitudinal modes* (Fig. 4.2). The longitudinal propagation

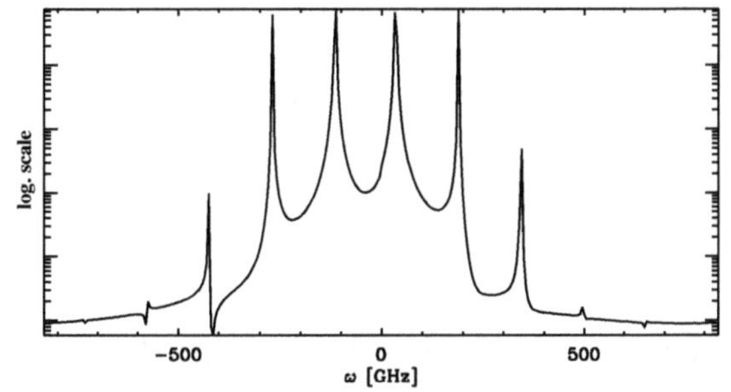

Fig. 4.2. Typical longitudinal mode spectrum of a free-running semiconductor laser (resonator length $L = 250$ μm)

dynamics plays an important role in the dynamic spectral characteristics of the laser. Figure 4.2 shows an example of a (temporally averaged) longitudinal mode spectrum of a free-running single-stripe semiconductor laser. Thereby the number and spectral distance $\Delta\omega$ between the modes depends via $\Delta\omega = c/(2n_l L)$ on both the length L of the laser cavity and the refractive index n_l of the active laser material.

Additionally, the longitudinal as well as the transverse modes influence each other via the strongly nonlinear coupling of the light fields with the spatio-temporal carrier dynamics in the active medium. Consequently, the temporal behavior and emission spectra that can be observed in experiments are the result of a complex interplay of transverse modes dynamically interacting with each other via the active medium and via the light fields propagating in the longitudinal direction of the resonator. Figure 4.3, showing

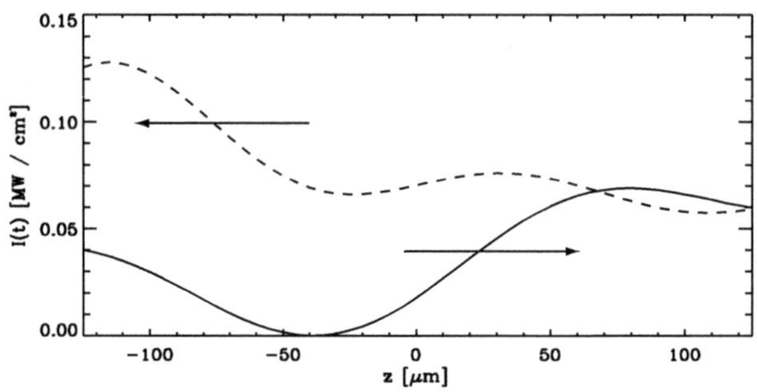

Fig. 4.3. Snapshot of the spatial distribution of forward (*solid line*) and backward (*dashed*) propagating light fields

a snapshot of the intensity profiles in the resonator, illustrates the spatial dependence of the light fields propagating in the forward (solid line) and backward (dashed) directions.

In this chapter we will present an effective multi-mode theory and simulation of large-area semiconductor lasers and laser arrays. The multi-mode Maxwell–Bloch equations consist of multi-mode wave equations and a two-level Bloch description for the active semiconductor medium. These equations represent a very practicable means to simulate the long-time behavior with realistic computational effort.

4.2 Multi-Mode Maxwell–Bloch Equations

For many applications such as frequency conversion and modulated laser systems not only the spatio-temporal dependence of the light and carrier distributions but particularly the long-time emission behavior like relaxation oscillation, dynamic instabilities and self-pulsations and modulation response are of fundamental interest. Generally, the high-speed dynamics of semiconductor lasers is determined by a complex interplay of ultrafast light field and carrier dynamics [1]. In order to theoretically describe and simulate these dynamics propagation effects and spatio-temporally varying mode competition have to be taken into account. Here we present a multi-mode Maxwell–Bloch description for the theoretical analysis and numerical simulation of the high-speed modulation dynamics of multi-segmented semiconductor lasers. The multi-mode Maxwell–Bloch equations consist of spatially dependent multi-mode wave equations and a two-level Bloch description for the active semiconductor medium. The Bloch equations for a two level homogeneously broadened medium model the complex material polarization and carrier density interacting with the propagating optical fields. The dynamic spatio-temporal interplay of longitudinal modes that is important for the long-time emission dynamics (relaxation oscillations, dynamic instabilities and self-pulsations) and, in particular, the spatio-temporal dynamics of modulated lasers is included via a multi-mode expansion of the fields.

In order to include the superposition and interaction of longitudinal modes in our theoretical description of a spatially extended laser structure we express the light fields, the polarization and the carrier density in terms of a multi-mode expansion:

$$E(z,t) = \mathrm{e}^{\mathrm{i}k_0 z}\sum_{n=0}^{\infty} E_n^+(z,t)\mathrm{e}^{2\mathrm{i}nk_0 z} + \mathrm{e}^{-\mathrm{i}k_0 z}\sum_{n=0}^{\infty} E_n^-(z,t)\mathrm{e}^{-2\mathrm{i}nk_0 z} \tag{4.1}$$

$$P(z,t) = \mathrm{e}^{\mathrm{i}k_0 z}\sum_{n=0}^{\infty} P_n^+(z,t)\mathrm{e}^{2\mathrm{i}nk_0 z} + \mathrm{e}^{-\mathrm{i}k_0 z}\sum_{n=0}^{\infty} P_n^-(z,t)\mathrm{e}^{-2\mathrm{i}nk_0 z} \tag{4.2}$$

$$N(z,t) = N_{(0)}(z,t) + \sum_{n=0}^{\infty} N_n(z,t)\mathrm{e}^{2\mathrm{i}nk_0 z} + \text{c.c.}. \tag{4.3}$$

Insertion of this expansion in the Maxwell–Bloch equations [2] leads to a hierarchy of equations for the polarization P and the carrier density N. Carrier diffusion and light diffraction lead to a smoothing of the grating structure induced in and by the counterpropagating light fields. In addition the spatio-temporal light field dynamics couples via the polariziation in the wave equation to the full (nonlinear) semiconductor properties and thus to the counterpropagating waves at every grid point of the laser. As a consequence we only have to consider the lowest order in the equations.

This leads to the following set of equations for the dynamics of the light fields propagating in the forward ('+') and backward ('−') directions in the laser

$$\frac{\partial}{\partial t}E^\pm + \frac{\partial}{\partial z}E^\pm = \mathrm{i}D_\mathrm{p}\frac{\partial^2}{\partial x^2}E^\pm - \mathrm{i}\eta E^\pm - \mathrm{i}k_0 E^\mp + \beta\Gamma P^\pm_{(0)}. \tag{4.4}$$

The diffraction coefficient is $D_\mathrm{p} = (2n_\mathrm{l}k_0)^{-1}$ with the vacuum wavenumber $k_0 = 2\pi/\lambda$. The wave-guiding properties derived from the effective index approximation are included in the parameter η. $\bar{\omega}$ denotes the frequency detuning between the frequency of the electron–hole pair and the light frequency. Via the polarization the light fields are locally coupled to carriers in the active medium.

The basic properties of the Bloch equations for a two level, homogeneously broadened medium to describe the complex material polarization and carrier density interacting with an optical field have been discussed in [2; 3]. Using the multi-mode expansions the Bloch equations read

$$
\begin{aligned}
\frac{\partial}{\partial t}P^{+}_{(0)} &= -\gamma_{\mathrm{p}}\left[\left(1+\mathrm{i}\frac{\bar{\omega}}{\gamma_{\mathrm{p}}}\right)+(\rho+\mathrm{i}\sigma)\,N\right]P^{+}_{(0)} \\
&\quad +\beta_2\left(\left(N_{(0)}+\mathrm{i}\alpha\right)E^{+}+N_{(1)}E^{-}\right) \\
\frac{\partial}{\partial t}P^{-}_{(0)} &= -\gamma_{\mathrm{p}}\left[\left(1+\mathrm{i}\frac{\bar{\omega}}{\gamma_{\mathrm{p}}}\right)+(\rho+\mathrm{i}\sigma)\,N\right]P^{-}_{(0)} \\
&\quad +\beta_2\left(\left(N_{(0)}+\mathrm{i}\alpha\right)E^{-}+N^{*}_{(1)}E^{+}\right) \\
\frac{\partial}{\partial t}P^{+}_{(1)} &= -\gamma_{\mathrm{p}}\left[\left(1+\mathrm{i}\frac{\bar{\omega}}{\gamma_{\mathrm{p}}}\right)+(\rho+\mathrm{i}\sigma)\,N\right]P^{+}_{(1)}+\beta_2 N_{(1)}E^{+} \\
\frac{\partial}{\partial t}P^{-}_{(1)} &= -\gamma_{\mathrm{p}}\left[\left(1+\mathrm{i}\frac{\bar{\omega}}{\gamma_{\mathrm{p}}}\right)+(\rho+\mathrm{i}\sigma)\,N\right]P^{-}_{(1)}+\beta_2 N^{*}_{(1)}E^{-} \\
\frac{\partial}{\partial t}N_{(0)} &= \Lambda + D_{\mathrm{f}}\nabla^{2}N_{(0)}-\gamma_{\mathrm{nr}}N_{(0)} \\
&\quad -2\left(E^{+}(P^{+}_{(0)}-\Lambda_0E^{+})^{*}+E^{-}(P^{-}_{(0)}-\Lambda_0E^{-})^{*}+\mathrm{c.c.}\right) \\
\frac{\partial}{\partial t}N_{(1)} &= -4D_{\mathrm{f}}k_z^{2}N_{(1)}-\gamma_{\mathrm{nr}}N_{(1)} \\
&\quad -2\left(E^{+}(P^{-}_{(0)}-\Lambda_0E^{-})^{*}+E^{-*}(P^{+}_{(0)}-\Lambda_0E^{+})\right. \\
&\quad \left. +E^{+*}P^{+}_{(1)}+E^{-}P^{-*}_{(1)}\right).
\end{aligned}
\tag{4.5}
$$

The time-dependent calculation of the light field and carrier dynamics on the basis of the set of equations, (4.4) and (4.5), allows for a simulation of the long-time behavior of spatially extended semiconductor laser with explicit consideration of the interplay of a number of longitudinal modes. This effective multi-mode theory will be applied in the following to the description and simulation of multi-stripe semiconductor lasers.

4.3 Transverse Coupling and of a Twin-Stripe Laser

In this section we will present selective results of numerical simulations of transversely coupled lasers. For a discussion and interpretation of the basic effects we will consider the twin-stripe laser (Fig. 4.4). The twin-stripe is not only the simplest model system for a multi-stripe laser configuration that can be used for a theoretical investigation of the complex interaction between adjacent semiconductor laser oscillators but also has – due to its modulation response – high technological significance. Figure 4.4 schematically shows the typical geometry of a twin-stripe laser. The active area consists of two stripes that are arranged parallel to each other on the device. The stripes have a lateral width w and a lateral distance (edge-to-edge) s, L is the length of the device. Typically the lateral extension of the individ-

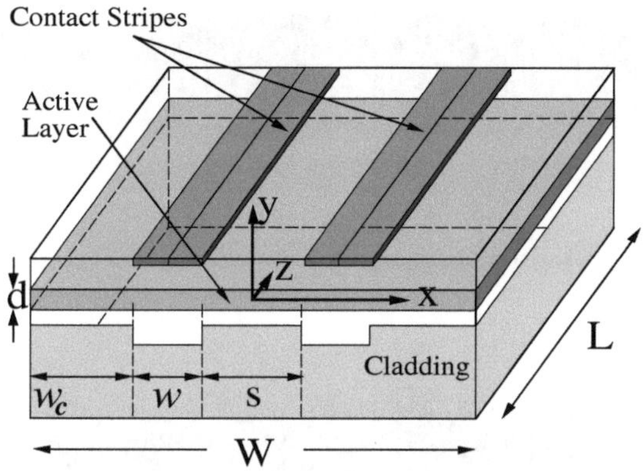

Fig. 4.4. Scheme of the model twin-stripe laser: two stripes (width w) are arranged parralel to each other on the laser structure. Their lateral (edge-to-edge) distance is s, L is the length of the laser

ual stripes is similar to the width of single-stripe lasers, i.e. approximately 3–5 μm. The length of the laser varies from a few hundred μm up to the mm-regime. The lateral distance of the stripes represents the most critical parameter. Its value (3–20 μm) is most relevant for the transverse coupling of the stripes and consequently for the spatio-temporal dyanamics of the laser.

For a visualization of the influence of the stripe separation Fig. 4.5 shows

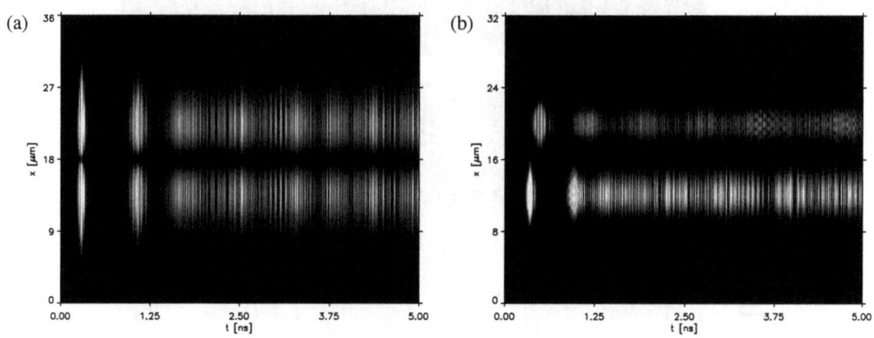

Fig. 4.5. Temporal behavior of the intensity at the output facet of a twin-stripe laser with width w =4 μm of the individual stripes and a lateral separation of 4 μm (**a**) and 8 μm (**b**). The injection current density is $j = 1.5 j_{\mathrm{thr}}$

the temporal behavior of the light field at the output facet of twin-stripe lasers with stripe separations of 4 μm and 8 μm. A small stripe separation (Fig. 4.5(a)) leads to a strong coupling between the stripes. As a result the two stripes may (depending on injection current and wave-guiding proper-

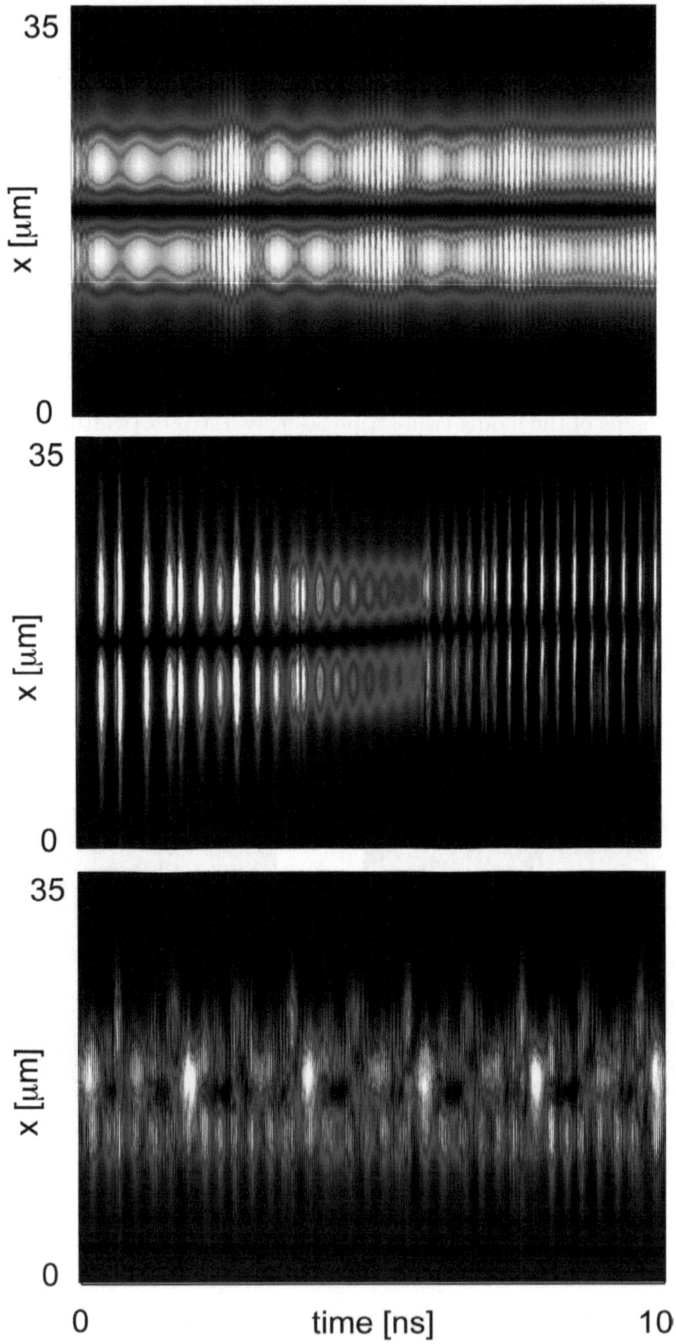

Fig. 4.6. Time dependence of the emitted intensity for a twin-stripe laser ($w =$ 4 μm). The injection current density is (from top to bottom) $j = 1.50 j_{\text{thr}}$, $j = 2.75 j_{\text{thr}}$ and $j = 3.00 j_{\text{thr}}$

ties) lock to each other and show a synchronized emission. The characteristic temporal behavior of the coupled stripes is then strongly dependent on the amount of external pumping via the electrical current. In the structure with larger stripe separation (Fig. 4.5(b)) the laser stripes show a very individual and decoupled dynamics. The dependence of the light field dynamics on current is shown in Fig. 4.6 for a stripe separation of 4 μm. The injection current density in the upper diagram is $j = 1.50 j_{\text{thr}}$, $j = 2.75 j_{\text{thr}}$ in the graph displayed in the center and $j = 3.00 j_{\text{thr}}$ in the graph displayed in the bottom of Fig. 4.6. The calculated temporal dependence of the intensity at the output facet shows a very typical behavior: the value of the injection current density significantly determines the amount of transverse coupling of the two stripes. As a consequence one can observe characteristic regimes. For moderate injection current the laser shows a rather uniform emission. With increasing pump current spatio-temporal instabilities and the formation of spontaneous periodic intensity pulsations arise which can be attributed to the complex nonlinear interaction of multiple transverse modes. The critical threshold current for the onset of pulsations thereby strongly depends on the specific geometry and material properties of the twin-stripe laser. High carrier injection leads to increased carrier diffusion and light–matter coupling resulting in a non-vanishing light field contribution between the stripes.

4.4 High-Frequency Modulation of Segmented Contact Semiconductor Lasers

One of the limiting factors in high-speed applications of semiconductor lasers is the well-known cut-off frequency in response to high-frequency injection-current modulation [4]. This is closely related to the internal carrier dynamics, i.e. the time scales of interband or intraband relaxation and scattering. They are generally accepted to be mainly material dependent. It would thus be of great technological relevance to find and explore avenues of circumventing these constraints, e.g. by mode-locking [5], multi-section lasers [6] or nonlinear intracavity interaction [7].

Encouraged by recent investigations on modulated semiconductor laser arrays [8] we here present results of spatio-temporal simulations of the high-frequency modulation characteristics of semiconductor lasers that have been modified with respect to the realization of their current contacts (e.g. a twin-stripe laser). We show that, indeed, lateral segmentation of the contact(s) with proper asymmetric application of the injection current may lead to a more than five-fold increase of the modulation bandwidth. The simulations on the basis of Maxwell–Bloch equations reveal that the increased high-speed modulation is closely associated with the coupled lateral and longitudinal multi-mode dynamics of the laser.

In order to analyze the response of the laser to current modulation we have varied the modulation frequency of the pump term and calculated the resulting spatio-temporal behavior of the nearfield and the carrier inversion. The simulation results depicted in Fig. 4.7 show the temporal behavior of the current-modulated two-segment laser (with average current density $j = 1.3j_{\rm thr}$) for modulation frequencies of (a, d) 2 GHz, (b, e) 4 GHz and (c, f) 6 GHz. The upper and lower rows show the dependence of the emitted intensity and the corresponding current density on time, respectively. In the case of the 2 GHz modulation (Fig. 4.7(a, d)) one can observe a profound

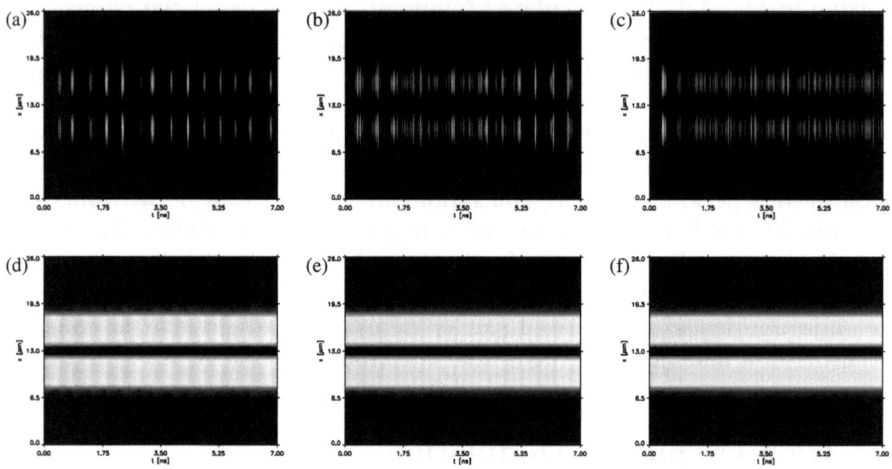

Fig. 4.7. Dependence of the nearfield intensity (**a**–**c**) and the carrier density (**d**–**f**) of a semiconductor laser with segmented contacts (segment separation 2 μm) on modulation frequency. The frequency of the current modulation is 2 GHz (**a**, **d**), 4 GHz (**b**, **e**) and 6 GHz (**c**, **f**). The (average) current density is $j = 1.3j_{\rm thr}$

and regular modulation of the carrier density. The modulations of the current are sufficiently slow so that the carrier inversion (Fig. 4.7(d)) can follow (i.e. within one period and before the start of the next peak). The resulting gain modulation leads to emission of a regular pulse train as displayed in Fig. 4.7(a). With increasing modulation frequency the pulses are emitted more and more out of phase. Irregular and spatio-temporally broadened pulses appear instead (Fig. 4.7(c)). This behavior is a direct consequence of the finite material dependent interaction times that govern the dynamics of the laser. Spontaneous and induced emission leading to spatio-temporal hole burning and the re-establishment of the gain via the pump current have characteristic time scales ranging from the picosecond up to the nanosecond regime. For high-speed modulation (i.e. more than 2 GHz) the carrier density can no longer follow fast enough due to the finite interaction times. As a consequence the modulation of the inversion is "smeared out" (Fig. 4.7(f)) leading to irregular pulse emission (Fig. 4.7(c)). Indeed, this cut-off frequency marks the relaxation oscillation and usually represents a modulation limit for a narrow stripe *single-contact* semiconductor laser. In spatially extended semiconductor lasers, however, the longitudinal and transverse dimensions generally enable the coexistence of numerous longitudinal and transverse modes. With suitable resonator design allowing segmented contact carrier injection and modulation it should thus be possible to directly influence the lateral coupling and transverse mode dyanamics of a given laser structure and modulate the laser with a beat frequency associated with these modes.

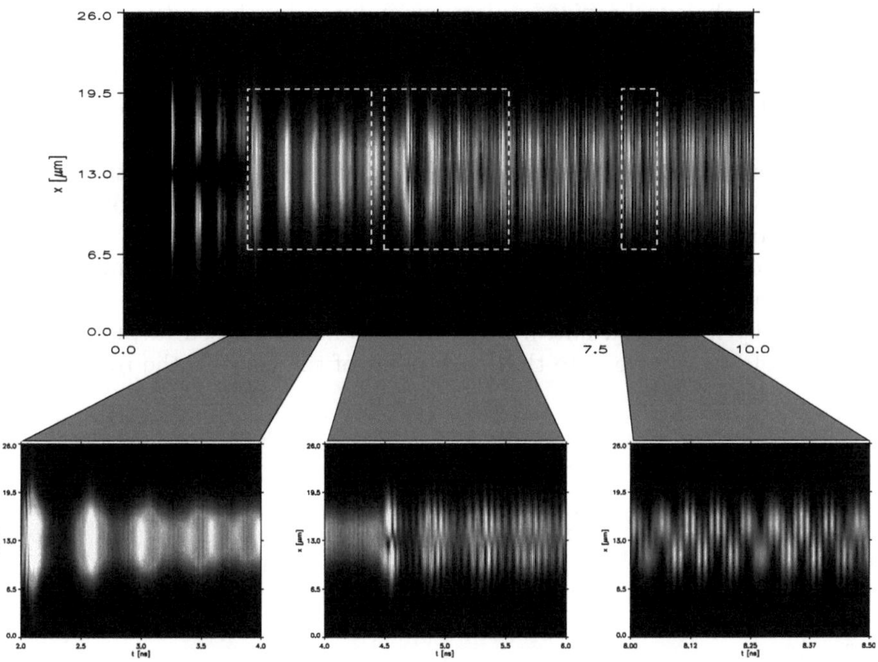

Fig. 4.8. Temporal behavior of the nearfield intensity of a segmented contact semiconductor laser (separation between the contact segments: 2 μm) for a modulation of 8 GHz. The high-speed modulated current is applied to one of the segments while the second segment is pumped with continuous current. The average injection current is $j = 2.00j_{\mathrm{thr}}$ in both contact segments

Indeed, as the example in Fig. 4.8 demonstrates, a spatially inhomogeneous high-speed modulation leads to a pronounced response of the two-segment laser at a modulation frequency of 8 GHz that is far beyond the former cut-off frequency of 2 GHz. The average injection current density is $j = 2.0j_{\mathrm{thr}}$ in both stripes. The sinusoidal modulation is only applied to the lower stripe. The modulation of the carrier inversion in the lower stripe is via the light diffraction and the nonlinear interaction of the evanescent waves partially transfered to the second (upper) stripe. The resulting laterally inhomogeneous pump profile leads to an increased lateral migration of the light fields that ends up in the excitation of a transverse eigenmode of the laser. The time-window displayed in Fig. 4.8 and, in particular, the magnified time-traces depicted below it demonstrate a sequence of different characteristic regimes experienced by the laser until a quasi-stationary regime is reached. Immediately after turning on the current (0...2 ns) the stripes show individual pulsations. After some time the coupling of lateral modes in combination with the light propagation leads to increasing interaction. First, at 2...4 ns, lateral modes merge and show an emission behavior similar to a single laser stripe. Then the synchronized pulsations induced by the mutual interaction of light and matter in combination with the laterally inhomogeneous carrier injection leads to the gradual formation of a transverse dynamics (4...6 ns). In its final state (8...8.5 ns), the laser shows a regular transverse migration of the light that can be identified with a higher-order transverse mode. The time for two left-right changes of the intensity (belonging to the positive and negative phase of the respective light field) thereby corresponds to the frequency of the modulation (8 GHz).

Figure 4.9 shows the result of a systematical variation of the modulation frequency and subsequent calculation of the response of the laser (temporally and spatially integrated data). With all parameters chosen identical to the situation of Fig. 4.8 we note that each point of the curve is the result of the

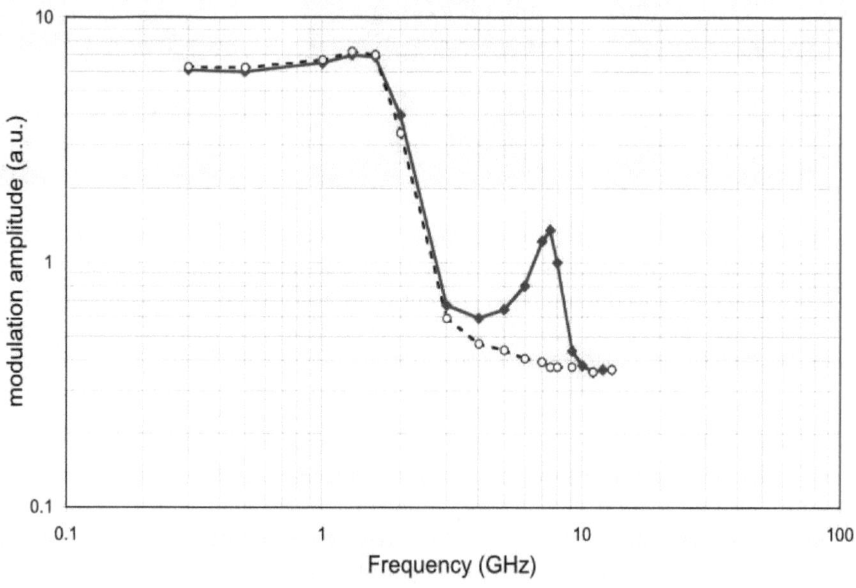

Fig. 4.9. Modulation response of a semiconductor laser with homogeneous (*circles, dashed line*) and segmented (*diamonds, full line*) contacts

full spatio-temporally resolved calculation of the light fields within the frame of the multi-mode Maxwell–Bloch equations. The modulation curve shows a first peak near 2 GHz corresponding to the typical laser-internal interaction times determined by the build-up and decay of inversion and intensity. Increasing the modulation frequency then leads to a dramatic decrease of the response. This effect is caused by limitations from the dynamics of the inversion that cannot follow the instantaneous injection current within a modulation period. In the high-frequency regime, near 8 GHz, a second maximum can be observed. This directly corresponds to the excitation of a transverse

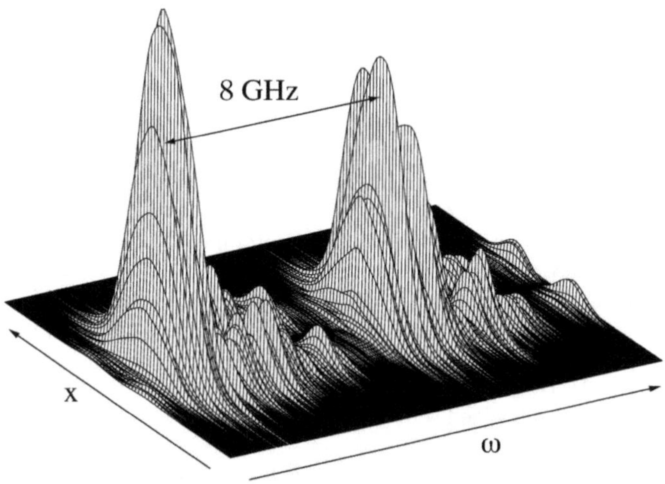

Fig. 4.10. Mode-splitting of a longitudinal mode occurring during high-frequency modulation.

mode (cf. Fig. 4.8) and high-frequency out-of-phase oscillation. In the emission spectrum this resonance is imprinted in a splitting of longitudinal modes (Fig. 4.10).

4.5 Conclusion

The aim of this chapter is to show that the combination of an effective two-level description of the active laser medium with a multi-mode wave equation (both explicitly dependent on transverse and longitudinal space coordinates) allows a realistic representation of typical spatio-spectral properties and modulation characteristics of transversely coupled semiconductor lasers. Numerical results on the basis of this theory visualize the dynamics of light fields and carriers and provide insight into the coupling and interplay of spatial and temporal degrees of freedom. For a given laser structure and set of parameters the multi-mode Maxwell–Bloch equations allow efficient calculation of temporal emission characteristics and of current-modulated lasers in view of a direct comparison with experimental results. Moreover, we have discussed the high-speed modulation dynamics of segmented contact semiconductor lasers. Simulations predict the existence of a high-frequency modulation response near 8 GHz and allow determining the origin of the observed behavior. It can be attributed to the dynamic lateral coupling of the stripes as well as the spatio-temporal coupling of transverse with longitudinal degrees of freedom. The multi-mode Maxwell–Bloch theory thus represents a fundamental basis for the control of the mode dynamics and for exploration and design of the high-speed modulation characteristics of semiconductor lasers.

References

1. W. W. Chow, S. W. Koch, and M. Sargent III. *Semiconductor-Laser Physics.* Springer-Verlag, Berlin, 1994.
2. M. Mnkel, F. Kaiser, and O. Hess. Spatio-temporal dynamics of multi-stripe semiconductor lasers with optical feedback. *Physics Lett. A*, 222:67–75, 1996.
3. J. M. Regalado, S. Balle, and N.B. Abraham. Spatio-temporal dynamics of gain-guided semiconductor laser arrays. *IEEE Journ. Quant. Electr.*, 32:257–276, 1996.
4. R. Olshansky, P. Hill, V. Lanzsiera, and W. Pawazinik. Frequency response of 1.3 μm InGaAsP high speed semiconductor lasers. *IEEE J. Quantum. Electron.*, 23:1410–1418, 1987.
5. E. Goutain, J. C. Renaud, M. Krakowski, D. Rondi, R. Blondeau, and D. Decoster. 30 Ghz bandwidth, 1.55 μm MQW-DFB laser diode based on a new modulation scheme. *Electron. Lett.*, 32:896, 1996.
6. K. Y. Lau. Narrow-band modulation of semiconductor lasers at millimeter wave frequencies (¿ 700 ghz) by mode locking. *IEEE J. Quantum Electron.*, 26:250–261, 1990.
7. R. Nietzke, W. Elsäßer, A.N. Baranov, and K. Wünstel. Generation of 170 ghz optical sidebands of a single-mode semiconductor laser by nonlinear intracavity interaction. *Appl. Phys. Lett.*, 58:554–556, 2001.
8. G. Carpintero, H. Lamela, M. Leones, C. Simmendinger, and O. Hess. Fast modulation scheme for a two laterally coupled laser diode array. *Appl. Phys. Lett.*, 78:4097–4099, 2001.

5 Polarization Fluctuations and Dynamics of Vertical-Cavity Surface-Emitting Lasers

Vertical-Cavity Surface-Emitting Lasers (VCSELs) have recently attracted intensive technological, experimental and theoretical effort. VCSELs have a very short resonator length of the order of the optical wavelength assuring single longitudinal mode operation. At the same time, both transverse (x and y) directions are equally large (typically 3–30 μm). To provide sufficient gain the mirror reflectivities have to be high and are technologically realized by dielectic multi-layers.

One of the major advantages that VCSELs have over convential in-plane edge-emitting semiconductor lasers is their highly symmetric geometry around the axis of laser light emission. However, due to the transverse symmetry the polarization of the emitted light is highly sensitive to the microscopic carrier and light field dynamics, anistropies in the crystal structure or strain and optical anistropies in the mirrors. Consequently VCSELs may exhibit polarization instabilities in the input-output characteristics that are the limitting factors in polarization-sensitive applications. Furthermore, the tranverse coupling induced by carrier diffusion and light diffraction leads to complex mode competition, frequency beating and filamentation. Depending on the particular VCSEL geometry and the device parameters (such as pumping current and operating temperature) very different types of behavior may arise. In particular, it is the interplay of transverse carrier and light field dynamics as well as the influence of thermal interactions and spontaneous fluctuations that act in concert to determine the overall spatio-temporal dynamics of the system. The theoretical description of spatially extended VCSELs consequently has to consider two transverse dimensions and two polarizations. Here, we will first base our discussion on a general and mesoscopic four band Maxwell–Bloch description (Sect. 5.1) that takes into account the spin-resolved spatio-spectral carrier dynamics of the carriers along with the spatio-temporal dynamics of the polarized optical fields and the temperature dynamics. In the second part of this chapter (Sect. 5.2) we will analyze the polarization fluctations on the basis of a simplified rate equation model.

5.1 Spatially Resolved Polarization and Temperature Dynamics in Quantum Well Vertical-Cavity Surface-Emitters

The multi-mode dynamics of VCSELs and the polarization state of the optical field are influenced by several mechanisms of physical and technological nature. A deeper understanding of them is of great importance to categorize their origin and is instrumental in the design and development of future devices.

The cavity properties associated with the geometric structure itself and the physical properties of the passive layers have an effect on the propagation and the confinement of the optical field as well as on the transport of carriers through the passive material of the device. Recently, significant design improvements for VCSEL devices have led to more efficient lasers in terms of low threshold current, good beam quality and higher output power. These are, e.g. the doping of the Bragg mirrors to lower the electrical resistance or the introduction of an oxide aperture resulting in better carrier and better optical field confinement. In the active area – usually realized by several

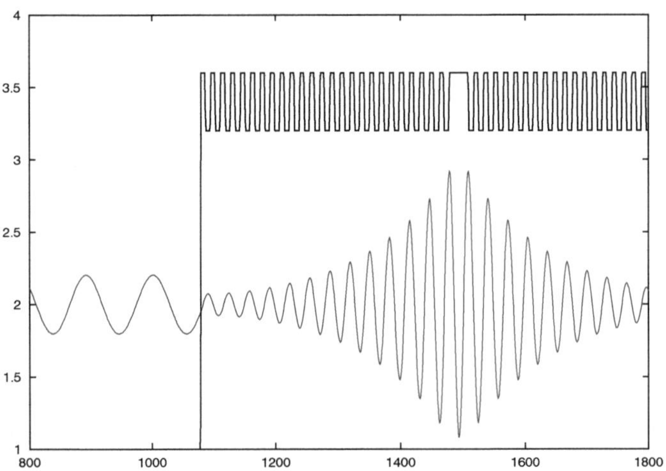

Fig. 5.1. Refractive index and the confined optical field mode plotted over the longitudinal z coordinate for a VCSEL structure with a λ-defect zone. In the model equations this direction will be approximated by a mean-field approach in which the detailed longitudinal cavity structure (e.g. the refractive index $n(z)$) will be treated as effective parameters (n_{eff})

specifically designed quantum wells and governed by the interaction between carriers and light field – the coherent amplification necessary for laser activity takes place. Many of the fundamental processes such as carrier generation or carrier scattering and their dependence on quantum well design or external parameters are not directly accessible in experiments and consequently have to be studied in a theoretical framework.

During recent years experimental and theoretical research concentrated on the subject of polarization behavior and transverse optical field modes. As common VCSEL structures bear a rotational symmetry around the axis of emission the two possible polarization states of the light field should be equally preferred. In contrast, it has been demonstrated that in many cases one of the polarization states dominates and polarization switching occurs due to a change of operation conditions such as pumping current or device temperature. Among the possible mechanisms responsible for this behavior cavity anisotropies and strong amplitude-phase coupling have been discussed. Here we will simulate and analyze the light field and polarization dynamics on the basis of a Maxwell–Bloch description. The results of our simulations help to identify the physical effects that determine the multi-mode field dynamics of modern quantum well VCSELs. For this we will compute and include wavenumber-dependent scattering rates in the equations describing the carrier dynamics.

5.1.1 Light-Field Dynamics and Carrier Transport

Maxwell's wave equations in mean-field approximation (i.e. averaging over the propagation direction depicted in Fig. 5.1) for the field amplitude $\boldsymbol{E}$ read:

$$\frac{1}{v_{\mathrm{l}}}\frac{\partial}{\partial t}\boldsymbol{E} = \mathrm{i}D_{\mathrm{p}}\nabla_{\perp}^{2}\boldsymbol{E} - \gamma_{\mathrm{res}}\boldsymbol{E} + \mathrm{i}\kappa\,\Gamma_{z}\,\boldsymbol{P} + \Theta\boldsymbol{E}\,. \tag{5.1}$$

The components of the vector $\boldsymbol{E}$ are the two possible circular polarization states (E^{+}, E^{-}). Γ_z is the longitudinal confinement factor, $v_{\mathrm{l}} = c/n_{\mathrm{eff}}$ the velocity of light in the medium, $D_{\mathrm{p}} = 1/(2K_{\mathrm{eff}})$ is the diffraction constant and $\kappa = K_{\mathrm{eff}}(2n_{\mathrm{eff}}^{2})^{-1}$ is the coupling factor for the active media polarization. n_{eff} and K_{eff} are the effective refractive index and the effective wavenumber which are averages over a cavity roundtrip weighted with the longitudinal field envelope. Finally, the tensor Θ of optical anisotropy accounts for optical birefrigence and dichroism of the cavity.

To represent carrier diffusion in the active layer we use a phenomenological diffusion equation for the density of carriers (electron–hole pairs) confined in the quantum well,

$$\frac{\partial}{\partial t}N = D_{\mathrm{f}}\nabla_{\perp}^{2}N - \gamma_{\mathrm{nr}}N + \Lambda + G - \Gamma_{\mathrm{sp}}\,. \tag{5.2}$$

Thereby we have introduced the transverse ambipolar diffusion constant D_{f}, the pumping rate

$$\Lambda = \frac{\eta}{ed}\,j_0\,, \tag{5.3}$$

(with current density j_0), the coherent carrier generation rate G as well as terms for the non-radiative recombination rate $\gamma_{\mathrm{nr}}N$, and the spontaneous emission rate Γ_{sp}. To implement a specific laser system one now has to define the functional dependence of the polarization P on light fields E and carrier density N. This will be done in the next section for a so-called spin resolved semiconductor four band system.

5.1.2 Bloch Equations for Quantum Well Vertical-Cavity Surface-Emitting Lasers

Equations of Motion

Today multi quantum well structures are widely used as gain media for VCSELs. As a consequence of the reduced dimensionality of the quantum wells the degeneracy along the z-axis is lifted and we are left with several discrete electronic sublevels for the band structure. For the sake of simplicity we will assume isotropy and only take into account ground states (i.e. the lowest bands with a subband index of one). Assuming cubic lattice symmetry (as for GaAs) the atomic quantum number of the angular momentum J can be considered as a sufficently good quantum number as long as the wavevectors $|\boldsymbol{k}|$ are not too far away from the band edge. In this case it is possible to associate corresponding atomic eigenstates with them. For GaAs quantum well structures $\mathbf{k}\cdot\mathbf{p}$ theory [1] shows that the situation is as follows: the s-like conduction band ($J = 1/2$) consists of two energetically degenerated eigenbands $|J = 1/2, J_z = \pm 1/2 >$. The p-like valence band energetically separates into the heavy–hole, the light–hole, and the split-off band. Moreover, as the E_z component is small compared to $E_\perp$, selection rules are $J_z = \pm 1$. The resulting four band model includes the two s-like electron bands ($|J = 1/2, J_z = \pm 1/2 >$) and the two p-like heavy–hole bands ($|J = 3/2, J_z = \pm 3/2 >$). Selection rules for optical transitions then separate the carrier reservoir into carriers that are involved in right and left polarized light for $J_z = \pm 1$. A four level model for VCSEL structures has been developed by San Miguel et al. [2]. Due to an inherent phenomenological treatment of various semiconductor characteristica a number of extensions e.g. to a four band model [3] have recently been proposed. For our approach we will particularly focus on the effect of scattering rates and the coupling to temperature fields.

In order to describe coherent laser dynamics the carrier–light interaction part has to be added to the non-interacting component of the Hamiltonian which is the sum of the potential energies for the four different bands shown in Fig. 5.2. Starting from Heisenberg's equation of motion and using Wigner's representation $n_k^{\mathrm{e},\pm}(\boldsymbol{r})$ for the electron densities, the hole densities $n_k^{\mathrm{h},\pm}(\boldsymbol{r})$ and the interband polarization $p_k^{\pm}(\boldsymbol{r})$ one can extend the Maxwell–Bloch equations to explicitly represent the four bands. Note that for notational reasons we will suppress the spatial coordinates (although they are fully taken

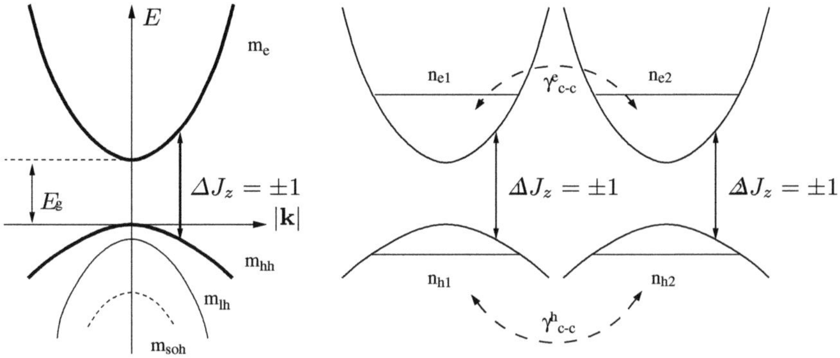

Fig. 5.2. Energetic scheme of a semiconductor quantum well structure indicating possible transitions that are represented in the four band split-density model. $n_{e1}, n_{e2}, n_{h1}, n_{h2}$ denote the electron (e) and hole (h) reservoirs

into account in the model and subsequent simulations). The four band Bloch equations for VCSELs then read:

$$\frac{\partial}{\partial t} p_k^{\pm} = -\left(\mathrm{i}\,\Delta\omega_k + \gamma_{\mathrm{p},k}\right) p_k^{\pm} + \frac{\mathrm{i}}{\hbar}\, \boldsymbol{d}_{\mathrm{cv}}^{\pm}\, \boldsymbol{E}^{\pm} \left(n_k^{\mathrm{e},\pm} + n_k^{\mathrm{h},\pm} - 1\right) \tag{5.4}$$

$$\frac{\partial}{\partial t} n_k^{\mathrm{e},\pm} = g_k^{\pm} - w_{\mathrm{sp},k}^{\pm} + \Lambda_k^{\mathrm{e/h}} \left(1 - n_k^{\mathrm{e},\pm}\right) \tag{5.5}$$

$$-\gamma_{\mathrm{c-c},k}^{\mathrm{e}} \left(n_k^{\mathrm{e},\pm} - f_k^{\mathrm{e}}\right) - \gamma_{\mathrm{c-LO},k}^{\mathrm{e}} \left(n_k^{\mathrm{e},\pm} - f_k^{\mathrm{e}}\right) - \gamma_{\mathrm{nr}}\, n_k^{\mathrm{e},\pm} \tag{5.6}$$

$$\frac{\partial}{\partial t} n_k^{\mathrm{h},\pm} = g_k^{\pm} - w_{\mathrm{sp},k}^{\pm} + \Lambda_k^{\mathrm{h}} \left(1 - n_k^{\mathrm{h},\pm}\right) \tag{5.7}$$

$$-\gamma_{\mathrm{c-c},k}^{\mathrm{h}} \left(n_k^{\mathrm{h},\pm} - f_k^{\mathrm{h}}\right) - \gamma_{\mathrm{c-LO},k}^{\mathrm{h}} \left(n_k^{\mathrm{h},\pm} - f_k^{\mathrm{h}}\right) - \gamma_{\mathrm{nr}}\, n_k^{\mathrm{h},\pm}\,, \tag{5.8}$$

where

$$\Delta\omega_k = \frac{\hbar k^2}{2m_r} + \omega_{\mathrm{g}} - \omega_{\mathrm{op}}\,, \tag{5.9}$$

$$g_k^{\pm} = -2\,\mathrm{Re}\left[\frac{\mathrm{i}}{\hbar}\, \boldsymbol{d}_{\mathrm{cv}}^{\pm}\, \boldsymbol{E}^{\pm} p_k^{\pm *}\right] \tag{5.10}$$

denote the frequency detuning between band energy and optical transition frequency and the stimulated carrier generation rate, respectively. The loss of free carriers due to spontaneous emission and the pump term are

$$w_{\mathrm{sp},k}^{\pm} = \gamma_{\mathrm{sp},k}\, n_k^{\mathrm{e},\pm} n_k^{\mathrm{h},\pm}, \tag{5.11}$$

$$\Lambda_k^{\mathrm{e/h}} = \Lambda f_k^{\mathrm{e/h}} \left[\frac{1}{A \cdot d} \sum_{\mathbf{k},\sigma=\pm} f_k^{\mathrm{e/h}} \left(1 - n_k^{\mathrm{e/h},\sigma}\right)\right]^{-1}, \tag{5.12}$$

respectively. $\boldsymbol{d}_{\mathrm{cv}}^{\pm}$ is the dipole matrix element at the wavenumber k_{op} of the optical transition with frequency $\omega_{\mathrm{op}} = c/n\, k_{\mathrm{op}}$, and $\boldsymbol{E}^{\pm}$ is the optical field amplitude as introduced in (5.1). Further we have to quantify the gap energy $\hbar\omega_{\mathrm{g}}$ and the reduced mass m_{r} of an electron–hole pair at $k = 0$. $f_k^{\mathrm{e/h}} = f_k^{\mathrm{e/h}}(N, T)$ are the equilibrium Fermi–Dirac functions.

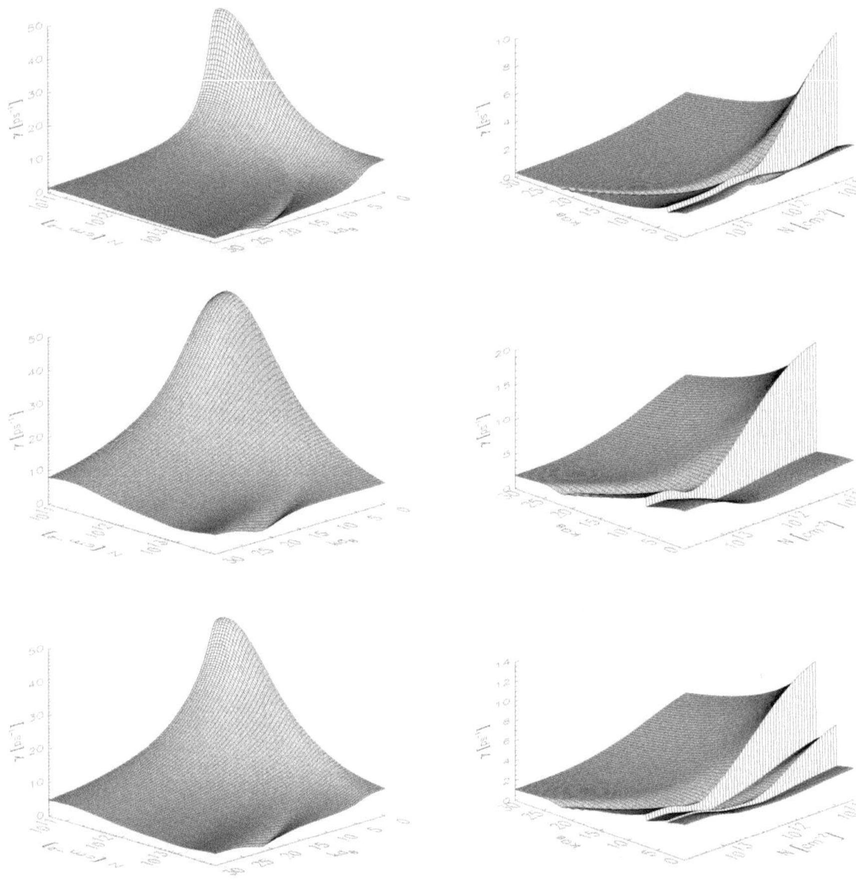

Fig. 5.3. The dephasing rates of the interband polarization are plotted against density (N) and momentum ($\boldsymbol{k}$) for the carrier–carrier (*left*) and the carrier–phonon (*right*) interaction. The polarization decay rates at the right (Fröhlich interaction) show threshold momenta of phonon creation. Both electron and the hole relaxation rates contribute to the polarization decay rate. This leads to two threshold "edges". Due the differences in effective masses the lower (higher) momentum edge pertains to electrons (holes). Under laser conditions typical scattering times near the band-edge are 50 fs for the Coulomb and 200 fs for the Fröhlich interaction

Equations (5.4), (5.6) and (5.8) take into account the dynamic impact of all relevant scattering processes by introducing corresponding phenomenological terms: the nonradiative recombination ($\gamma_{\rm nr}$) of carriers, the dephasing of polarization due to carrier–carrier scattering and the scattering of carriers with LO phonons ($\gamma_{{\rm p},k}$) as well as the redistribution of the carrier distributions $n_k^{\rm e/h}$ to equilibrium Fermi–Dirac functions $f_k^{\rm e/h}$ by carrier–carrier scattering ($\gamma_{{\rm c-c},k}^{\rm e/h}$) and carrier–phonon scattering ($\gamma_{{\rm c-LO},k}^{\rm e/h}$) processes. It is the scattering rates $\gamma_{{\rm c-c},k}^{\rm e/h}$ and $\gamma_{{\rm c-LO},k}^{\rm e/h}$ that describe the frequency (i.e. k-)dependent mixing of carriers belonging to the two reservoirs '+' and '−'. The associated time scale $\tau_{{\rm c-c},k}^{\rm e/h}$ is significantly smaller (typically 50 fs) than the time scale of any other process affecting the angular momentum state of the carriers. Therefore we do not introduce such additional processes. By summing up the microscopic quantities using the appropriate two dimensional density of state function $D(k)$ one obtains the respective macroscopic quantities,

$$G = \frac{1}{A \cdot d} \sum_{\pm,\mathbf{k}} g_k^{\pm} \tag{5.13}$$

$$W_{\rm sp} = \frac{1}{A \cdot d} \sum_{\pm,\mathbf{k}} w_{{\rm sp},k}^{\pm} \tag{5.14}$$

$$\boldsymbol{P}^{\pm} = \frac{1}{A \cdot d} \sum_{\mathbf{k}} \boldsymbol{d}_{\rm cv}^{\pm} \, p_k^{\pm} \,, \tag{5.15}$$

where G is the total coherent carrier generation rate, $W_{\rm sp}$ is the total carrier loss rate by spontaneous emission and $\boldsymbol{P}^{\pm}$ is the total polarization of the active medium. The four band Bloch equations for VCSELs together with the partial differential equations (5.1) and (5.2) for the electric field amplitudes $E^{\pm}$ and the total inversion N constitute a closed set of equations that can be numerically solved in time and space in a self-consistent way.

Scattering Processes

The four band Bloch equations for VCSELs describe the kinetic part and the coherent interaction of the charge carriers with the optical laser field (single-particle Hamiltonians) as well as the mean-field part of many-body Hamiltonians. The strong influence of many-body interaction is one of the prominent features of semiconductur laser material. The many-body interactions first of all comprise the Coulomb interaction among carriers, leading to screening effects and thermalization of non-equilibrium distribution functions due to scattering. The second important type of interaction is the interaction with phonons, in particular with polar optical phonons (Fröhlich interaction), resulting in energy transfer between the subsystems of the carriers and the crystal lattice [4]. Here due to their central significance for the optical field polarization of VCSELs the scattering processes are discussed in more detail.

Disregarding memory effects (Markov approximation) and neglecting non-linear terms in the polarization we have to add adequate Boltzmann collision terms that take into account these scattering processes:

$$\frac{\mathrm{d}}{\mathrm{d}t} n_k^{\mathrm{e/h}}|^{\mathrm{sc.}} = -\Gamma_k^{\mathrm{e/h,out}} n_k^{\mathrm{e/h}} + \Gamma_k^{\mathrm{e/h,in}} \left(1 - n_k^{\mathrm{e/h}}\right) . \tag{5.16}$$

Both, the out- and in-scattering from and into a k-state are decay processes. In equation (5.4) this loss of coherence is expressed by

$$\frac{\mathrm{d}}{\mathrm{d}t} p_k^{\pm}|^{\mathrm{sc.}} = -\frac{1}{2}\left(\Gamma_k^{\mathrm{e,out}} + \Gamma_k^{\mathrm{e,in}} + \Gamma_k^{\mathrm{h,out}} + \Gamma_k^{\mathrm{h,in}}\right) p_k^{\pm} , \tag{5.17}$$

where the Γ scattering matrices are derived on the basis of Fermi's golden rule. For the Coulomb interaction they read [5] ($\circ = \mathrm{e, h}$)

$$\begin{aligned} &\mathbf{k} \to \mathbf{k} - \mathbf{q} : \\ \Gamma_k^{\mathrm{e,out}} &= \sum_{\mathbf{k}',\mathbf{q},\circ} \frac{2\pi}{\hbar} \delta\left(E_k^{\mathrm{e}} - E_{k-q}^{\mathrm{e}} + E_{k'}^{\circ} - E_{k'+q}^{\circ}\right) |V(q)|^2 n_{k'}^{\circ} \\ &\quad \left(1 - n_{k'+q}^{\circ}\right)\left(1 - n_{k-q}^{e}\right) , \end{aligned} \tag{5.18}$$

$$\begin{aligned} &\mathbf{k} - \mathbf{q} \to \mathbf{k} : \\ \Gamma_k^{\mathrm{e,in}} &= \sum_{\mathbf{k}',\mathbf{q},\circ} \frac{2\pi}{\hbar} \delta\left(-E_k^{\mathrm{e}} + E_{k-q}^{\mathrm{e}} - E_{k'}^{\circ} + E_{k'+q}^{\circ}\right) |V(q)|^2 \\ &\quad \left(1 - n_{k'}^{\circ}\right) n_{k'+q}^{\circ} n_{k-q}^{e} , \end{aligned} \tag{5.19}$$

where the Delta function guarantees energy conservation. $V(q)$ is the strength of the interaction. The bare 2D Coulomb potential is weakened due to plasma screening described by the static Lindhard formula and contains additional corrections required for the description of quantum well systems. The last terms in the products represent the populations of initial states and the probabilities of unfilled final states (Pauli-principle).

In the case of the Fröhlich interaction there are two possible scattering processes involving either the emission or the absorption of a phonon,

$$\begin{aligned} &\mathbf{k} \to \mathbf{k} - \mathbf{q} : \\ \Gamma_k^{\mathrm{e,out}} &= \sum_{\mathbf{q},q_z} \sum_{\pm} \frac{2\pi}{\hbar} \delta\left(E_k^{\mathrm{e}} - E_{k-q}^{\mathrm{e}} \pm \hbar\omega_{\mathrm{LO}}\right) |\gamma(q,q_z)|^2 \\ &\quad \left(n_{q,q_z} + \frac{1}{2} \mp \frac{1}{2}\right)\left(1 - n_{k-q}^{\mathrm{e}}\right) , \end{aligned} \tag{5.20}$$

$$\begin{aligned} &\mathbf{k} - \mathbf{q} \to \mathbf{k} : \\ \Gamma_k^{\mathrm{e,in}} &= \sum_{\mathbf{q},q_z} \sum_{\pm} \frac{2\pi}{\hbar} \delta\left(-E_k^{\mathrm{e}} + E_{k-q}^{\mathrm{e}} \pm \hbar\omega_{\mathrm{LO}}\right) |\gamma(q,q_z)|^2 \\ &\quad \left(n_{q,q_z} + \frac{1}{2} \mp \frac{1}{2}\right) n_{k-q}^{\mathrm{e}} , \end{aligned} \tag{5.21}$$

respectively. n_{q,q_z} denotes the phonon distribution function (that may be approximated with a Bose–Einstein function). Due to the high carrier den-

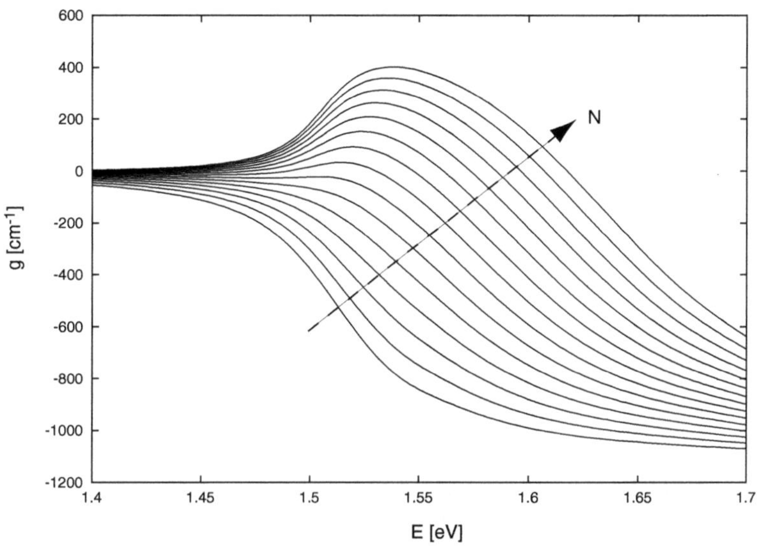

Fig. 5.4. Small-signal gain vs. energy for increasing carrier densities $N \in [2 \cdot 10^{11}\mathrm{cm}^{-2}, 3 \cdot 10^{12}\mathrm{cm}^{-2}]$ at room temperature (T=300 K)

sities in typical quantum well lasers we assume that the distributions show only small deviations from the corresponding quasi-equilibrium Fermi–Dirac functions. We therefore linearize (5.16) and (5.17) in the vicinity of the equilibrium state:

$$\frac{\mathrm{d}}{\mathrm{d}t} n_k^{\mathrm{e/h}}|^{\mathrm{sc.}} = -\gamma_k^{\mathrm{e/h}} \left(n_k^{\mathrm{e/h}} - f_k^{\mathrm{e/h}} \right) \tag{5.22}$$

$$\frac{\mathrm{d}}{\mathrm{d}t} p_k^{\pm}|^{\mathrm{sc.}} = -\gamma_{\mathrm{p},k}\, p_k^{\pm} \,. \tag{5.23}$$

These equations illustrate the significance of scattering: the carrier distributions relax towards their quasi-equilibrium state and the polarization is destroyed. The corresponding scattering rates are expressed in terms of the scattering matrices:

$$\gamma_k^{\mathrm{e/h}} = \Gamma_k^{\mathrm{e/h,out}} + \Gamma_k^{\mathrm{e/h,in}}\Big|^{\mathrm{eq}} , \tag{5.24}$$

$$\gamma_{\mathrm{p},k} = \frac{1}{2} \left(\gamma_k^{\mathrm{e}} + \gamma_k^{\mathrm{h}} \right) \,. \tag{5.25}$$

In Fig. 5.3 the polarization relaxation rates for the Coulomb and the Fröhlich interaction are shown as functions of carrier density and momentum (here: for a quantum well width of 5 nm).

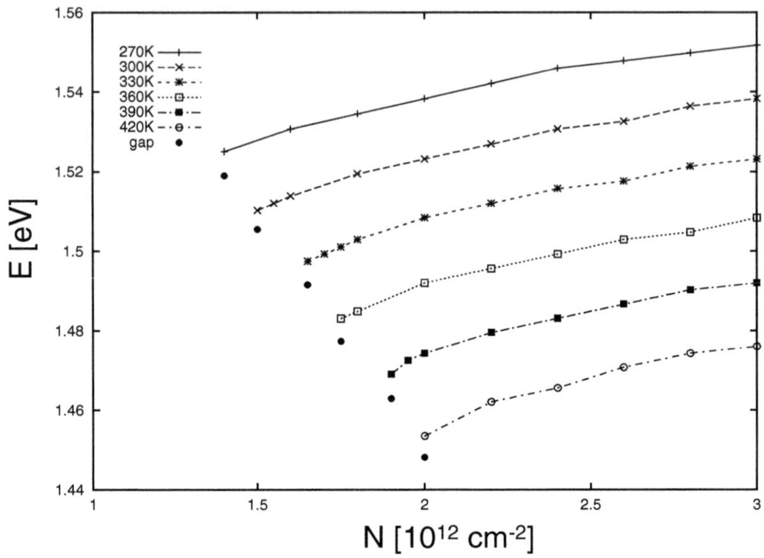

Fig. 5.5. Dependence of the energy of the gain maximum on the carrier density for various values of the plasma temperature

The coupling of the light field with the quantum well gain medium may be described in terms of an induced susceptibility. In many cases of interest the different time scales of the microscopic and macroscopic variables allow us to assume that the microscopic dynamics follows the macroscopic fields adiabatically. The complex susceptibility can then be calculated as follows:

$$\mathrm{Re}[\chi^{\pm}(N,T)] + \mathrm{i}\,\mathrm{Im}[\chi^{\pm}(N,T)] = \frac{|\boldsymbol{d}_{\mathrm{cv}}^{\pm}|^2}{\epsilon_0 \hbar d \cdot A} \sum_{\mathbf{k}} \frac{-\Delta\omega_k - \mathrm{i}\gamma_{\mathrm{p},k}}{(\Delta\omega_k)^2 + (\gamma_{\mathrm{p},k})^2} \times \left(f_k^{\mathrm{e}} + f_k^{\mathrm{h}} - 1\right) . \tag{5.26}$$

Figure 5.4 shows the density dependence of the small-signal gain as obtained from the imaginary part of the (spatially averaged) complex susceptibility (5.26). Adabatically eliminating the polarization dynamics leads to $P = \chi E$. To obtain a static relation between the field and the polarization we eliminate the time dependence in the wave equation leading to

$$\frac{\partial \mid \boldsymbol{E} \mid}{\partial z} = g \mid \boldsymbol{E} \mid \tag{5.27}$$

with

$$g = -\frac{k\Gamma}{2n^2\varepsilon_0}\mathrm{Im}[\chi(N,T)] . \tag{5.28}$$

These relations form the basis for an exploration of the spectral properties of quantum well gain media. The following examples apply to a GaAs single quantum well structure.

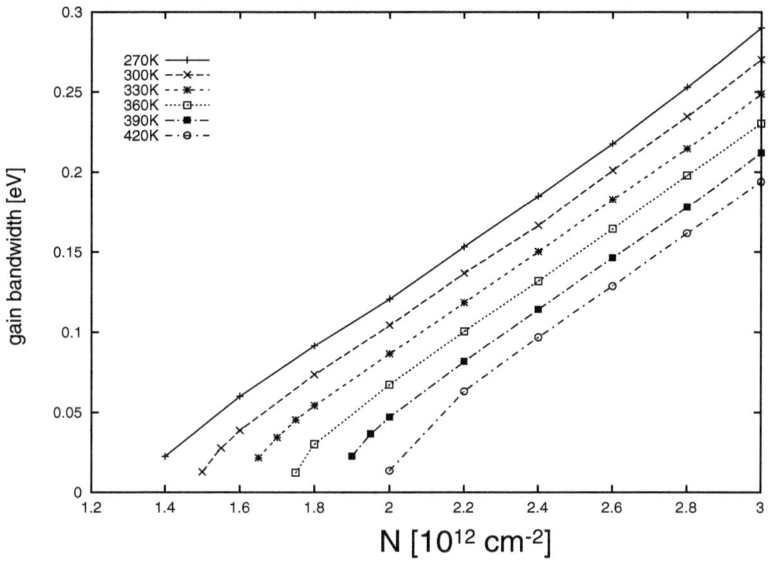

Fig. 5.6. Full gain width for $g > 0$

The spectral dependence of the small-signal gain on energy is shown in Fig. 5.4 for various values of the carrier density at temperature $T = 300K$. With increasing density the gain changes from absorption to amplification. At the same time its maximum is spectrally shifted towards higher energies. Relative to the bandgap correction due to a possible lattice deformation this carrier-induced shift of the energy of the gain maximum is plotted against the carrier density in Fig. 5.5 for various temperatures ranging from 270 K up to 420 K.

The probability for electron–hole recombination is determined by the carrier distributions (filled according to the Pauli-principle) as well as the density of states. As a consequence the average energy of a plasma exciton (electron–hole pair) shifts to higher energies if the carrier density increases. Furthermore, a rise in temperature energetically smoothes the electron and hole distributions, leading to the additional shift that is evident from Fig. 5.5. In addition to the energy of the recombining electron–hole pairs it is also the width and the maximum of the gain that is affected by the particular shape of the carrier distributions as well as the band shapes and their degree of filling. With increasing carrier density the spectral dependence of the bands determines the energy range available for electron–hole recombination. The approximately parabolic shape of the bands thus leads to a larger spectral gain range with increasing carrier density. For a similar reason a rise in temperature increases the spectral range of carriers available for recombination. This leads to an additional gain-broadening. Figure 5.6 shows the carrier density and temperature dependence of the gain bandwidth (defined by $g > 0$).

At the same time with increasing temperature the spectral broadening of the gain leads to the effective decrease of the gain maximum shown in Fig. 5.7. Moreover, as Fig. 5.7 shows, for a fixed temperature an increasing

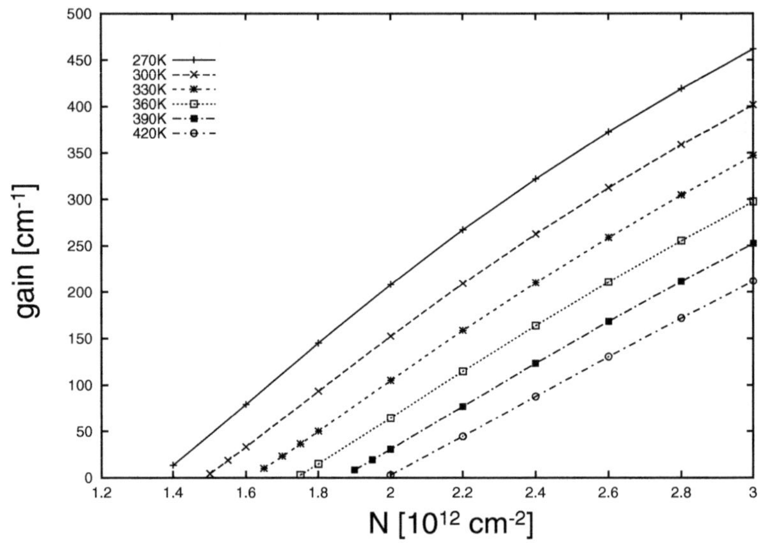

Fig. 5.7. Dependence of the gain maximum on carrier density for various temperatures

carrier density leads to an increasing gain (maximum) due to a larger number of possible electron–hole pairs available for recombination.

5.1.3 Spatio-Temporal Dynamics

Having set up the four band polarization and temperature dependent Maxwell–Bloch model we will now discuss selected results of corresponding spatio-temporal simulations. All simulations are carried out for a quantum well GaAs based VCSEL structure emitting at 810 nm and using typical parameters for this material system. The time series Fig. 5.8 and Fig. 5.9 cover a time window of 900 ps during which the relaxation oscillation can be observed. We note that for such relatively short time scales the temperature is still far away from a global equilibrium state.

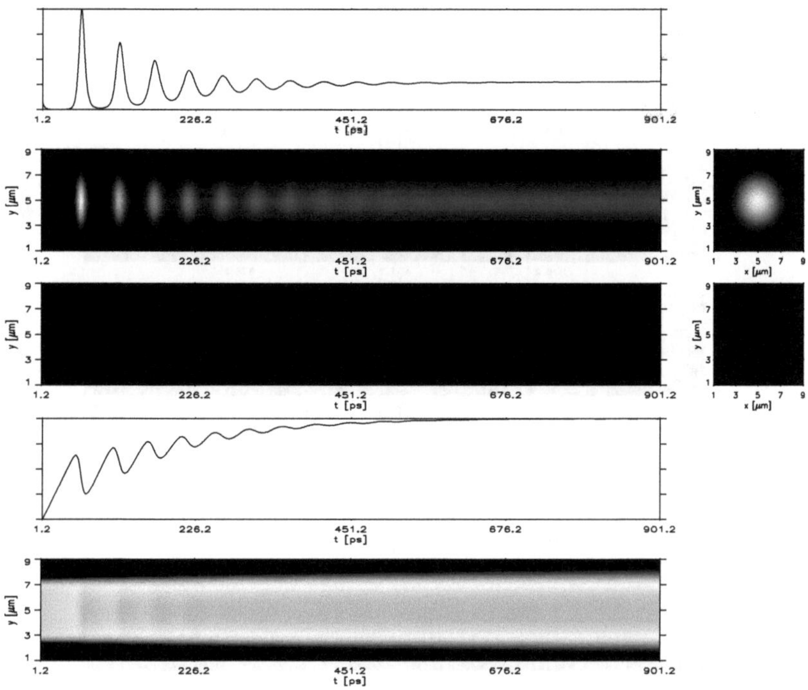

Fig. 5.8. Relaxation oscillation of a VCSEL with 6 μm aperture. Temperatures are constant and homogeneous. From top to bottom: total intensity, x-polarized field intensity $|E_x|^2$, y-polarized field intensity $|E_y|^2$, total inversion and carrier density N

The sample simulation in Fig. 5.8 shows the relaxation oscillation of a small VCSEL with a 6 μm aperture. The time series for the total intensity, the x-polarized field intensity $|E_x|^2$, the y-polarized field intensity $|E_y|^2$, the total inversion and the carrier density N are shown for 900 ps and illustrate the typical single mode relaxation behavior. The pictures on the right are the averaged mode patterns showing only the ground mode. Due to birefringence symmetry is broken and the y-polarized mode is entirely suppressed.

Broader VCSELs, such as the 11 μm sample device simulated in Fig. 5.9 reveal the typical irregular relaxation oscillations. This behavior can be attributed to the appearance of higher order modes that arise after the first 100 ps. The temporally averaged intensity distributions (Fig. 5.9, right) demonstrate the coexistence of x and y-polarized modes. They are a consequence of an interplay of higher-order modes and arise although we have assumed the very same anisotropies as in Fig. 5.8.

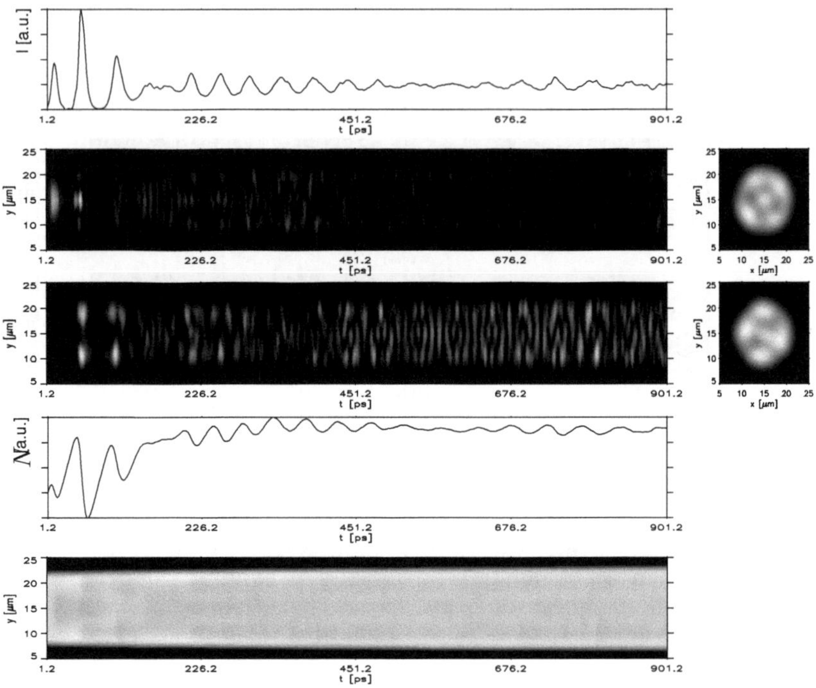

Fig. 5.9. Relaxation oscillation of a VCSEL with 11 μm aperture. Temperatures are constant and spatially homogeneous

As Fig. 5.9 shows, the polarization behavior of a VCSEL depends sensitively on the interplay of spatial and temporal degrees of freedom in the light field and carrier dynamics. An additional complexity appears due to the simultaneous influence of temperature dynamics. To elucidate this inherent coupling in the following we will focus on the dynamics in the post-relaxation oscillation regime. We will compare the polarization dynamics of a VCSEL whose temperature is artificially fixed to the case where the temperatures are calculated dynamically. The series depicted in Fig. 5.10 and Fig. 5.11 pertain to a 11 μm VCSEL. In both cases the full spatio-temporal temperature dynamics was considered up to T_0. In Fig. 5.10 this temperature profile was then frozen to its value at T_0. The figures show (from top to bottom) the dynamics (along a cut through the center of the VCSEL) of the x and y-polarized optical field intensity, the carrier density, and the electron temperature.

Comparison of Fig. 5.10 with Fig. 5.11 then highlights the influence of dynamic thermal effects on the spatio-temporal emission characteristics of VCSELs: in the case of a static temperature profile the emission characteristics follow the internal anisotropy such that one of the polarizations is suppressed (Fig. 5.10). A dynamically varying carrier temperature profile,

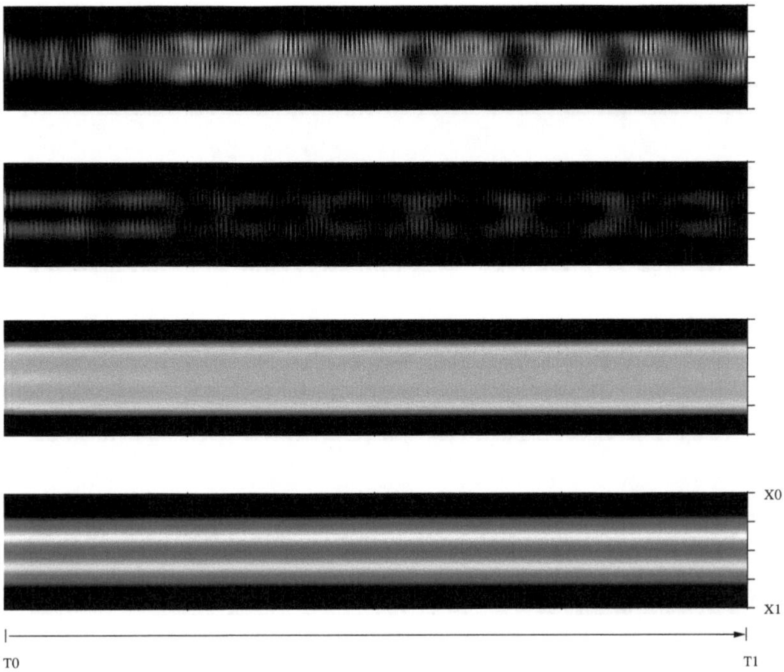

Fig. 5.10. Post-relaxation-oscillation dynamics of a VCSEL with 11 μm aperture during a time window of 2.5 ns. The figures show (from top to bottom) $|E_x|^2$, $|E_y|^2$, N and T_{p}. The plasma temperatures are artificially frozen to the spatial profile at T_0

however, effectively decreases the influence of this anisotropy. As a consequence, both polarization fields oscillate at approximately the same intensity (Fig. 5.11).

After having seen the strong influence of polarization and temperature dynamics on the VCSEL in the following we will concentrate on analyzing the effect and influence of fluctuations on the emission dynamics of VCSELs based on a simplified (phenomenological) rate equation approach.

5.2 Polarization fluctuations in VCSELs

5.2.1 The 'Split-Density' Rate Equation Model

In the following we will use a simplified rate equation level of description to discuss the effects of polarization fluctuations on the emission characteristics of VCSELs. In this rate equation model [2] we will base our discussion

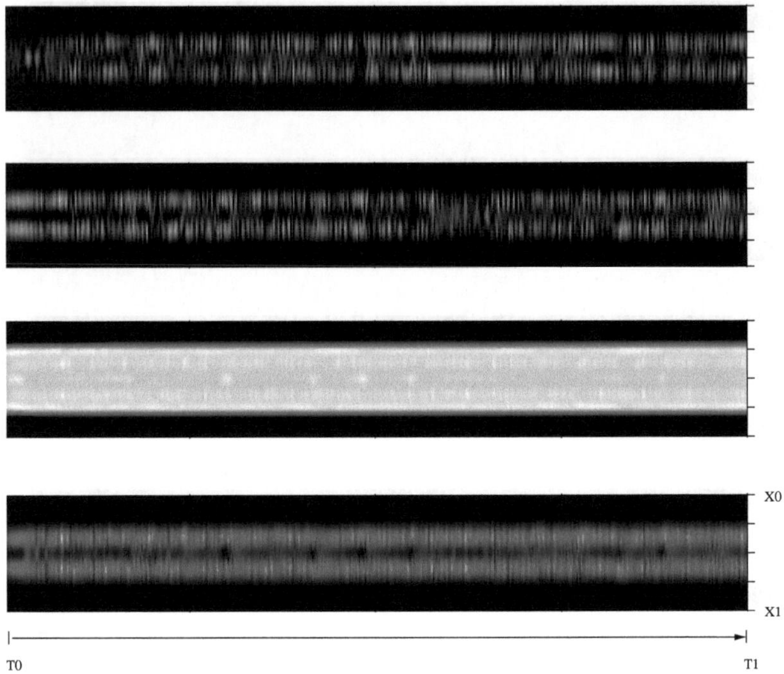

Fig. 5.11. Post-relaxation oscillation dynamics of a VCSEL with 11 μm aperture during a time-window of 2.5 ns. The figures show (from top to bottom) $|E_x|^2$, $|E_y|^2$, N and T_{p}. The coupled spatio-temporal light field and temperature dynamics are explicitly taken into account

on an approximation of the optical band-to-band transitions with transitions occurring between discrete energy levels. Further, we will assume that the carrier system is split (according to angular momentum) into separate density pools ('split-density model'). These simplifying assumptions are illustrated in Fig. 5.12. Photons with an angular momentum of $+1$ (right circular polarized light) may only be emitted by electron–hole pairs with a total angular momentum of $+1$ and photons with an angular momentum of -1 (left circular polarized light) may only be emitted by electron–hole pairs with a total angular momentum of -1. If only the bands closest to the fundamental gap are considered, electrons have an angular momentum of $\pm 1/2$ and holes have an angular momentum of $\pm 3/2$ along the axis of quantization perpendicular to the quantum wells. Consequently, electrons with an angular momentum of $+1/2$ may only recombine with holes of an angular momentum of $-3/2$, emitting left circular polarized light with an angular momentum of -1. Electrons with an angular momentum of $-1/2$ may only recombine with holes of an angular momentum of $+3/2$, emitting right circular polarized light with an angular momentum of $+1$. A schematic illustration of these transitions is

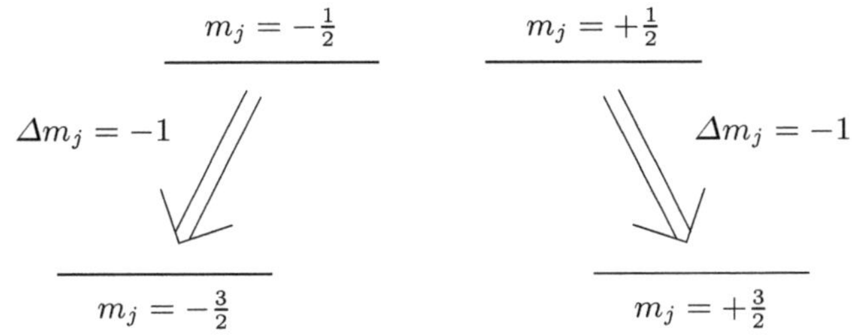

Fig. 5.12. Illustration of the angular momentum changes in the transitions between electronic states of the lowest energy conduction bands and electronic states of the highest energy heavy–hole valence bands at the center of the Brillouin zone ($\mathbf{k} = 0$)

shown in Fig. 5.12. Due to conservation of angular momentum, the carrier density is thus split into two carrier density pools. In terms of the carrier densities N_r and N_l coupled to the right (r) and left (l) circularly polarized light fields E_r and E_l, respectively, the equations read

$$\begin{aligned}
\frac{\mathrm{d}}{\mathrm{d}t}N_\mathrm{r} &= \frac{j}{2} - \left(\frac{1}{\tau} + \frac{1}{2}\Gamma_\mathrm{S}\right) N_\mathrm{r} + \frac{1}{2}\Gamma_\mathrm{S} N_\mathrm{l} \\
&\quad - \frac{\beta}{\tau_\mathrm{sp}}(N_\mathrm{r} - N_\mathrm{T}) \mid E_\mathrm{r} \mid^2 \\
\frac{\mathrm{d}}{\mathrm{d}t}N_\mathrm{l} &= \frac{j}{2} - \left(\frac{1}{\tau} + \frac{1}{2}\Gamma_\mathrm{S}\right) N_\mathrm{l} + \frac{1}{2}\Gamma_\mathrm{S} N_\mathrm{r} \\
&\quad - \frac{\beta}{\tau_\mathrm{sp}}(N_\mathrm{l} - N_\mathrm{T}) \mid E_\mathrm{l} \mid^2 \\
\frac{\mathrm{d}}{\mathrm{d}t}E_\mathrm{r} &= -(\kappa_0 + \mathrm{i}\omega_0)E_\mathrm{r} - \frac{1}{2}(s - \mathrm{i}\Omega)E_\mathrm{l} \\
&\quad + \frac{\beta}{2\tau_\mathrm{sp}}(1 - \mathrm{i}\alpha)(N_\mathrm{r} - N_\mathrm{T})E_\mathrm{r} \\
\frac{\mathrm{d}}{\mathrm{d}t}E_\mathrm{l} &= -(\kappa_0 + \mathrm{i}\omega_0)E_\mathrm{l} - \frac{1}{2}(s - \mathrm{i}\Omega)E_\mathrm{r} \\
&\quad + \frac{\beta}{2\tau_\mathrm{sp}}(1 - \mathrm{i}\alpha)(N_\mathrm{l} - N_\mathrm{T})E_\mathrm{l},
\end{aligned} \tag{5.29}$$

where β is the spontaneous emission factor describing the coupling between the gain medium and the laser mode, N_T ist the effective carrier density at dephasing, $\Gamma_\mathrm{S}|$ is the spin flip rate and α is the linewidth enhancement factor. Note that at the high carrier densities typically present in semiconductor lasers above laser threshold this spin flip scattering rate is significantly larger than e.g. the carrier diffusion rate. The parameter s describes the dichroism of the cavity mirrors and Ω represents the birefringence of the active medium inside the cavity. In practical VCSELs dichroism and birefringence are typically attributed to strain effects in the semiconductor medium. Both usually stabilize light emission with linear polarization along the [110] crystaline axes [5; 6]. Note that the sign of s and Ω may be both positive or negative.

5.2.2 Spin Dynamics

Since dichroism and birefringence stabilize one of the linear polarization directions, it is convenient to formulate the rate equations in terms of the amplitudes E_x and E_y of the corresponding orthogonal linear polarizations. The spin polarization parameter Δ describes the carrier density difference and is defined as $\Delta = N_\mathrm{r} - N_\mathrm{l}$; the total carrier density is given by $N_\mathrm{tot} = N_\mathrm{r} + N_\mathrm{l}$. The rate equations then read

$$\begin{aligned}
\frac{\mathrm{d}}{\mathrm{d}t} N_\mathrm{tot} &= j - \frac{N_\mathrm{tot}}{\tau} \\
&\quad - \frac{\beta}{2\tau_\mathrm{sp}} (N_\mathrm{tot} - 2N_\mathrm{T}) \left(| E_x |^2 + | E_y |^2 \right) \\
&\quad - \frac{\beta}{2\tau_\mathrm{sp}} \Delta \left(E_x^* E_y + E_y^* E_x \right) \\
\frac{\mathrm{d}}{\mathrm{d}t} \Delta &= - \left(\frac{1}{\tau} + \Gamma_\mathrm{S} \right) \Delta \\
&\quad - \frac{\beta}{2\tau_\mathrm{sp}} \Delta \left(| E_x |^2 + | E_y |^2 \right) \\
&\quad - \frac{\beta}{2\tau_\mathrm{sp}} (N_\mathrm{tot} - 2N_\mathrm{T}) \left(E_x^* E_y + E_y^* E_x \right) \\
\frac{\mathrm{d}}{\mathrm{d}t} E_x &= - \left(\kappa_0 + \frac{s}{2} + \mathrm{i}\omega_0 - \mathrm{i}\frac{\Omega}{2} \right) E_x \\
&\quad + \frac{\beta}{4\tau_\mathrm{sp}} (1 - \mathrm{i}\alpha) (N_\mathrm{tot} - 2N_\mathrm{T}) E_x \\
&\quad + \frac{\beta}{4\tau_\mathrm{sp}} (1 - \mathrm{i}\alpha) \Delta E_y \\
\frac{\mathrm{d}}{\mathrm{d}t} E_y &= - \left(\kappa_0 - \frac{s}{2} + \mathrm{i}\omega_0 + \mathrm{i}\frac{\Omega}{2} \right) E_y \\
&\quad + \frac{\beta}{4\tau_\mathrm{sp}} (1 - \mathrm{i}\alpha) (N_\mathrm{tot} - 2N_\mathrm{T}) E_y \\
&\quad + \frac{\beta}{4\tau_\mathrm{sp}} (1 - \mathrm{i}\alpha) \Delta E_x .
\end{aligned} \tag{5.30}$$

Experiments investigating the polarization dynamics and stability of VCSELs suggest that Γ_S is the fastest relevant time scale [7; 6]. It is therefore reasonable to simplify the dynamics by adiabatically eliminating the spin polarization of the gain medium Δ using the approximation that Δ is instantaneously given by

$$\Delta \approx -\frac{\beta}{2\Gamma_{\mathrm{S}}\tau_{\mathrm{sp}}}(N_{\mathrm{tot}} - 2N_{\mathrm{T}})\left(E_x^* E_y + E_y^* E_x\right). \tag{5.31}$$

The rate equations are thus reduced to a single carrier density coupled to the two light field modes by a nonlinear interaction. Since the nonlinearity is a consequence of the hole burning in the angular momentum distribution of the carrier system, the nonlinear effects all depend on the interference term $E_x^* E_y + E_y^* E_x$ between the linear polarizations. The relative magnitude of the nonlinearities is a function of the ratio between the spontaneous emission rate into the laser modes and the spin relaxation. It is convenient to define a rate χ which characterizes the nonlinearity with respect to the cavity loss rate $\kappa_y = \kappa_0 - s/2$,

$$\chi = \frac{\beta \kappa_y}{2\Gamma_{\mathrm{S}}\tau_{\mathrm{sp}}}. \tag{5.32}$$

With this definition the rate equations of the nonlinear single density model derived by adiabatical elimination of Δ read

$$\begin{aligned}
\frac{\mathrm{d}}{\mathrm{d}t}N_{\mathrm{tot}} &= j - \frac{N_{\mathrm{tot}}}{\tau} \\
&\quad -\frac{\beta}{2\tau_{\mathrm{sp}}}(N_{\mathrm{tot}} - 2N_{\mathrm{T}})\left(| E_x |^2 + | E_y |^2\right) \\
&\quad +\frac{\chi\beta}{2\kappa_y\tau_{\mathrm{sp}}}(N_{\mathrm{tot}} - 2N_{\mathrm{T}})\left(E_x^* E_y + E_y^* E_x\right)^2 \\
\frac{\mathrm{d}}{\mathrm{d}t}E_x &= -(\kappa_0 + \frac{s}{2} + \mathrm{i}\omega_0 - \mathrm{i}\frac{\Omega}{2})E_x \\
&\quad +\frac{\beta}{4\tau_{\mathrm{sp}}}(1 - \mathrm{i}\alpha)(N_{\mathrm{tot}} - 2N_{\mathrm{T}})E_x \\
&\quad -\frac{\chi\beta}{4\kappa_y\tau_{\mathrm{sp}}}(1 - \mathrm{i}\alpha)(N_{\mathrm{tot}} - 2N_{\mathrm{T}})\left(E_x^* E_y + E_y^* E_x\right)E_y \\
\frac{\mathrm{d}}{\mathrm{d}t}E_y &= -(\kappa_0 - \frac{s}{2} + \mathrm{i}\omega_0 + \mathrm{i}\frac{\Omega}{2})E_y \\
&\quad +\frac{\beta}{4\tau_{\mathrm{sp}}}(1 - \mathrm{i}\alpha)(N_{\mathrm{tot}} - 2N_{\mathrm{T}})E_y \\
&\quad -\frac{\chi\beta}{4\kappa_y\tau_{\mathrm{sp}}}(1 - \mathrm{i}\alpha)(N_{\mathrm{tot}} - 2N_{\mathrm{T}})\left(E_x^* E_y + E_y^* E_x\right)E_x.
\end{aligned} \tag{5.33}$$

These rate equations provide a complete description of the nonlinear polarization dynamics of VCSELs.

5.2.3 Fluctuation Dynamics of the Non-Lasing Polarization

The stationary solution describing the case of stable linear polarization along the y axis is described by

$$
\begin{aligned}
N_{\text{tot}} &= 2N_{\text{T}} + \frac{2\tau_{\text{sp}}(2\kappa_0 - s)}{\beta} \\
E_x &= 0 \\
E_y &= \sqrt{I_0}\, \mathrm{e}^{-\mathrm{i}\bar{\omega}t},
\end{aligned}
\tag{5.34}
$$

where $\bar{\omega} = \omega_0 + \Omega/2 + \alpha\kappa_0 - \alpha s/2$ and I_0 is the total laser mode intensity given in terms of photon number inside the cavity. The linearized fluctuation dynamics of the non-lasing polarization E_x for this stationary solution is given by a one dimensional complex Langevin equation which reads

$$
\begin{aligned}
\frac{\mathrm{d}}{\mathrm{d}t} E_x = &-(s + \mathrm{i}\bar{\omega} - \mathrm{i}\Omega)E_x \\
&-\chi I_0(1 + \mathrm{i}\alpha)\left(\mathrm{e}^{-\mathrm{i}2\bar{\omega}t} E_x^* + E_x\right) + q(t).
\end{aligned}
\tag{5.35}
$$

The nonlinearity χ couples the fluctuations in the non-lasing polarization mode to the lasing mode oscillations. Although the carrier density dynamics has been eliminated, the fluctuations are still phase locked to the laser light. The description of this phase locking is best illustrated by defining

$$
E_x = (E_{\parallel} - \mathrm{i}E_{\perp})\mathrm{e}^{-\mathrm{i}\bar{\omega}t}
\tag{5.36}
$$

in analogy to the fluctuation dynamics of arrays. The two dimensional real Langevin equation then reads

$$
\frac{\mathrm{d}}{\mathrm{d}t}\begin{pmatrix} E_{\parallel} \\ E_{\perp} \end{pmatrix} = \begin{pmatrix} -s & -\Omega - \alpha\chi I_0 \\ \Omega & -s - \chi I_0 \end{pmatrix} + \begin{pmatrix} q_{\parallel} \\ q_{\perp} \end{pmatrix}.
\tag{5.37}
$$

The quantum noise terms $q_{\parallel}$ and $q_{\perp}$ are given by

$$
\begin{aligned}
\langle q_{\parallel}(t) q_{\parallel}(t+\tau)\rangle &= \eta\kappa_y\, \delta(\tau) \\
\langle q_{\perp}(t) q_{\perp}(t+\tau)\rangle &= \eta\kappa_y\, \delta(\tau),
\end{aligned}
\tag{5.38}
$$

where $\eta \geq 1$ is the excess noise factor due to incomplete inversion above threshold. There are only two types of solutions for this equation. If the nonlinearity is weaker than the birefringence, the dynamical matrix has complex eigenvalues and the spectrum shows a pair of sidebands to the laser line. If the nonlinear effects overdamp the oscillation, the non-lasing mode is phase locked to the laser line and the spectrum consists of two Lorentz lines at the laser frequency $\omega = 0$.

5.2.4 Amplified Spontaneous Emission Spectra of the Non-Lasing Polarization

The two functions of the light field fluctuations in the non-lasing mode may be determined by calculating the eigenvalues and the eigenvectors of the dynamical matrix $\boldsymbol{S}$ given by

$$\boldsymbol{S} = \begin{pmatrix} s & \Omega + \alpha\chi I_0 \\ -\Omega & s + \chi I_0 \end{pmatrix}$$

$$= \sum_i \lambda_i \mathbf{a}_i \otimes \mathbf{b}_i^\dagger . \tag{5.39}$$

If the pair of eigenvalues is complex the solution of the eigenvalue problem reads

$$\begin{aligned} \lambda_\pm &= s + \frac{\chi I_0}{2} \pm \mathrm{i}\omega_\mathrm{R} \\ \mathbf{a}_\pm^\dagger &= \frac{1}{\sqrt{2}}\left(1,\ \frac{\chi I_0 \mp \mathrm{i}\,2\omega_\mathrm{R}}{2\Omega}\right) \\ \mathbf{b}_\pm^\dagger &= \approx \frac{1}{\sqrt{2}}\left(1,\ \frac{\chi I_0 \mp \mathrm{i}\,2\omega_\mathrm{R}}{2\Omega + 2\alpha\chi I_0}\right), \end{aligned} \tag{5.40}$$

where the resonant frequency ω_R is given by

$$\omega_\mathrm{R} = \Omega\sqrt{1 + \frac{\alpha\chi I_0}{\Omega} - \left(\frac{\chi I_0}{2\Omega}\right)^2} . \tag{5.41}$$

The two time correlations of $E_\parallel$ and $E_\perp$ can be determined with this solution. The results read

$$\begin{aligned} \langle E_\parallel(t)E_\parallel(t+\tau))\rangle &= \frac{\eta\kappa_y}{2s+\chi I_0}\left(\frac{1+\frac{\alpha\chi I_0}{2\Omega}}{1+\frac{\alpha\chi I_0}{\Omega}}\right)\mathrm{e}^{-(s+\chi I_0/2)\tau}\cos(\omega_\mathrm{R}\tau) \\ \langle E_\parallel(t)E_\perp(t+\tau)\rangle &= \frac{\eta\kappa_y}{2s+\chi I_0}\left(\frac{1+\frac{\alpha\chi I_0}{2\Omega}}{1+\frac{\alpha\chi I_0}{\Omega}}\right)\mathrm{e}^{-(s+\chi I_0/2)\tau} \\ &\quad\times\left(\frac{\chi I_0}{2\Omega}\cos(\omega_\mathrm{R}\tau) - \frac{\omega_\mathrm{R}}{\Omega}\sin(\omega_\mathrm{R}\tau)\right) \\ \langle E_\perp(t)E_\parallel(t+\tau)\rangle &= \frac{\eta\kappa_y}{2s+\chi I_0}\left(\frac{1+\frac{\alpha\chi I_0}{2\Omega}}{1+\frac{\alpha\chi I_0}{\Omega}}\right)\mathrm{e}^{-(s+\chi I_0/2)\tau} \\ &\quad\times\left(\frac{\chi I_0}{2\Omega}\cos(\omega_\mathrm{R}\tau) + \frac{\omega_\mathrm{R}}{\Omega}\sin(\omega_\mathrm{R}\tau)\right) \\ \langle E_\perp(t)E_\perp(t+\tau)\rangle &= \frac{\eta\kappa_y}{2s+\chi I_0}\left(1+\frac{\alpha\chi I_0}{2\Omega}\right)\mathrm{e}^{-(s+\chi I_0/2)\tau}\cos(\omega_\mathrm{R}\tau). \end{aligned} \tag{5.42}$$

This result describes the fluctuation properties of the non-lasing polarization. In order to obtain the spectrum, the total complex two time correlation of the field amplitude E_x may be determined as

$$\begin{aligned} \langle E_x^*(t)E_x(t+\tau)\rangle &= \langle E_\parallel(t)E_\parallel(t+\tau)\rangle + \langle E_\perp(t)E_\perp(t+\tau)\rangle \\ &\quad + \mathrm{i}\langle E_\perp(t)E_\parallel(t+\tau)\rangle - \mathrm{i}\langle E_\parallel(t)E_\perp(t+\tau)\rangle \end{aligned} \tag{5.43}$$

$$= \frac{I_{\rm SB}}{2}\left(1+\frac{2\omega_{\rm R}}{2\Omega+\alpha\chi I_0}\right)e^{-(s+\chi/2)\tau}e^{+i\omega_{\rm R}\tau}$$
$$+\frac{I_{\rm SB}}{2}\left(1-\frac{2\omega_{\rm R}}{2\Omega+\alpha\chi I_0}\right)e^{-(s+\chi/2)\tau}e^{-i\omega_{\rm R}\tau}, \qquad (5.44)$$

where $I_{\rm SB}$ is the total fluctuation intensity given by

$$I_{\rm SB} = \frac{2\eta\kappa_y}{2s+\chi I_0}\,\frac{(1+\frac{\alpha\chi I_0}{2\Omega})^2}{1+\frac{\alpha\chi I_0}{\Omega}}. \qquad (5.45)$$

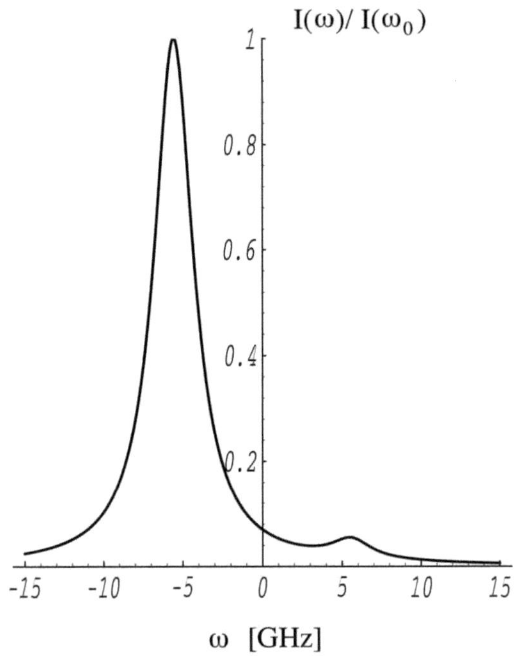

Fig. 5.13. Spectrum of the non-lasing polarization for $\alpha = 2$, $s = 0.5$ GHz, $\Omega = 4$ GHz and $\chi I_0 = 2$ GHz

Note that for $\chi = 0$ this intensity is equal to the intensity of amplified spontaneous emission expected at a gain value of $\kappa_x - s = \kappa_y$ for a spontaneous emission rate of $2\eta\kappa_y$. Weak nonlinearities modify this intensity only slightly. The spectrum corresponding to the two time correlations given in (5.44) reads

$$I_x(\omega) = \frac{I_{\rm SB}}{2\pi}\left(1+\frac{2\omega_{\rm R}}{2\Omega+\alpha\chi I_0}\right)\frac{s+\chi I_0/2}{(s+\chi I_0/2)^2+(\omega+\omega_{\rm R})^2}$$
$$+\frac{I_{\rm SB}}{2\pi}\left(1-\frac{2\omega_{\rm R}}{2\Omega+\alpha\chi I_0}\right)\frac{s+\chi I_0/2}{(s+\chi I_0/2)^2+(\omega-\omega_{\rm R})^2}. \qquad (5.46)$$

A typical spectrum is shown in Fig. 5.13. In this example the frequency of the lines has been shifted from $\Omega = 4$ GHz to a frequency of about 6 GHz, yet the mirror image peak above the lasing mode frequency is still very weak.

For weak nonlinearities only the lowest order modifications of the amplified spontaneous emission spectrum caused by the nonlinearity χI_0 are relevant. The approximated spectrum reads

$$\begin{aligned} I_x(\omega) \approx\ & \frac{\eta\kappa_y}{2\pi s}\left(1-\frac{\chi I_0}{2s}\right)\frac{s+\chi I_0/2}{(s+\chi I_0/2)^2+(\omega+\Omega+\alpha\chi I_0/2)^2} \\ & + \frac{\eta\kappa_y}{2\pi s}\left(1+\alpha^2\right)\left(\frac{\chi I_0}{4\Omega}\right)^2\frac{s+\chi I_0/2}{(s+\chi I_0/2)^2+(\omega-\Omega-\alpha\chi I_0/2)^2}. \end{aligned} \tag{5.47}$$

The nonlinearity shifts the frequency by $\alpha\chi I_0/2$, increases the linewidth (full width half maximum) by χI_0 and reduces the total intensity of the low frequency peak by a fraction of $\chi I_0/2s$. At $\chi I_0 = 0$ there is only a low intensity peak corresponding to the cavity resonance of the non-lasing polarization. However, the nonlinearity causes the appearance of a high-frequency sideband at a frequency of ω_R above the laser frequency $\bar{\omega}$. The intensity ratio between the high-frequency sideband and the main peak at low frequency for $\chi I_0 \ll \Omega$ is approximately given by

$$\frac{I_x(+\omega_R)}{I_x(-\omega_R)} = \left(1+\alpha^2\right)\left(\frac{\chi I_0}{4\Omega}\right)^2. \tag{5.48}$$

These characteristics of the non-lasing polarization spectrum of quantum well VCSELs may provide an experimental test of the split-density model [7; 8; 9]. Indeed, the prediction of a triplet spectrum [9] has experimentally been confirmed [10].

5.2.5 Fluctuations of the Stokes Parameters

The major qualitative effect caused by the nonlinearity is the appearance of the high-frequency sideband in the spectrum. However, the intensity of this sideband is only quadratic in χI_0. A promising method of observing the nonlinearity predicted by the split-density model may be the measurement of two time correlations in the intensity fluctuations of the Stokes parameters [9].

The normalized Stokes parameters are defined as

$$P_1 = \frac{E_x^* E_x - E_y^* E_y}{E_x^* E_x + E_y^* E_y} \tag{5.49}$$

$$P_2 = \frac{\mathrm{i}E_x^* E_y - \mathrm{i}E_y^* E_x}{E_x^* E_x + E_y^* E_y} \tag{5.50}$$

$$P_3 = \frac{E_x^* E_y + E_y^* E_x}{E_x^* E_x + E_y^* E_y} \tag{5.51}$$

and represent intensity differences between orthogonally polarized light field modes normalized by the total intensity. In particular, P_3 is equal to the normalized intensity difference between right circular polarized light and left circular polarized light. P_3 therefore represents the ellipticity of the light field. P_2 is a measure of the normalized intensity difference between the linear polarizations at angles of $\pi/4$ to the x and y axes. P_2 therefore represents the polarization direction. If $E_x = 0$, P_1 is equal to -1 while P_2 and P_3 are equal to zero. Small fluctuations in the Stokes parameters P_2 and P_3 are approximately proportional to the quadrature components $E_\parallel$ and $E_\perp$ of the non-lasing polarization. The relations between the linearized field fluctuations and the Stokes parameters read

$$P_2 \approx \frac{2}{\sqrt{I_0}} E_\perp$$
$$P_3 \approx \frac{2}{\sqrt{I_0}} E_\parallel . \tag{5.52}$$

The two quadrature components of the non-lasing mode thus describe fluctuations in the ellipticity and in the polarization direction of the light field. It should be possible to experimentally observe the two time correlation functions given by (5.42) by streak camera measurements of the correlated intensity fluctuations in the circular polarizations and in the linear polarizations at angles of $\pm\pi/4$ to the stable polarization axis. Alternatively, the power spectra of these intensity fluctuations may be obtained using fast photodiodes. Since the dichroism s and the birefringence Ω are in the regime of $10^9 s^{-1}$, a time resolution of about 100 ps should be sufficient to resolve the dynamics.

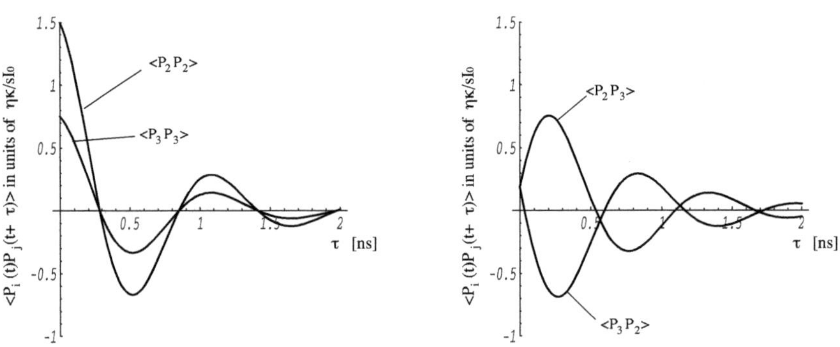

Fig. 5.14. Two time correlations of the polarization fluctuations $\langle P_i(t)P_j(t+\tau)\rangle$ for $\alpha = 2$, $s = 0.5$ GHz, $\Omega = 4$ GHz and $\chi I_0 = 2$ GHz. The *left figure* shows the fluctuations in polarization direction $\langle P_2(t)P_2(t+\tau)\rangle$ and the fluctuations in ellipticity $\langle P_3(t)P_3(t+\tau)\rangle$. The *right figure* shows the correlations $\langle P_2(t)P_3(t+\tau)\rangle$ and $\langle P_3(t)P_2(t+\tau)\rangle$ between polarization direction and ellipticity

On the basis of our analysis, two important predictions for experiments may be obtained. The first prediction is that the fluctuations in polarization direction are stronger than the fluctuations in ellipticity. The ratio is given by

$$\frac{\langle P_2(t)P_2(t+\tau)\rangle}{\langle P_3(t)P_3(t+\tau)\rangle} = 1 + \frac{\alpha}{\Omega}\chi I_0. \tag{5.53}$$

The second prediction is that the fluctuations of ellipticity and of polarization direction have a component oscillating in phase as seen in the correlation of ellipticity and polarization direction at $\tau = 0$,

$$\langle P_2(t)P_3(t)\rangle = \frac{2\eta\kappa_y}{sI_0}\frac{\alpha}{2\Omega}\chi I_0. \tag{5.54}$$

Both effects are a direct consequence of the hole burning dynamics described by the split-density model. The linewidth enhancement factor α indicates that the physical reason for the asymmetry in the relaxation dynamics of split-density fluctuations in ellipticity and fluctuations in the polarization direction is caused by the birefringence induced by the carrier density difference Δ associated with a non-zero ellipticity P_3. This birefringence rotates the polarization direction. Therefore, fluctuations in P_3 cause correlated fluctuations in P_2. By investigating the Stokes parameter dynamics, the physical processes described by the split-density model are thus revealed in much greater detail than by investigation of the total frequency spectrum.

Figure 5.14 shows the two time correlations of the Stokes parameters which correspond to the spectrum shown in Fig. 5.13. Note that the fluctuations in polarization direction are twice as strong as the fluctuations in ellipticity. It should therefore be much easier to measure this difference than it is to observe the low intensity sideband on the high-frequency side of the laser line shown in Fig. 5.13.

5.2.6 Spectra at the Spin Flip Transition

One of the interesting features of the polarization dynamics of VCSELs which initially motivated the investigation of nonlinearities in this context is the polarization flip observed in some devices as the intensity is increased. This polarization flip has been interpreted as a consequence of the nonlinear modification of stability due to the α factor by Martin-Regalado and coworkers soon after the introduction of the split-density model for VCSEL polarization [11]. According to this interpretation, the spin flip is caused by the competition between the linear stabilization provided by the dichroism s and the nonlinear stabilization $\alpha\Omega$. The conditions for stability can be derived from the fluctuation dynamics of (5.37). For $s > 0$ and $\Omega < 0$ the stability condition is

$$\chi I_0 < \frac{\Omega^2 + s^2}{\mid \alpha\Omega \mid - s}. \tag{5.55}$$

For $s < 0$ and $\Omega < 0$ the stability condition is

$$\chi I_0 > \mid 2s \mid . \tag{5.56}$$

If the stable polarization switches the signs of s and Ω must be exchanged. Therefore the two stability conditions correspond to the stability conditions of x-polarization and of y-polarization, respectively. The model thus predicts bistability for

$$\mid 2s \mid < \chi I_0 < \frac{s^2 + \Omega^2}{\mid \alpha\Omega \mid - s} . \tag{5.57}$$

This bistability is one of the fundamental predictions of the split-density model. It indicates a hysteresis loop at the polarization switch defined only by the optical parameters α, Ω and s. Another significant prediction can be obtained for the frequency of the non-lasing fluctuations. The frequency ω_{R} for $\Omega < 0$ is given by

$$\omega_{\mathrm{R}}(\Omega < 0) = \Omega \sqrt{1 - 2\alpha \left| \frac{\chi I_0}{2\Omega} \right| - \left| \frac{\chi I_0}{2\Omega} \right|^2} . \tag{5.58}$$

This drop in the non-lasing mode frequency should be visible in the non-lasing mode spectrum. The frequency shift of the non-lasing mode should move the peak closer to the lasing mode peak until the two frequencies are locked. Switching is only possible after this frequency locking takes effect. If the frequency of the non-lasing mode peak is still shifted with respect to the laser peak just below the switching intensity, other mechanisms must be responsible for the switch.

The spectrum for intensities below the intensity at which frequency locking occurs is given by (5.46). Note that the frequency Ω is negative. However, it is possible to characterize the polarization switch more thoroughly if the dichroism s is much smaller than the nonlinearity χI_0. Since polarization switching requires nonlinearities χI_0 of the same order of magnitude as Ω, the assumption of negligible dichroism is valid if the dichroism is much weaker than the birefringence, i.e. if $s \ll \Omega$. However, this approximation does not hold close to zero intensity where polarization stability is exclusively caused by dichroism. For $s \approx 0$, the polarization switch occurs at $\alpha\chi I_0 = \Omega$. By defining a normalized intensity

$$\mu := \frac{\alpha\chi I_0}{\mid \Omega \mid} \tag{5.59}$$

the spectrum of the non-lasing mode can be written as

$$\begin{aligned} I_x(\omega) = \; & \frac{\alpha\eta\kappa_y}{\pi\Omega} \frac{(1-\mu/2)^2}{\mu - \mu^2} \left(1 + \frac{\omega_{\mathrm{R}}}{\Omega(1+\mu/2)} \right) \frac{\mid \mu\Omega/2\alpha \mid}{(\mu\Omega/2\alpha)^2 + (\omega + \omega_{\mathrm{R}})^2} \\ & + \frac{\alpha\eta\kappa_y}{\pi\Omega} \frac{(1-\mu/2)^2}{\mu - \mu^2} \left(1 - \frac{\omega_{\mathrm{R}}}{\Omega(1+\mu/2)} \right) \frac{\mid \mu\Omega/2\alpha \mid}{(\mu\Omega/2\alpha)^2 + (\omega - \omega_{\mathrm{R}})^2} . \end{aligned} \tag{5.60}$$

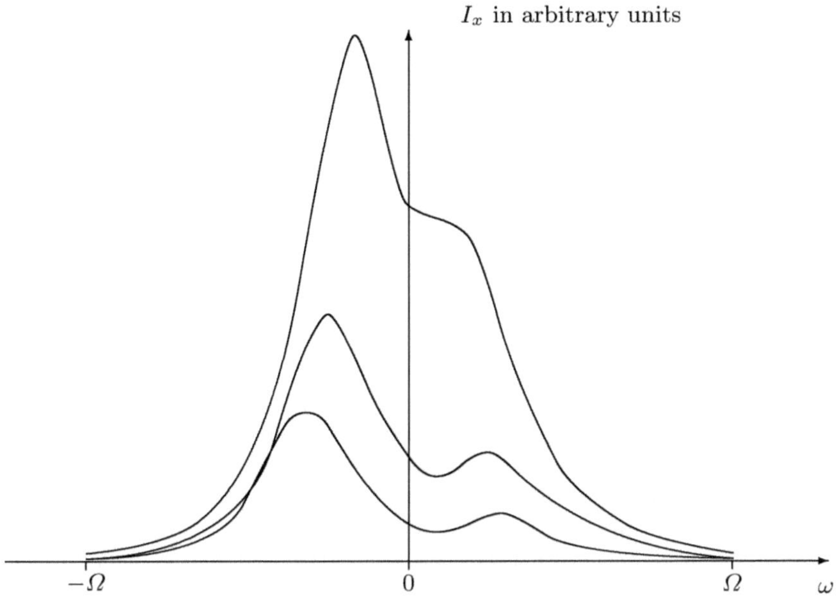

Fig. 5.15. Spectra of the non-lasing mode as the switching point is approached from the low intensity side. The only relevant device parameter is the linewidth enhancement factor α, which is equal to six for the spectra shown here. The scaled laser intensities for the spectra shown are $\mu = 8/9$ (*lower curve*), $\mu = 15/16$ (*intermediate curve*) and $\mu = 35/36$ (*upper curve*)

The frequency ω_{R} is given by

$$\omega_{\mathrm{R}} = \Omega\sqrt{1-\mu-\left(\frac{\mu}{2\alpha}\right)^2}. \tag{5.61}$$

For large values of α, frequency locking ($\omega_{\mathrm{R}} = 0$) occurs close to the switching point at $\mu = 1$. For typical linewidth enhancement factors α of three or higher, it is reasonable to approximate ω_{R} by

$$\omega_{\mathrm{R}} \approx \Omega\sqrt{1-\mu}. \tag{5.62}$$

The polarization switch then occurs as soon as the frequency of the non-lasing mode is locked to the lasing mode frequency. The spectrum close to the switching point may be approximated by considering only lowest order terms in $1-\mu$. The spectrum then reads

$$\begin{aligned} I_x(\omega) = \; & \frac{\alpha\eta\kappa_y}{4\pi\Omega}\left(\frac{1+2\sqrt{1-\mu}}{1-\mu}\right)\frac{|\,\Omega/2\alpha\,|}{(\Omega/2\alpha)^2+(\omega+\Omega\sqrt{1-\mu})^2} \\ & + \frac{\alpha\eta\kappa_y}{\pi\Omega}\left(\frac{1-2\sqrt{1-\mu}}{1-\mu}\right)\frac{|\,\Omega/2\alpha\,|}{(\Omega/2\alpha)^2+(\omega-\Omega\sqrt{1-\mu})^2}. \end{aligned} \tag{5.63}$$

The evolution of this spectrum close to the polarization switch at $\mu = 1$ for $\alpha = 6$ is shown in Fig. 5.15. Note that the linewidth relative to the frequency scale is invariably given by Ω/α. The low frequency line and the high-frequency line can thus be resolved up to a laser intensity of $(1 - \mu) \approx \alpha^{-2}$. Closer to the spin flip transition the sidebands appear to merge into a single line which rises and narrows until the intensity of the non-lasing polarization becomes comparable to the lasing mode intensity.

5.3 Conclusion

In this chapter we have discussed the polarization dynamics and fluctuations in vertical-cavity surface-emitting semiconductor lasers (VCSELs). In the first part we have introduced a four band Maxwell–Bloch model in which the momentum (spin) resolved carrier dynamics is described on the basis of a rate equation approximation. The detailed scattering rates for quantum wells as well as the gain characteristics based on them have been calculated. The microscopic properties have been self-consistently considered with effects and fields on the macroscopic level. In particular, they are the spatio-temporal dynamics of the optical fields, the carrier density, the energy densities and the temperatures. Results of numerical simulation based on this model demonstrate that static temperature profiles, especially for the plasma temperatures, do not include the full spatio-temporal polarization dynamics in common VCSEL structures. We note that even with careful approximation of the model equations it is only possible to consider steady-state properties of VCSELs. Their dynamic features (e.g. in high-frequency modulation) require a spatio-temporal treatment.

Bearing in mind these limits, in the second part of the chapter we have discussed the polarization fluctuations of VCSELs on the basis of a simplified split-density rate equation model. The polarization fluctuations show features typical of the split-density model, which can be used as test criteria for whether the model is valid in a given device or not. Since the split density represents the effect of quantum well confinement on the polarization properties of VCSELs, this effectively tests the quantum well structure in the active region. The quantity that depends strongly on the size of the quantum wells is the spin relaxation rate. Experimental results on this rate are also of interest in connection with calculations [12; 13] and luminesence experiments [14] carried out to investigate spin flip scattering in quantum wells.

The rate equations presented here are formulated in a very general way and may also be applied to cases with more exotic anisotropies and time scales such as discussed in [11], which show switching and/or include a magnetic field. In all these cases, the investigation of noise adds additional predictions for experiment to the results and thereby increases our understanding of the physics involved in the polarization properties of VCSELs.

References

1. E. O. Kane. *Handbook on Semiconductors I*, chapter KP Method. North Holland, Amsterdam, 1982.
2. M. San Miguel, Q. Feng, and J.V. Moloney. Light-polarization dynamics in surface-emitting semiconductor lasers. *Phys. Rev. A*, 52:1728–1739, 1995.
3. D. Burak, J. V. Moloney, and R. Binder. Microscopic theory of polarization properties of optically anisotropic vertical-cavity surface-emitting lasers. *Phys. Rev. A*, 61:053809, 2000.
4. O. Hess and T. Kuhn. Maxwell-Bloch equations for spatially inhomogeneous semiconductor lasers I: Theoretical description. *Phys. Rev. A*, 54:3347–3359, 1996.
5. W. W. Chow, S. W. Koch, and M. Sargent III. *Semiconductor-Laser Physics*. Springer-Verlag, Berlin, 1994.
6. A. K. J. van Doorn, M. P. Exter, and J. P. Woerdman. Elasto-optic anisotropy and polarization orientation of vertical-cavity surface emitting lasers. *Appl. Phys. Lett.*, 69:1041–1043, 1996.
7. A. K. J. van Doorn, M.P. van Exter, A. M. van der Lee, and J.P. Woerdman. Coupled-mode description for the polarization state of a vertical-cavity semiconductor laser. *Phys. Rev. A*, 55:1473–1484, 1997.
8. H. van der Lem and D. Lenstra. Saturation-induced frequency shift in the noise spectrum of a birefringent vertical-cavity surface-emitting laser. *Opt. Lett.*, 22:1698–1700, 1997.
9. H. F. Hofmann and O. Hess. Spontaneous-emission spectrum of the nonlasing supermodes in semiconductor laser arrays. *Opt. Lett.*, 23:391–393, 1998.
10. S. F. Pereira, M. P. van Exter, and J. P.Woerdman. Triplet structure in the spontaneous-emission spectrum of two coupled vertical-cavity semiconductor lasers. *Phys. Rev. A*, 62:043811, 2000.
11. J. Martin-Regalado, M. S. Miguel, N. B. Abraham, and F. Prati. Polarization switching in quantum-well vertical-cavity surface-emitting lasers. *Opt. Lett.*, 21:351–353, 1996.
12. T. Uenoyama and L. J. Sham. Hole relaxation and luminescence polarization in doped and undoped quantum wells. *Phys. Rev. Lett.*, 64:3070–3073, 1990.
13. R. Ferreira and G. Bastard. 'spin'-flip scattering of holes in semiconductor quantum wells. *Phys. Rev. B*, 43:9687–9691, 1991.
14. A. Vinattieri, Jagdeep Shah, T. C. Damen, D. S. Kim, L. N. Pfeiffer, M. Z. Maialle, and L. J. Sham. Exciton dynamics in gaas quantum wells under resonant excitation. *Phys. Rev. B*, 50:10868–10879, 1994.

6 Nonlinear Pulse Propagation and Solitons in Active Semiconductor Media

Due to the dynamic interplay of light and matter involving a hierarchy of different time (fs ... ns) and length scales (nm ... µm) the semiconductor laser represents a highly complex media in which dynamic nonlinear phenomena may occur. The ultrashort time dynamics in the charge carrier plasma of a semiconductor laser shows a strong dependence on amplitude and spatio-spectral properties of the optical fields. Simulations of propagation and amplification of one or more light pulses thus are of particular interest for visualization and analysis of the physical interactions within the active medium.

Furthermore, the simulation of pulse propagation is of high technological relevance. Due to its large amplification bandwidth the semiconductor laser is particularily appropriate for the generation and amplification of ultrashort pulses with peak powers exceeding the constant output power of the respective continuous systems. They have far-reaching applications as light source for imaging systems in medicine and biology [1], high-speed data transfer[2], data storage [1], optical switching [3; 4], temporal resolved detection of physical processes [5] as well as nonlinear frequency conversion [6].

In recent years the impressive progress in technology and the large application bandwidth has lead to many experimental and theoretical investigations [2; 7; 8; 9; 10; 11; 12; 13; 14; 15; 16; 17; 18; 19; 20; 21; 22]. In most theoretical descriptions of pulse propagation, however, microscopic carrier and polarization dynamics are not considered and counterpropagation effects are neglected. Thereby both the spatial variations and the spectral dependence represented by the band structure of the medium are disregarded. Typically a transformation of the variables to the travelling frame of the propagating pulse is made thereby neglecting the dynamic interaction of the light field with the nonlinear matter. But it is, in particular, the microscopic spatio-temporal interaction of the counterpropagating light fields with the microscopic distribution of the electron–hole system and the nonlinear interband polarization which is responsible for the dynamic spatial and spectral shaping of a light pulse propagating in a semiconductor medium. The dynamic physical processes affecting the properties of a light pulse are e.g. spatio-spectral hole burning, carrier heating and thermalization leading to dynamic spatial and spectral gain saturation, induced index dispersion and self phase modulation. This is of particular importance for spatially extended inhomogeneous media, where severe nonlinearites and instabilities can emerge [23].

In this chapter we will discuss the spatio-temporal dynamics of an InGaAs broad-area semiconductor laser amplifier with injection of one or more ultrashort light pulses. We will base our simulations on the Maxwell–Bloch–Langevin equations presented and discussed in Chap. 3. Section 6.1 presents results of our numerical simulations of the propagation of a picosecond pulse in a broad-area semiconductor laser with particular focus on the spatio-temporal dynamics within the active layer. In Sect. 6.2 we will consider the simultaneous and oblique injection of two optical pulses in the charge carrier plasma of a broad-area semiconductor laser and analyze the nonlinear mechanisms responsible for the formation of a spatio-spectral soliton.

6.1 Pulse Propagation in Nonlinear Semiconductor Media

The dynamic interactions between light and matter have a strong influence on the physical properties of a light pulse propagating in the active area of a semiconductor laser. The reduction of the spatio-spectral inversion by induced recombination and the dynamics of the interband polarization representing spatio-spectral gain and index (see Chap. 3) lead to saturation, self-focusing and diffraction. Together they affect both the amplitude and phase of a propagating light field. In the following we will present results of our numerical modelling of the pulse propagation in semiconductor lasers. As an example we consider a light pulse with a duration of 7.5 ps propagating in the active area of a large-area semiconductor laser (length 2 mm, width 100 μm). We will visualize the dynamic pulse amplification and reshaping during the pulse propagation with snapshots of spatial distributions, e.g. intensity, carrier density, gain and refractive index, in the active area and discuss the underlying physical processes.

6.1.1 Dynamic Pulse Shaping

Figure 6.1 shows snapshots of the intensity distribution (left) and the spatial carrier density (right) during the propagation of an optical pulse with duration 7.5 ps (full width at half maximum, FWHM) in a semiconductor laser amplifier (length 2 mm, width 100 μm). The injection current density has been chosen so that the carrier density in the active area of the amplifier is significantly above the transparency value N_t ($N \approx 3.0\ N_t$). The time between successive plots is 4 ps. Figures 6.1(a) and (d) represent the spatial distributions of the intensity (a) and the carrier density (d) immediately after the injection of the pulse. The lateral spatial shape of the injected light field is a Gaussian distribution with a width (FWHM) of 100 μm. The temporal

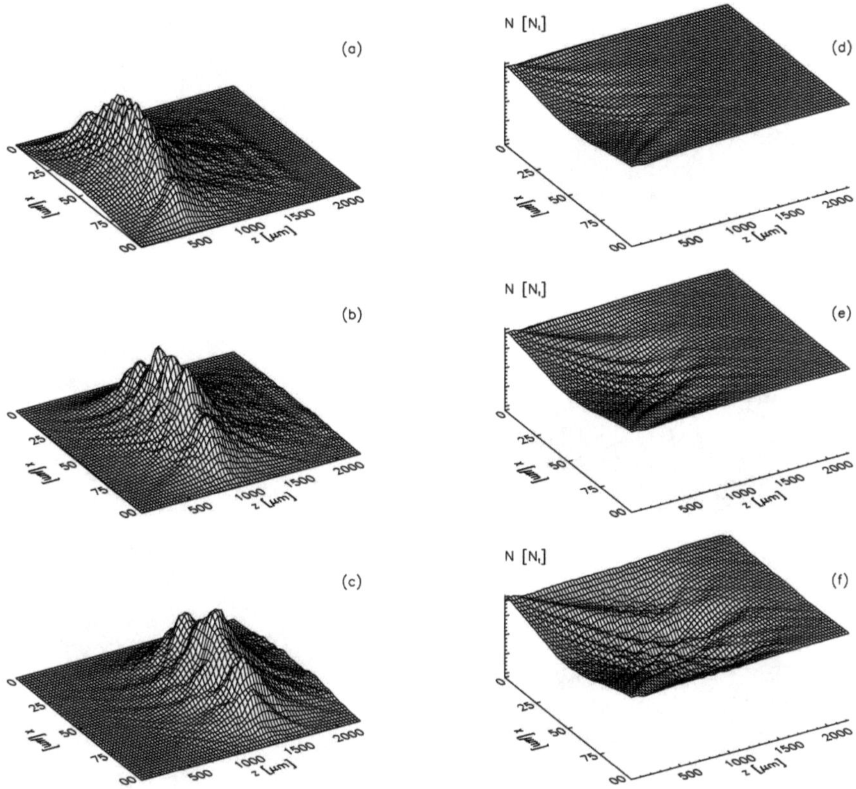

Fig. 6.1. Pulse propagation in the linear gain regime: snapshots of the intensity (*left*) and carrier density (*right*) in the active area of a broad-area laser amplifier 4 ps (**a**, **d**), 8 ps (**b**, **e**) and 12 ps (**c**, **f**) after optical injection of a light pulse with duration 7.5 ps (FWHM) and a total input power of $P_s = 0.02\ P_s$

shape of the pulse is a double-side exponential function in accordance with the output pulse shape observed in experiments [24]. Due to the low input power of $P_{in} = 0.02\ P_s$ (where P_s denotes the power required for a complete saturation of the gain) no saturation of the gain occurs (small-signal regime).

Immediately after injection the light pulse starts to interact with the nonlinear medium. As a consequence, its leading part is modulated whereas the trailing part still maintains its initial shape. The charge carrier plasma established by the injection current is significantly excited by the propagating pulse via spatio-spectral hole burning and carrier heating. Due to the finite "response time" of the carrier system given by the carrier relaxation times characteristic oscillations occur which lead to spatial and temporal modulations in the distributions of the intensity and carrier density. In particular, the modifications in the carrier density induced by the leading part of the pulse (i.e. the spatial reduction via induced recombination as well as thermal

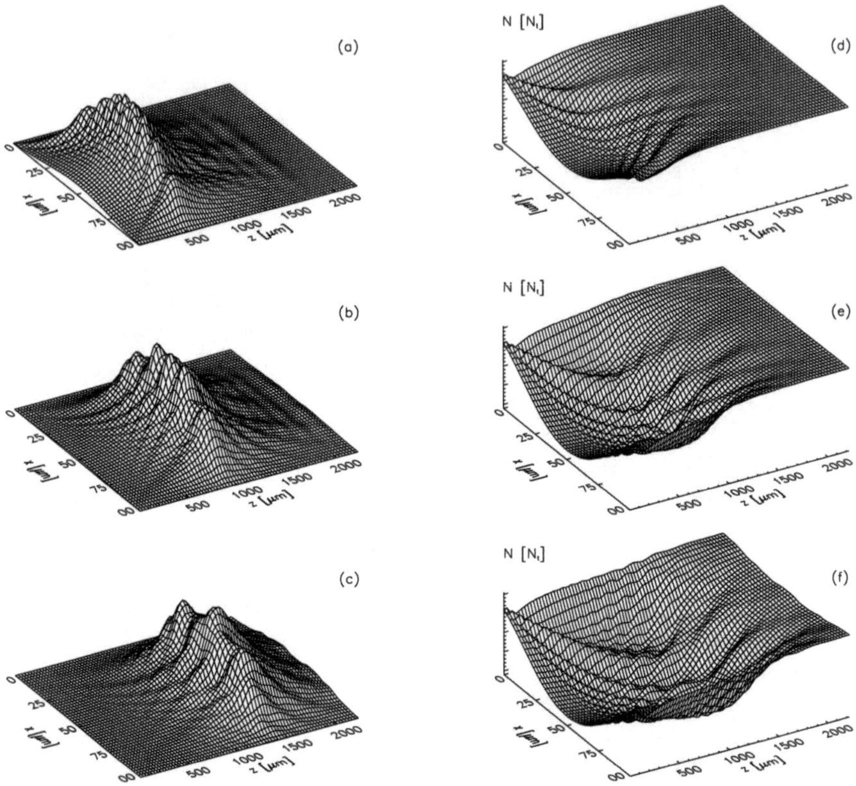

Fig. 6.2. Pulse propagation in the saturated gain regime: snapshots of the intensity (*left*) and the carrier density (*right*) in the active area of a broad-area laser amplifier 4 ps (**a**, **d**), 8 ps (**b**, **e**) and 12 ps (**c**, **f**) after optical injection of a pulse with duration 7.5 ps (full width at half maximum) and input power $P_{\text{in}} = 2.0\ P_{\text{s}}$

excitation of carriers in the bands) affect the amplification of the trailing part of the pulse as well as the temporal behavior of the amplified spontaneous emission (ASE) arising in the active area after the passing of the pulse. The trailing part of the pulse thereby increases the amplitude of an already existing excitation of the carriers whereas the leading part of the pulse experiences the unaffected carrier system. As a consequence the magnitude of the induced changes per time step is larger during the interaction of the charge carrier system with the leading part of the pulse while the trailing part of the pulse in general shows less intensity modulation. In addition to these propagation effects, Fig. 6.1 shows the interplay of transverse effects such as carrier diffusion and light diffraction. In particular, the formation of characteristic spatio-temporal modulations in the leading part of the pulse can be transfered to the trailing part of the pulse and to the spontaneous emission arising behind the pulse (Fig. 6.1(a–c)) via the light propagation and the nonlinear coupling between light and carrier system. Consequently structures and modulations also appear in the "back side" of the pulse (i.e. the part of the pulse facing the input facet). The optical patterns emerging in the intensity distributions are transfered to the respective carrier density distributions via induced recombination (Fig. 6.1(d–f)). Since the amplifier is driven in the small signal regime (i.e. low input power) the carrier injection via the pump rate rebuilds the gain on a picosecond time scale so that the spatial area in which reduction of inversion occurs more or less covers the area determined by the propagating pulse light field. Due to the different time scales of the dynamics of the carrier density and of the optical fields this spatial hole burning is slightly shifted compared to the intensity distributions of the same time instance (see e.g. Fig. 6.1(a) and (d)). Additionally, a "trench" in propagation direction can be seen in Fig. 6.1(f) that reflects the slow dynamics of the carrier density. Near the output the induced reflected wave leads to spatio-temporal interference of counterpropagating light pulses. Depending on the reflectivity of the back facet the part of the pulse reflected back into the active area can lead to multiple pulse formation [2].

If the input power is of the order of the saturating power, P_s, (gain saturation regime) efficient saturation of the amplifier is achieved for a given injection current (in the example $j = 2.0 \cdot 10^7$ A/m^2) and the characteristic spatial distributions change significantly. This situation is shown in Fig. 6.2. The strong reduction of the carrier density by the leading part of the pulse (Fig. 6.2(d)) yields a reduction of the gain available for the trailing part of the pulse. As a consequence, the amplitude of the leading part of the pulse is increased to a higher degree than the trailing part which shows a steep decrease. The initially symmetric pulse (Fig. 6.2(b)) is strongly distorted during propagation. Near the back facet of the laser the dynamic interaction between the forward propagating light field and the carrier distribution lead to additional counterpropagating effects even before the output facet is entirely reached by the pulse. Due to the strong saturation achieved by the high input power the carrier density keeps at a low value after the passing of the pulse (Fig. 6.2(f)). The re-establishment of the initial inversion via the injection current then occurs on long time scales of typically a few hundred ps (i.e. many resonator round trips). The gain and refractive index changes correlated to these density variations cause self focusing which in the lateral direction lead to a spatial distortion (Fig. 6.2(c)) i.e. the side parts of the pulse (i.e. near the lateral boundaries) stay behind the pulse center.

6.1.2 Spatial Amplification and Wave-Guiding

The dynamic coupling of the light field of the pulse and the carrier density occurs via the term in the nonlinear polarization that depends on the microscopic carrier distributions of the charge carriers. The induced gain (i.e. the inverse generation rate of the diffusion equation (2.26)) and the induced refractive index dynamics are displayed in Fig. 6.3 for the same operation

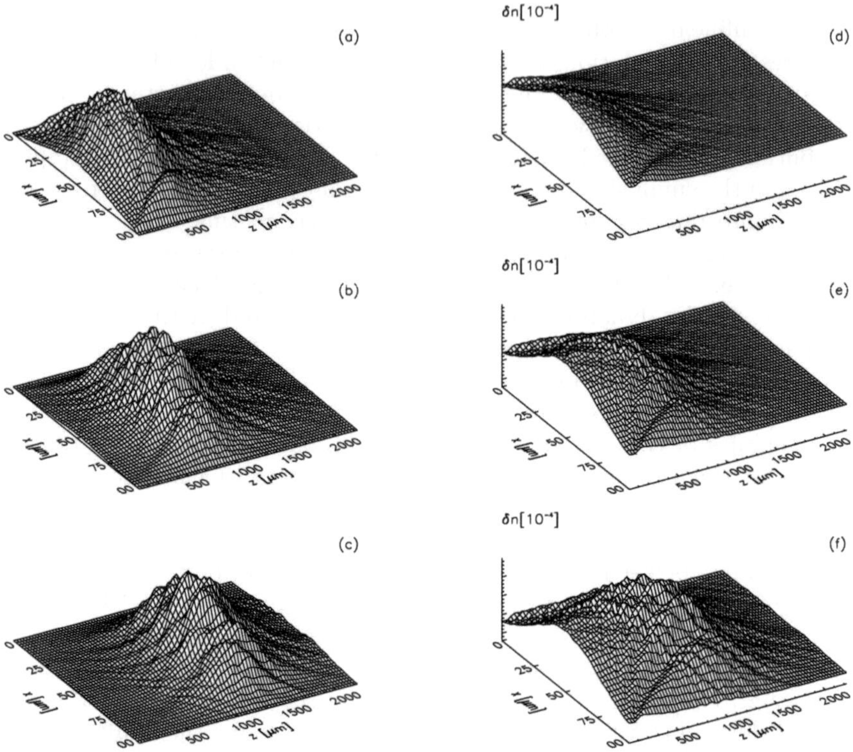

Fig. 6.3. Pulse propagation in the linear gain regime: snapshots of the gain (*left*) and the induced refractive index (*right*) in the active area of a broad-area laser amplifier 4 ps (**a**, **d**), 8 ps (**b**, **e**) and 12 ps (**c**, **f**) after optical injection of a light pulse with duration 7.5 ps (FWHM) and a total input power of $P_{\text{in}} = 0.02\ P_{\text{s}}$

conditions ($N \approx 3.0\ N_{\text{t}}$, power of the injected pulse 0.02 P_{s}, i.e. small signal regime) and the same time instances as in Fig. 6.1. Due to the negligible saturation effects the spatial gain distribution (Fig. 6.3(a–c)) is mostly determined by the intensity of the pulse which "shapes" the gain experienced by the pulse (see e.g. Fig 6.3(a) and 6.1(a)). At every time instance a sufficient inversion is available for the amplification process so that the temporal pulse shape (Fig. 6.1(a–c)) is maintained. The induced refractive index profile (Fig. 6.3(d–f)) depends on the distribution of electrons and holes [25] and consequently shows a qualitatively similar behavior as the carrier density (Fig. 6.1(d–f)). The spatially dependent reduction of the carrier density by the light pulse leads to an increase in the induced refractive index. Due to the slow spatial dynamics of the carrier density (ps... ns) (Fig. 6.3(d–f)) the variations in the index distribution are temporally delayed and "smeared-out" compared to the intensity distribution (Fig. 6.1(a–c)). In the situation displayed in Fig. 6.3 the power of the injected light pulse and consequently

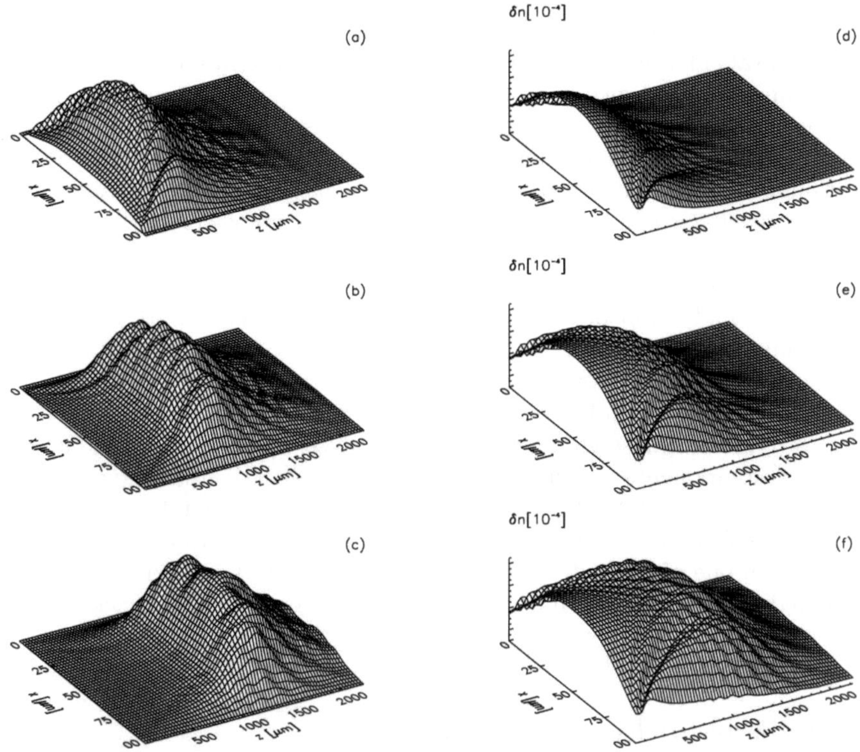

Fig. 6.4. Pulse propagation in the saturated gain regime: snapshots of the gain (*left*) and the induced refractive index (*right*) in the active area of a broad-area laser amplifier 4 ps (**a**, **d**), 8 ps (**b**, **e**) and 12 ps (**c**, **f**) after optical injection of a pulse with duration 7.5 ps (full width at half maximum) and input power $P_{\text{in}} = 2.0\ P_{\text{s}}$

the amount of induced refractive index change is low so that no significant phase and wavelength changes occur during the pulse propagation. As a consequence also the spectral properties of the pulse before and after passing the amplifier are nearly identical.

If we look at the temporal properties of the spatial gain and index formation during the propagation of an intense light pulse we can observe a very different behavior. Figure 6.4 shows the spatial distributions of the gain (Fig 6.4(a–c)) and the induced refractive index (Fig. 6.4(d–f)) for the situation of Fig. 6.2, i.e. a laser amplifier operated in the gain saturation regime. The spatial distribution of the gain (Fig 6.4(a–c)) is significantly modified by the propagating light pulse. The carrier density is preferentially reduced in the lateral center (due to the laterally Gaussian shaped injected pulse). As a result the gain “re-shapes” and shifts with increasing propagation towards the borders (table-top shape)(Fig 6.4(c)) which can lead to a spatial enlargement of the injected light pulse. Additionally, the saturation of the gain by the intense light pulse leads with increasing propagation to a longitudinal and temporal enlargement of the gain compared to the intensity distribution of the same time instance (Fig. 6.2(a)).

The spatially varying refractive index profile which is induced at the same time (Fig. 6.4(d–f)) is responsible for dynamic self focusing and thus an additional influence on beam propagation. In particular, the different time scales of the changes in the carrier density and the intensity distribution lead to a temporal delay between the changes in gain and refractive index. This leads to nonlinear phase changes that are correlated with dynamic frequency changes. For high input power levels and high injection current the increase in the dynamic refractive index may thus lead to a shift of the central wavelength of the picosecond pulse towards lower energies (i.e. longer wavelengths) [26]. Furthermore, the dependence of the refractive index on the carrier density leads to dynamic phase changes within the pulse envelope. As a result new frequency components are created resulting in spectral broadening and self phase modulation [14; 22; 27; 28]. The spatial gain and the induced refractive index establish a strong coupling between intensity and carrier density so that variations in one of these distributions are (with spatial and temporal delay) transfered to the other distributions. The result is a positive feedback between light and matter.

6.2 Solitons in Partially Coherent Active Semiconductor Media

The charge carrier plasma in an active broad-area semiconductor waveguide is a highly nonlinear medium in which the dynamic light–matter interaction occurs on time scales ranging from the femtosecond to the nanosecond regime [29]. Due to the simultaneous incoherent spatial transport of carriers and coherent dynamics of interband polarization as well as the counterpropagation of light the internal processes linked to amplification, absorption and relaxation occur on both spatial and temporal scales. The discussion of the previous section has demonstrated that the injection of an ultrashort optical pulse into this partially coherent semiconductor leads to considerable modification of the pulse properties during its passage [28]. This strong modification in shape and duration can be seen directly in Fig. 6.5. The characteristic sequence of snapshots of the optical intensity directly visualize the strong distortion a typical picosecond pulse (duration 2 ps) has to suffer within active semiconductor media.

On the other hand, recently a first demonstration has been given that the simultaneous and oblique injection of two optical pulses in the charge carrier plasma of a broad-area semiconductor laser may lead to a spatio-spectral wave mixing and self-trapping process [30]. Immediately this leads us to ask if it is possible to use the nonlinear light–matter interactions for general light stabilization? Is it possible to tailor the pulses such that even this vividly

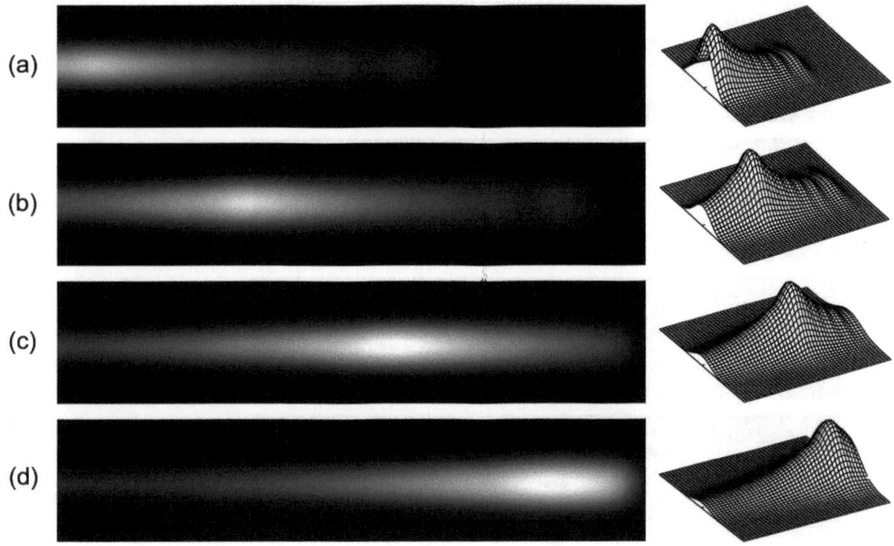

Fig. 6.5. Grey-scale (*left*) and corresponding surface (*right*) intensity plots of a picosecond pulse (duration 2 ps full width at half maximum) propagating in the active charge carrier plasma in a broad-area semiconductor laser

varying, partially coherent active semiconductor environment may support solitons similar to the Kerr-like nonlinear media [31; 32]? What would be the intrinsic nonlinear mechanisms responsible for their formation?

Figure 6.6 demonstrates a successful soliton formation. Two light pulses (duration 2 ps full width at half maximum, transverse width 20 μm, total input power 150 mW, wavelength: $\lambda = 790$ nm) are injected into the active layer of a wide semiconductor waveguide structure (length 1 mm, width 100 μm, electrically pumped slightly above threshold, maximum of the gain curve: ≈ 820 nm) at an angle α of 18° with respect to the facet normal. The sequence of microscopically calculated intensity distributions directly visualizes the trapping process. With the oblique injection (Fig. 6.6 (a)) the geometrical point of intersection of the pulses would be located at $z \approx 400$ μm. The next time step (Fig. 6.6 (b)) demonstrates the beginning of their interference. Instead of the usual strong temporal distortion of a single, but otherwise identical pulse displayed in Fig. 6.5 the optical light pattern which has been formed by the self-induced molding of the two pulses is spatially and temporally stable (Fig. 6.6 (c)). From there on the pulse is robust with respect to spatial and spectral deformation and propagates towards the output facet (Fig. 6.6 (d)) instead of being separated into two pulses that follow the original geometrical light rays.

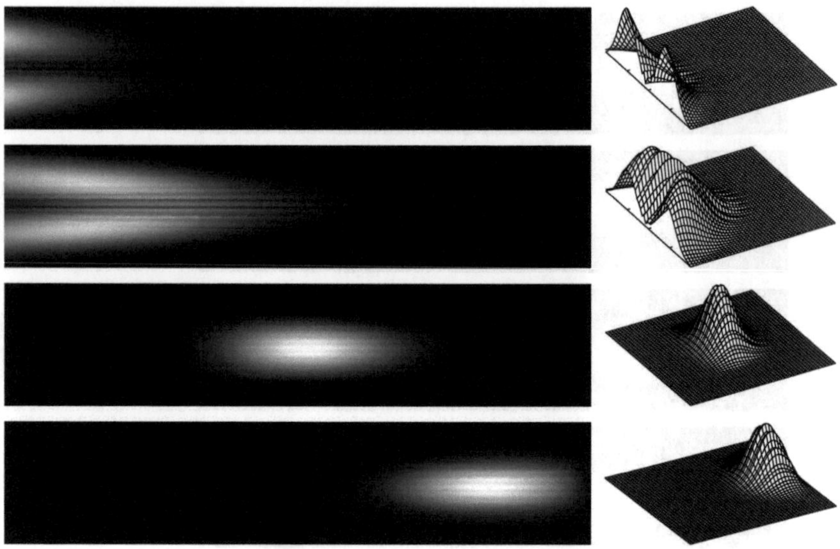

Fig. 6.6. Formation and propagation of a spatial optical soliton in the partially coherent active semiconductor media in a broad-area laser. The pumping conditions and pulse parameters are identical to Fig. 6.5

The investigation of optical solitons is usually performed on the basis of a Kerr-type model. These models approximately describe the internal nonlinear processes (e.g. the balance between diffraction and nonlinearity) and frequently allow a derivation of analytic expressions describing the solitons. In active semiconductor media, however, the nonlinearities and dispersion arise not only from a spatially varying gain and index distribution. Instead, they are governed by a multitude of (partially coherent and incoherent) dynamic intraband and interband processes within the charge carrier system with time scales ranging from the femtosecond to the nanosecond regime. We will thus base our analysis of the coupled optical and semiconductor spatio-spectral dynamics on the Maxwell–Bloch–Langevin approach ((3.1) in Chap. 3) taking into account the spatio-temporal light field dynamics, microscopic (carrier–carrier and carrier–phonon) scattering processes [25; 33], the interplay of spontaneous and stimulated emission as well as light and dipole fluctuations relevant for and in the soliton formation. Here we note that any further simplification in the description of the nonlinear active medium will exclude vital processes necessary in the soliton formation, as the following discussion shows.

In the following we will present results of our modelling of the spatio-spectral mixing and interference of light pulses in an inverted semiconductor laser medium. We will discuss the multistage dispersion compensation involv-

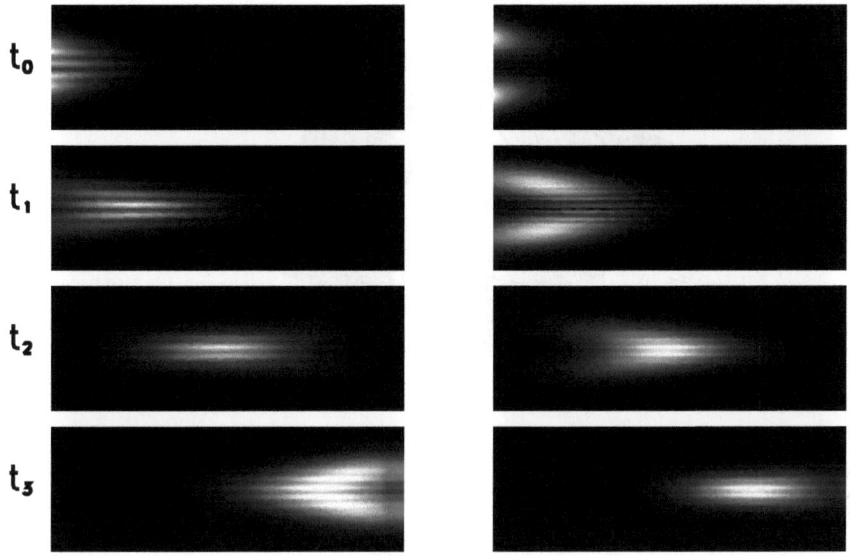

Fig. 6.7. Dependence of soliton formation on the pulse properties and propagation direction. *Left column*: $\alpha = 8°$, $\delta x = 16$ µm (peak-to-peak). *Right column*: $\alpha = 18°$, $\delta x = 30$ µm (peak-to-peak)

ing a combination of coherent and incoherent nonlinear physical processes that are altogether associated with and needed for soliton formation. We will see that as a characteristic feature the solitons show a pronounced anisotropy whose origin may be attributed to the combination of spatial effects (carrier transport, diffraction and self-focusing) and the microscopic dynamics of the nonlinear active semiconductor medium reflecting the spatio-spectral gain and refractive index dynamics.

6.2.1 Spatio-Temporal Refractive Index and Gain Dynamics

How strongly the successful formation of a soliton depends on the particular properties of the pulses may be anticipated when we vary the conditions of Fig. 6.6 slightly. In Fig. 6.7 two similar light pulses as in Fig. 6.6 have a transverse width of 15 µm. In the sequence shown in the left column the angle α is set to 8° and the two pulses are injected at a lateral distance (peak to peak) of $\delta x = 16$ µm. The right column shows a sequence with a somewhat steeper angle of injection ($\alpha = 18°$) and injection at $\delta x = 30$ µm. In the first case (left column) both pulses interfere strongly with each other from the start. However, in spite of a clear formation of interference fringes the interaction length is not long enough for pulse trapping and subsequent pulse coagulation. The snapshots of the right column on the other hand show

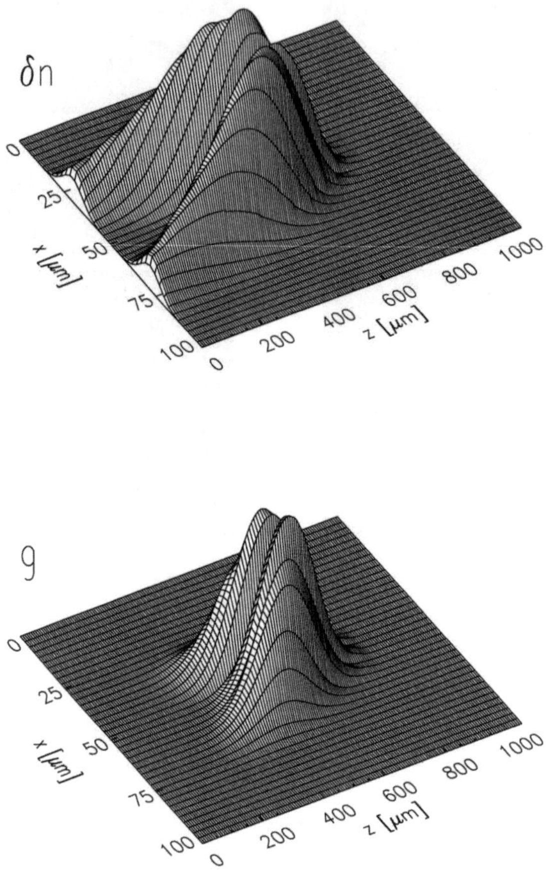

Fig. 6.8. Spatial distribution of the induced refractive index δn and gain g after the molding of the two light pulses in the broad-area semiconductor laser. Positive values denote high refractive index and high amplification

that for the particular conditions the light trapping has been successful and a self-induced stable pulse is being formed.

Generally, the formation of a soliton requires a spatial (and temporal) balancing of the dispersive effects by a form of nonlinearity. For a time instance after the interference of the light pulses in their geometrical point of intersection Fig. 6.8 (a) and (b) show the spatial distributions of the refractive index (real part of $p^{\pm}$) and the spatial gain (imaginary part of $p^{\pm}$), respectively. The off-resonant pulse injection (i.e. near the absorption regime of the semiconductor gain curve) restricts to a large degree the reduction of the carrier density to the spatial area of high intensity, i.e. the overlap region of the two pulses. Moreover, the spatial distribution of the refractive index is increased at locations of high intensity and low density. The dynamics of the carrier distributions is determined by fast incoherent scattering processes between electrons and holes in the valence and conduction bands of the semiconductor ($\approx$ 200 fs) and by slow variations ($\approx$ ps ... ns) originating from the carrier injection via the current, the (spontaneous and induced) recombination as well as carrier diffusion. The radiative recombination in the charge carrier plasma leads to coherent dynamics of the interband polarization (with a dephasing time of a few hundred fs) which via the wave equation is transferred to the propagating light fields. The typical time scales of the light field given by the diffraction length and typical round trip times in the optical cavity are in the picosecond regime. Due to this difference in time scales the spatial distribution of the refractive index visualized in Fig. 6.8(a) shows a light-induced "hill" in the intersection area and a smooth decay in the trailing part of the pulses (left side in Fig. 6.8(b)) due to the gain saturation. Additionally, a comparison of the lateral distributions at $z = 0$ μm and $z \approx 500$ μm demonstrates the nonlinear interaction occurring during the spatial overlap of the two light pulses. At $z = 0$ μm the light fields can still be distinguished in the index distribution. Due to the amplification in the inverted active medium the amplitude of the index increases in the z direction. The spatial overlap in combination with the light diffraction and carrier diffusion then lead at $z \approx 500$ μm to an index distribution which is enlarged in the lateral (x) direction. As a consequence the light fields are laterally confined by this dynamic "waveguide" which acts as an "attractive force" and leads to molding of the pulses. In particular, the trailing parts of the pulses experience the "fingerprints" of the leading parts, i.e. the two light "trenches" in the first half of the sample increase the guiding of the diffracting light by leading the lateral outer parts via the index distribution towards the lateral center of the waveguide structure.

The guiding of the pulses can also be seen in the spatial distribution of the gain shown in Fig. 6.8(b). The gain reflects the recombination rate and thus determines the intensity distribution of the following time step. High values in Fig. 6.8(b) represent high amplification. The spatial gain depends on the intensity and on the inversion. Since the time scale of the carrier density is much slower than the light field dynamics the gain can remember and "save" the past intensity distributions. Thereby it develops a stabilization of the light field via spatially dependent amplification.

6.2.2 Dynamic Spatio-Spectral Dispersion Compensation

The interaction of the pulses with the nonlinear medium and their mixing with the spontaneously emitted light thus leads to a spatial localization of the gain and refractive index. However, the spatial distributions shown in Fig. 6.8 alone are not sufficient for the creation and stabilization of the solitons. The "macroscopic" variations in e.g. gain and refractive index depend via the carrier density on the microscopic electron–hole plasma and are thus composed

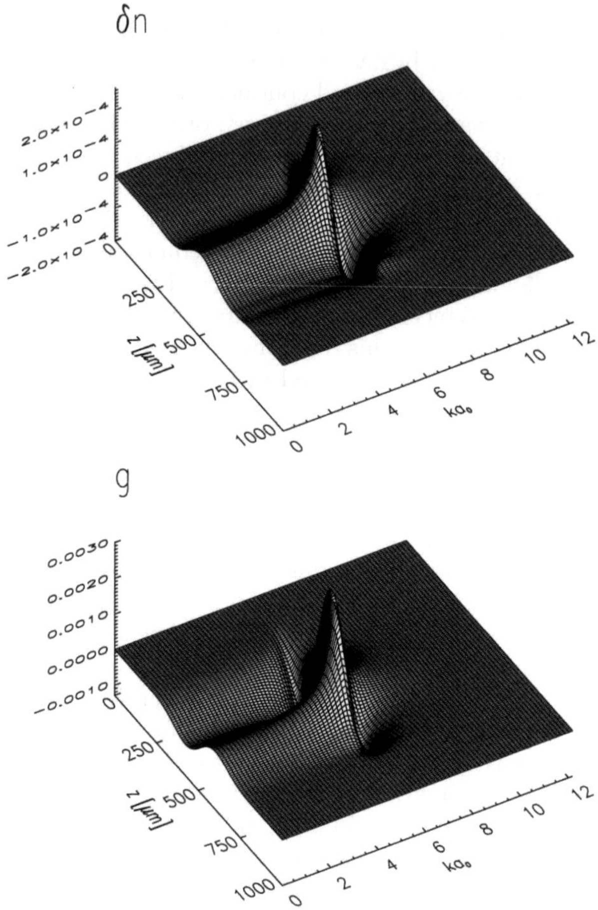

Fig. 6.9. Longitudinal cut of the microscopic refractive index δn and the microscopic spectral gain g after the molding of the two light pulses in the broad-area semiconductor laser. Positive values denote high refractive index and high amplification

of microscopic changes. The "balance" between defocusing and focusing effects in the laser-active medium can only be observed in the "microscopic" spectral distributions of electrons, holes and the interband polarization. A convenient way of analyzing the complex spatio-spectral processes is therefore to focus on the microscopic Wigner distributions [30]. As an example, Fig. 6.9 shows snapshots of longitudinal cuts of the "microscopic refractive index" and the "microscopic gain". The snapshots were taken in the lateral center of the waveguide structure after the molding of the two pulses. The coherent injection of the pulses in the absorption regime of the charge carrier plasma leads to spatial and spectral hole burning at $ka_0 \approx 4$, i.e. significantly above the bandgap. At the same time the carrier injection accumulates heated carriers of higher energy (at $ka_0 \approx 6$), implying a decrease of the spectral refractive index. The relaxation of the excited carriers via incoherent carrier–carrier as well as carrier–phonon scattering occurs on a time scale of a few hundred fs. Thus, together with the "unaffected" carrier inversion at the bandgap a characteristic spectral dispersion and spectral wave-guiding can build up.

The self-focusing and guiding of the light fields occurring during the propagation within the active medium is thus significantly increased by the "spectral guiding" observed in the microscopic carrier dynamics. The "microscopic gain" (i.e. the active nonlinearity) shows a similar behavior. The spatio-spectral hole burning created by the pulses increases the recombination rate and thus the gain (high values in Fig. 6.9). The spatio-spectral region of high gain thereby is (in spatial and spectral dimensions) surrounded by absorbing regions (negative values in Fig. 6.9). It is the combination and interplay of spatio-spectrally confined nonlinear gain, index dispersion and the spatio-spectral wave mixing of optical pulses that – as a consequence of coherent and incoherent interactions in the active nonlinear charge carrier plasma – lead to the self-induced formation and stabilization of a propagating spatio-spectral soliton.

6.3 Conclusion

In conclusion we have seen the effects of the dynamics of the spatial gain and refractive index distributions in determining the beam propagation and the optical properties of a pulse passing a semiconductor laser. They are a direct consequence of the microscopic gain and refractive index dispersion discussed in Chap. 3. Measurable output parameters such as power, shape, duration and spectrum of the amplified pulse are the result of both the complex microscopic carrier dynamics and the spatio-temporally dynamics of gain and index.

Simulations predict formation of propagating anisotropic and spatio-spectral solitons in the active charge carrier plasma inside broad-area semiconductor lasers and visualize the multi-stage cascade of nonlinear internal self-organizing processes that are the physical origin of soliton formation. Although considerably more complex in their origin, the propagating spatial optical solitons in the partially coherent active semiconductor media are thus reminiscent of walking solitons found in simpler and coherent quadratic nonlinear optical systems [34] and await a particle-like description of solitons [35].

References

1. G. P. Agrawal. *Semiconductor lasers: past, present, and future.* AIP Press, Woodbury, 1995.

2. P. J. Delfyett, L. T. Florez, N. Stoffel, T. Gmitter, N. C. Andreadakis, Y. Silberberg, J. P. Heritage, and G. A. Alphonse. High-power ultrafast laser diodes. *IEEE J. Quant. Electr.*, 28:2203–2219, 1992.
3. P. W. Smith. Applications of all-optical switching and logic. *Phil Trans. R. Soc. Land.*, 313:159–165, 1984.
4. S. Scholz, O. Hess, and R. Rühle. Dynamic cross-waveguide optical switching with a nonlinear photonic band-gap structure. *Opt. Expr.*, 3:28–34, 1998.
5. S. L. Shapiro. *Ultrashort Light Pulses.* Springer-Verlag, Berlin, 1977.
6. L. Goldberg and D. A. V. Kliner. Tunable UV generation at 286 nm by frequency tripling of a high-power mode-locked semiconductor laser. *Opt. Lett.*, 20:1640–1642, 1995.
7. G. P. Agrawal and N. A. Olsson. Self-phase modulation and spectral broadening of optical pulses in semiconductor laser amplifiers. *IEEE J. Quant. Electr.*, 25:2297–2305, 1989.
8. T. Saitoh and T. Mukai. Gain saturation characteristics of traveling-wave semiconductor laser amplifiers in short optical pulse ampification. *IEEE J. Quant. Electr.*, 26:2086–2094, 1990.
9. Y. Lai, K. L. Hall, E. P. Ippen, and G. Eisenstein. Short pulse gain saturation in InGaAsP diode laser amplifiers. *IEEE Photon. Technol. Lett.*, 2:711–713, 1990.
10. M. Trippenbach and Y. B. Band. Effects of self steepening and self-frequency shifting on short-pulse splitting in dispersive nonlinear media. *Phys. Rev. A*, 57:4791–4803, 1998.
11. C. Headley III and G. P. Agrawal. Simultanous amplification and compression of picosecond optical pulses during raman amplification in optical fibers. *J. Opt. Soc. Am. B*, 10:2383–2389, 1993.
12. V. I. Tolstikhin, M. Willander, and A. N. Mamaev. Picosecond pulse shaping using dynamic carrier heating in a gain-switched semiconductor laser. *J. Appl. Phys.*, 78:2955–2958, 1995.
13. M. Artoni and R. Loudon. Quantum theory of optical-pulse propagation through an amplifying slab. *Phys. Rev. A*, 57:622–627, 1998.
14. K. Sutkus, K. Shum, R. R. Alfano, and P. J. Delfyett. Effect of carrier heating on the wavelength chirp of ultrashort laser pulses in semiconductor traveling wave amplifiers. *IEEE Photon. Technol. Lett.*, 6:372–374, 1994.
15. A. Mecozzi and J. Mørk. Saturation induced by picosecond pulses in semiconductor optical amplifiers. *J. Opt. Soc. Am. B*, 14:761–770, 1997.
16. G. P. Agrawal and N. A. Olsson. Amplification and compression of weak picosecond optical pulses by using semiconductor-laser amplifiers. *Opt. Lett.*, 14:500–502, 1989.
17. K. L. Hall, J. Mark, E. P. Ippen, and G. Eisenstein. Femtosecond gain dynamics in InGaAsP optical amplifiers. *Appl. Phys. Lett.*, 56:1740–1742, 1990.
18. A. Dienes, J. P. Heritage, M. Y. Hong, and Y. H. Chang. Time- and spectral-domain evoluation of subpicosecond pulses in semiconductor optical amplifiers. *Opt. Lett.*, 17:1602–1604, 1992.
19. C. T. Hultgren and E. P. Ippen. Ultrafast refractive index dynamics in AlGaAs diode laser amplifiers. *Appl. Phys. Lett.*, 59:635–637, 1991.
20. A. V. Uskov, J. R. Karin, J. E. Bowers, J. G. McInerney, and J. L. Bihan. Effects of carrier cooling and carrier heating in saturation dynamics and pulse

propagation through bulk semiconductor absorbers. *IEEE J. Quant. Electr.*, 34:2162–2171, 1998.

21. R. A. Indik, R. Binder, M. Mlejnek, J. V. Moloney, S. Hughes, A. Knorr, and S. W. Koch. Role of plasma cooling, heating, and memory effects in subpicosecond pulse propagation in semiconductor amplifiers. *Phys. Rev. A*, 53:3614–3620, 1996.
22. N. A. Olsson and G. P. Agrawal. Spectral shift and distortion due to self-phase modulation of picosecond pulses in 1.5 μm optical amplifiers. *Appl. Phys. Lett.*, 55:13–15, 1989.
23. O. Hess and T. Kuhn. Spatio-temporal dynamics of semiconductor lasers: Theory, modeling and analysis. *Prog. Quant. Electr.*, 20:85–179, 1996.
24. L. Goldberg, D. Mehuys, and D. Welch. High-power mode-locked compound laser using a tapered semiconductor amplifier. *IEEE Photon. Technol. Lett.*, 6:1070, 1994.
25. E. Gehrig and O. Hess. Nonequilibrium spatio-temporal dynamics of the Wigner-distributions in broad-area semiconductor lasers. *Phys. Rev. A*, 57:2150–2163, 1998.
26. G. P. Agrawal. *Nonlinear Fiber Optics.* Academic Press, San Diego, CA, 1995.
27. B Dagens, S. Balsamo, and I. Montrosset. Picosecond pulse amplification in AlGaAs flared amplifiers. *IEEE J. Quant. Electr.*, 3:233–244, 1997.
28. E. Gehrig, D. Woll, M. Tremont, A. Robertson, R. Wallenstein, and O. Hess. Saturation behavior and self phase modulation of picosecond pulses in single stripe- and tapered semiconductor laser amplifiers. *J. Opt. Soc. Am. B*, 17:1452–1456, 2000.
29. W. W. Chow, S. W. Koch, and M. Sargent III. *Semiconductor-Laser Physics.* Springer-Verlag, Berlin, 1994.
30. E. Gehrig and O. Hess. Pulse trapping and nonequilibrium spatiotemporal wave mixing in broad-area semiconductor lasers. *J. Opt. Soc. Am. B*, 15:2861–2867, 1998.
31. I. Carvalho, Y. H. Coskun, D. N. Christodoulides, M. Mitchell, and M. Segev. Coherence properties of multimode incoherent spatial solitons in noninstantaneous kerr media. *Phys. Rev. E*, 59:1193, 1999.
32. T. H. Coskun, D. N. Christodoulides, Y.-R. Kim, Z. Chen, M. Soljacic, and M. Segev. Bright spatial solitons on a partially incoherent background. *Phys. Rev. Lett.*, 84:2374, 2000.
33. O. Hess and T. Kuhn. Maxwell-Bloch equations for spatially inhomogeneous semiconductor lasers I: Theoretical description. *Phys. Rev. A*, 54:3347–3359, 1996.
34. L. Torner, D. Mazilu, and D. Mihalache. Walking solitons in quadratic nonlinear media. *Phys. Rev. Lett.*, 77:2455, 1996.
35. M. R. Belic, A. Stepken, and F. Kaiser. Spatial screening solitons as particles. *Phys. Rev. Lett.*, 84:83, 2000.

7 Spatio-Spectral Wave Mixing in High-Power Amplifier and Laser Systems

7.1 Introduction

Semiconductor lasers have proven to represent compact and efficient light sources in modern laser systems. Due to a narrow facet of typically 3 - 5 μm the single-stripe laser efficiently produces coherent light with good spatial and spectral quality. Catastrophic optical mirror damage (COMD), however, sets an upper limit to the output power that can be achieved in devices of this type. One therefore is conceiving alternative laser and amplifier geometries in which COMD is avoided. In large area lasers with rectangular or tapered geometry the width of the facet (typically 100 -250 μm) allows the generation of high output powers of several Watts. However the microscopic laser-internal dynamics gains influence in these systems and leads to a significant deterioration of the quality of the emitted radiation [1; 2; 4; 5; 6; 8; 7; 3]. In recent years the need for high output power and good spatial and spectral purity often required by applications in e.g. nonlinear optics or communication lead to the realization of specific laser amplifier configurations with improved beam quality [9; 10; 11; 12; 13; 14; 15; 16; 17; 18; 19; 20]. In these technologically extremely important systems a coherent light signal which is injected into the active region of an antireflection-coated large-area semiconductor laser is – via the induced recombination in the inverted medium – amplified thereby basically maintaining its spatial and spectral properties [10]. However, the large lateral dimensioning of the high-power laser amplifier not only increases the output power but also leads to complex spatio-temporal interactions and to the formation of optical filaments [6; 21; 22; 23; 24; 25; 26; 27; 28; 29; 30; 31].

The amplification processes that characterize a laser amplifier system is determined by dynamic microscopic changes in the electron and hole distributions in the semiconductor medium [22]. The dynamic interactions between the light fields and the active charge carrier plasma lead to spatial and spectral gain and refractive index variations which determine both power and spectral properties of the emitted radiation. In particular it is the microscopic processes such as spatio-spectral hole burning, saturation, dynamic spatio spectral wave mixing and carrier scattering that determine the overall behavior of large-area laser amplifiers with continuous or pulsed light injection.

In this chapter we present results of detailed numerical modelling of semiconductor laser amplifiers. The model is based on the Maxwell–Bloch–Langevin equations presented in Chap. 2 and Chap. 3 and includes all relevant dynamic microscopic and macroscopic processes responsible for the special emission properties of high-power amplifiers. Since the active area of high-power laser amplifiers is large compared to characteristic length scales given by e.g. carrier diffusion and diffraction, the longitudinal and transverse dependence of the optical fields as well as their counterpropagation are of particular importance.

7.2 Modelling of Optical Injection

The microscopic modeling allows one to consider an optical injection in a very general and detailed way. Not only the input power serves as an input parameter, but also the spatial and spectral shape and angle of incidence are self-consistently transferred to the internal dynamics.

The laterally dependent injected light field at the possible injection positions $z = 0$ or $z = L$ may conveniently be expressed via the boundary conditions

$$\begin{aligned} \boldsymbol{E}^{+}(x, z=0, t) &= -R_1 \boldsymbol{E}^{-}(x, z=0, t) + \boldsymbol{E}_{\text{inj}}(x, z=0, t) \\ \boldsymbol{E}^{-}(x, z=L, t) &= -R_2 \boldsymbol{E}^{+}(x, z=L, t) + \boldsymbol{E}_{\text{inj}}(x, z=L, t), \end{aligned} \tag{7.1}$$

where

$$\boldsymbol{E}_{\text{inj}}(x, z=0, L, t) = \boldsymbol{E}^{0}_{\text{inj}}(x, z=0, L, t) \times \exp\left[\mathrm{i} K_z \frac{(x-x_0)^2}{2R(z)}\right] \tag{7.2}$$

with

$$\begin{aligned} \boldsymbol{E}^{0}_{\text{inj}}(x, z=0, L, t) &= \mathcal{E}^{0}_{\text{inj}}(x, z=0, L, t) \\ &\quad \times T_{1,2} \exp\left[-\mathrm{i}\omega_{\text{inj}}(z=0, L)t\right] \exp\left[-\frac{(x-x_0)^2}{w_0^2}\right] \\ &\quad \times \exp\left[\frac{\mathrm{i}2\pi}{\lambda}(x-x_0)\sin\left(\alpha_{\text{inj}}(z=0, L)\right)\right] \\ R(z) &= z + \frac{z_0^2}{z} \end{aligned} \tag{7.3}$$

In (7.2) we have assumed a Gaussian shaped monofrequent optical injection. $T_{1,2} = 1 - R_{1,2}$ is the transmission of the front and back facet and $x - x_0$ is the transverse (i.e. parallel to the facets) distance from the lateral injection position x_0. $\omega_{\text{inj}}(z=0, L) = 2\pi c / \left(\lambda_{\text{inj}}(z=0, L)\right)$ and $\alpha_{\text{inj}}(z=0, L)$ are the frequency and the injection angle of the injected light field, respectively, $z_0 = \pi w_0^2 / \lambda_{\text{inj}}$ is the Rayleigh range, $R(z)$ is the radius of curvature of areas of constant phase and $2w_0$ the beam waist of the injected beam. Since propagation after entering the active layer occurs automatically via Maxwell's wave equation, no term $e^{ik_z z}$ which is usually a part of the formula for Gaussian beams has to be considered here. Equation (7.3) describes the transversely varying amplitude of the injected optical field after transmission in the active layer, including the phase change between different lateral positions due to the angle of incidence. Thus (7.2) very generally represents the dynamic spatio-temporally resolved injection of an optical beam or pulse and allows one to optimize the injection (e.g. width of the injected light beam, angle of incidence, frequency with respect to the gain bandwidth of the amplifier).

7.3 Continuous-Wave Laser Amplifiers

The Maxwell–Bloch–Langevin equations will in the following be applied to large-area semiconductor lasers with coherent injection of a continuous light field. We will thereby focus on a description and interpretation of relevant physical properties that are responsible for typical operation characteristics of high-power laser amplifier systems. Typical values of characteristic parameters can be obtained via an experimental characterization. In particular, they are the dependence of the output power of the amplifier on the injected input power, the spectral properties of the emitted radiation (particularly the contributions of the amplified signal and the amplified spontaneous emission) as well as the spatial properties of the emitted radiation (i.e. the intensity distribution in the nearfield and farfield). Experimental investigations have shown that these quantities are of high relevance for the quality of the laser source. In recent experiments they show a strong dependence on the material properties, the geometry of the active layer and the operation conditions (i.e. injection current, input power) and thus vary from system to system. However, due to the complex spatio-temporal carrier and light field dynamics occurring in the active area of a semiconductor laser an explicit correlation between measurable output quantities on the one hand and system properties on the other hand is difficult to obtain. The microscopic description of the carrier dynamics and the interband polarization as well as the inclusion of the spatial and spectral properties of the optical fields allows the derivation of the contributions ("signal" and "ASE") the emitted radiation is composed of as well as the calculation of the spatial beam properties of large-area laser amplifiers.

Up to now various amplifier systems have been realized (broad-area amplifiers in single-pass or double-pass configuration, amplifiers with tapered geometry). Due to its small signal gain and good wave-guiding properties the tapered amplifier has been in the focus of theoretical [26; 27; 32] and experimental [15; 33] investigations in recent years. In the following analysis we will thus concentrate on the tapered laser amplifier (Fig. 7.1) which consists of two parts, a single-stripe waveguide with a width of $\approx$ 3 to 5 μm and a

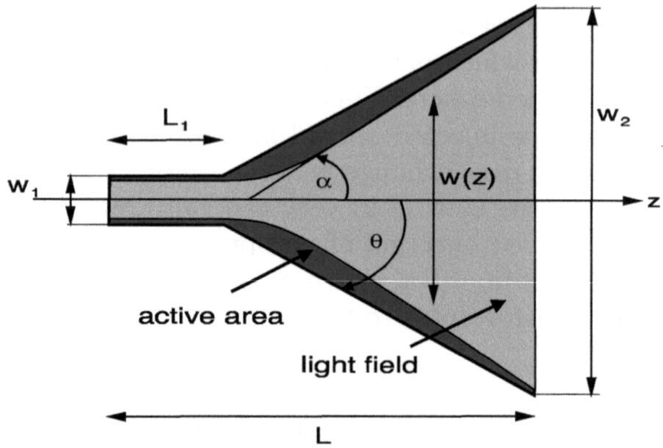

Fig. 7.1. Scheme of a large-area semiconductor laser with tapered geometry

tapered section in which the active area enlarges in the propagation direction so that the intensity at the output facet is kept below the threshold value for COMD. The facets of the tapered amplifiers are antireflection coated. For good beam quality, this facet reflectivity should be less than 10^{-4} [21]. Alternatively, the wave propagation in the resonator should be off-axis, i.e. the facets should be angled with respect to the resonator axis. The small transverse dimension of the waveguide at the input facet of the active area leads to a high small signal gain allowing efficient saturation of the inversion in the active layer for very moderate input powers of a few mWs. Typical lengths L_1 of the small waveguide are a few 100 μm at a total length of $L \approx$ 1 to 3 mm of the device.

In the following we analyze the propagation of a continuous-wave light beam in amplifiers of tapered geometry. Results of our modelling visualize the dynamic spatio-spectral beam shaping experienced by the propagating light and reveal the microscopic physical processes that are responsible for the particular amplitude and spatial shape of the light beam at the output facet.

7.3.1 Spatio-Spectral Wave Mixing

During the propagation of a coherent light signal in the active area of a semiconductor laser amplifier a multitude of physical interaction processes occur. In the amplifier configuration it is in particular the dynamic mixing of the injected light signal with light from spontaneous emission processes in the gain medium that has a strong influence on the amplification process. The nonlinear coupling of the two light fields, the signal and the amplified spontaneous emission ('ASE'), to the charge carrier plasma of the inverted medium induces a complex spatio-temporal carrier–light field dynamics that

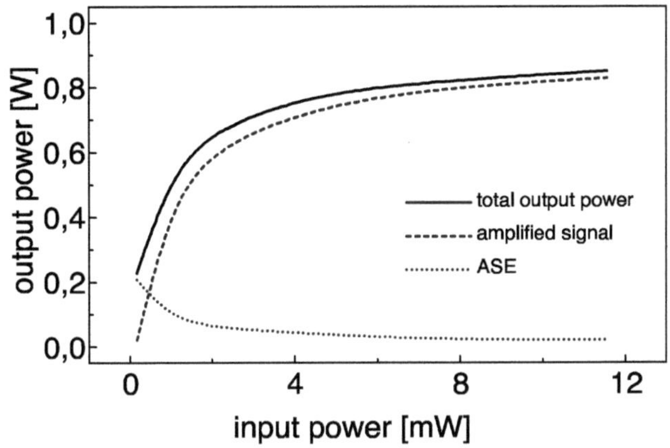

Fig. 7.2. Dependence of the total emitted power, the amplified signal and the ASE on input power

affects the spatial and spectral properties of the laser amplifier. The dynamic changes in the carrier system and in the light field distribution result from physical processes such as spatio-spectral hole burning, carrier relaxation, gain saturation, diffraction and self focusing.

In the following we show numerical results of a typical tapered amplifier (InGaAs/GaAs, gain-guided, total length $L = 2$ mm, length of the quantum well waveguide $L_1 = 500$ µm, width of the output facet $w_2 = 150$ µm, emission wavelength 925 nm). In particular, we will analyze its saturation behavior and the spectral composition of the amplified optical fields. Figure 7.2 shows the dependence of the amplified signal power and the ASE power contribution. They are via

$$\mathcal{P}_{\mathrm{sig,se}}(z=L) \propto \left(\mathrm{Re}[\boldsymbol{d}_{\mathrm{cv}}\boldsymbol{E}_{\mathrm{sig,se}}(z=L)]^2 + \mathrm{Im}[\boldsymbol{d}_{\mathrm{cv}}\boldsymbol{E}_{\mathrm{sig,se}}(z=L)]^2\right) \quad (7.4)$$

related to the dynamically calculated optical fields. The third curve in Fig. 7.2 represents the total output power which can be extracted from the total field via $\mathcal{P}_{\mathrm{ges}}(z=L) \propto \mathrm{Re}[U(z=L)]^2 + \mathrm{Im}[U(z=L)]^2$. The turning point of the curve as well the absolute value of the emitted power thereby depends on both the geometry of the high-power laser and the material properties (i.e. facet reflectivity, wave-guiding properties) and may vary from system to system.

The spatio-spectral inversion established by the injection current, i.e. the spatially varying distribution of the carriers at every wavenumber k within the bands, is reduced by both, the propagating signal light field and the amplified spontaneous emission (ASE). Together with the light field (interacting at every location and in each frequency interval with the respective carrier population) this inversion constitutes a spatially and spectrally varying gain. In our microscopic equations this is reflected in the term of the generation

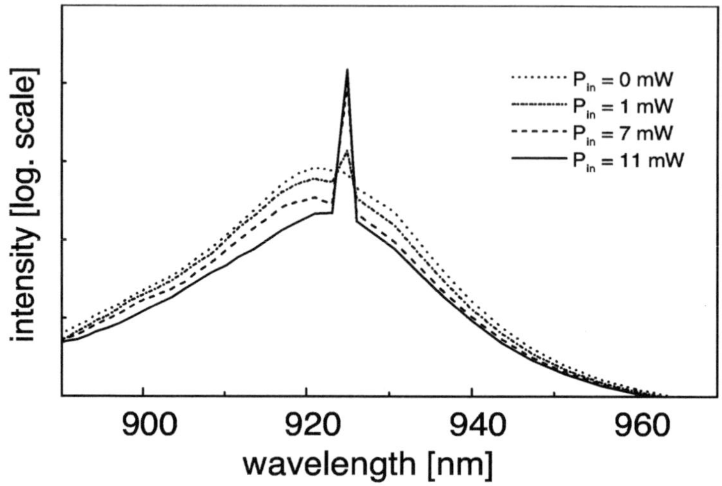

Fig. 7.3. Calculated emission spectra of a tapered amplifier against the injected power (logarithmic scale)

rate g of the semiconductor Bloch equations (3.5)). High amplitudes of the light fields (which contribute to the local field U) and a strong reduction of the carrier distribution $f^{e,h}$ (which is correlated with respective changes in the interband polarization p via the microscopic coupling of the Bloch equations) yield a high generation rate. The changes in the polarization then lead to corresponding changes in the light fields via the wave equation. The generation rate g consequently can be interpreted as spatio-spectral gain, i.e. the spatially and spectrally varying amplification by stimulated emission. Due to the light field dependence of the generation rate the gain available for the signal light field and for the ASE is determined by the amplitudes and the dynamic mixing of both field contributions at every location within the active area and at every time instance. Via the local field U constituted by both fields and via their microscopic coupling to the carrier distribution in the bands a continuous interaction between both light fields and between light fields and carriers occurs. Depending on both, the location in the active area (r) and the wavenumber k within the bands this process consequently can be interpreted as a dynamic spatio-spectral wave mixing. With increasing input power the gain available for the ASE is thus reduced by the dynamic gain competition and spatio-spectral wave mixing between signal and ASE. For a further understanding of the spectral composition of the amplified fields Fig. 7.3 shows (in logarithmic scale) calculated emission spectra of the tapered amplifier. The numerical modelling confirms the typical spectral properties of a high-power semiconductor laser amplifier. With increasing input power the amplified signal increases whereas the emitted spontaneous radiation is significantly reduced. The origins of this characteristic behavior are the microscopic spatio-spectral wave mixing processes mentioned above. Due to the spectral bandwidth of the injected signal the inversion is mostly reduced in a narrow frequency interval whereas the spontaneous emission occurs over the whole gain bandwidth of the amplifier. On the other hand, the induced recombination occurring preferentially in this spectral region confines the spectrally dependent gain. As a consequence the amplification is highest within the spectral bandwidth of the injected light field so that the signal maintains its good spectral purity. The reduction of the spectral inversion and the refilling of the spectral hole via the injection current and microscopic carrier–carrier as well as carrier–phonon scattering lead at the same time to a reduction of the gain available for the ASE. Thus a "spectral condensation" occurs, i.e. the ratio of amplified signal power and amplified spontaneous emission rises with increasing input power.

The microscopic calculation of the spatio-spectral coupling and wave mixing processes thus allows description and interpretation of the relevant physical interactions leading to the characteristic spectral emission properties of high-power laser amplifiers.

7.3.2 Optimization of the Tapered Geometry

The geometry of the active area of a tapered amplifier is of fundamental interest for technology needs and thus successful industrial application. In particular, the length of the waveguide and the angle of the tapered region play a major role for the spatially dependent light amplification and thus the power and the quality of the emitted radiation. Our microscopic description allows a systematic analysis of the influencing factors and parameters. It may thus significantly contribute to an optimization of tapered amplifiers.

Spatial gain

The spatial distribution of the gain in the active area G given by the spectrally integrated values of the generation rate g, $G(\mathbf{r},t) = \int \mathrm{d}^3k \; g(k,\mathbf{r},t) = \int \mathrm{d}^3k \frac{1}{\mathrm{i}\hbar} \left[\mathcal{U}(k,\mathbf{r},t) \; p^*(k,\mathbf{r},t) - \mathcal{U}(k,\mathbf{r},t)^* \; p(k,\mathbf{r},t)\right]$ depends (via $\mathcal{U}$) on the optical fields and (via the interband polarization p) on the spatial and spectral carrier inversion. It represents the spatially resolved carrier recombination by stimulated emission in the active area, i.e. local amplification. It includes the local inversion and light field distribution and thus is an adequate quantity for visualization of the physical effects that are responsible for the spatio-temporal dynamics and output quantities of tapered amplifiers. Its dependence on the light field distribution is particularly important to self-consistently capture (spatial and spectral) saturation effects. Via the dependence on the interband polarization the spatial gain includes linear and nonlinear gain contributions. In each spatial element the efficiency of the amplification process during propagation of a light signal depends on both inversion and intensity. As a consequence, the value of the spatial gain immediately at the output facet is determined by the transfer of carrier inversion into coherent light after the passage of the last segment Δz. It is thus a direct measure of the intensity at the output facet, i.e. the output power of the device. The calculation of the gain for a specific geometry can be used to extract information on the saturation properties of a semiconductor laser amplifier which lead to the measurable quantity of the output power. The theory can thereby visualize the physical interactions in the active area whereas experimental characterization is limited to the measurement of the output parameters.

In the following we will analyze the spatial gain distributions in tapered amplifiers (InGaAs/GaAs) with a length of 2 mm. We will vary the length of the waveguide section (200 µm, 500 µm, 750 µm, 1000 µm). The tapering angle of $\theta = 6°$ corresponds to the diffraction angle α of the injected beam. The emission wavelength of the amplifiers and of the injected beam is 925 nm. For the input power a typical value of 3 mW is chosen in the numerical simulation. The injection current density is $1.0 \cdot 10^7$ A/m^2. Figure 7.4 shows a snapshot of the spatial gain in the active area (left) and a longitudinal cut (right) taken approximately 1 ns after the optical injection for the various tapered geometries. Light shading corresponds to high gain. The propagating light beam injected into the narrow waveguide leads to induced recombination and thus to amplification of the light. At the same time the carrier density is reduced. Due to the dependence of the spatial gain on intensity and on the inversion an increase in the gain can be observed in the propagation direction. All distributions show a maximum at the end of the waveguide. This typical effect can be explained by the dynamic beam shaping induced by the tapered geometry. The intensity of the injected light field is increased during the propagation in the waveguide thereby increasing the spatial gain. Reaching the end of the waveguide, the preamplified radiation spreads in the lateral direction due to its diffraction. The sudden "lack of lateral confinement" leads to a reduction of the spatial intensity immediately behind the waveguide and thus to a corresponding lowering of the gain. With preceding signal propagation in the tapered region the intensity is increased by induced recombination thereby producing a second rise in the gain. Via the intensity dependence of the spatial gain not only the carrier inversion but also the spatially varying intensity at every position of the active area determines the light amplification in the following time step. In particular, Fig. 7.4 clearly elucidates the diffraction of the signal light fields. Near the output facet the spatially dependent shading visualizes the curvature of the phasefront induced by the diffraction of the light. The distributions of the spatial gain and of the longitudinal cuts visualized in the right side of Fig. 7.4 demonstrate that both the gradient of the gain in the propagation direction

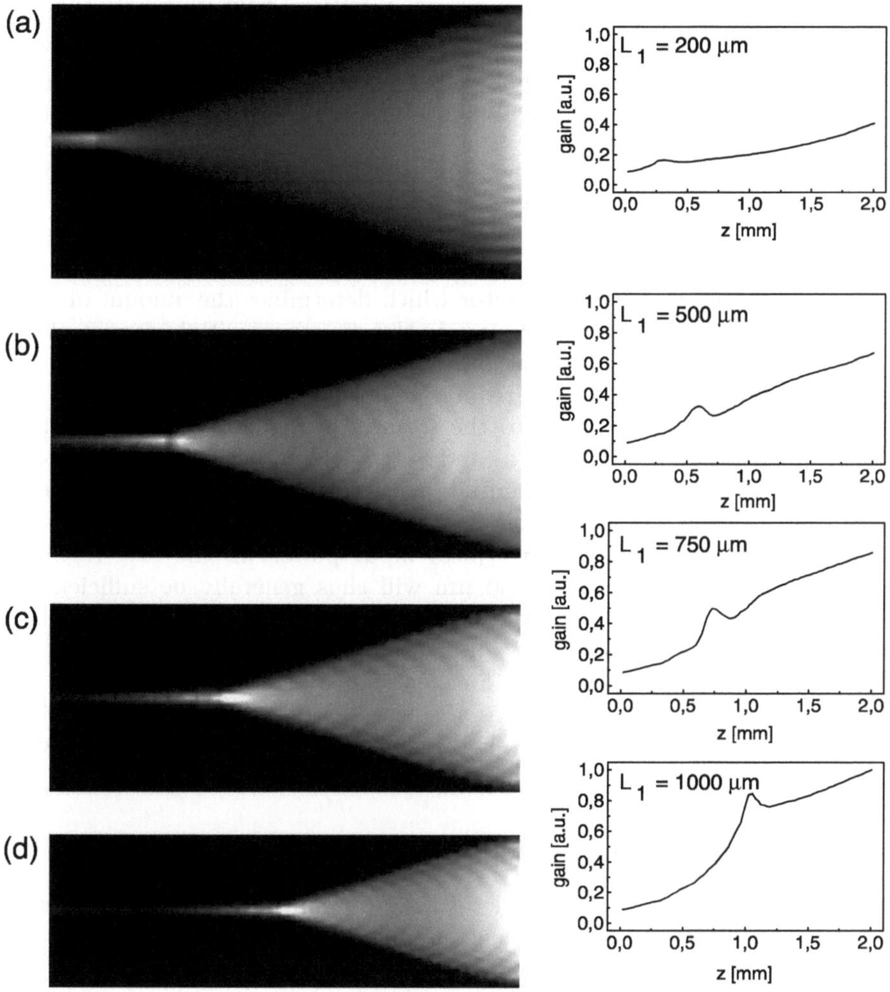

Fig. 7.4. Snapshots of the spatial gain in the active area of tapered amplifiers with different waveguide lengths and identical total length of the device (2 mm). The figures on the right show longitudinal cuts. The input power is 3 mW and the injection current density is $1.0 \cdot 10^7$ A/m^2. The length of the waveguide is (**a**) 200 µm, (**b**) 500 µm, (**c**) 750 µm, (**d**) 1000 µm.

and its end value at the output facet are strongly influenced by the length of the device (the scaling to arbitrary units is identical for all geometries).

The increase of the gain within the waveguide section approximately obeys an exponential function. Consequently, the waveguide is not saturated with low input powers of a few mWs that are typically used in experimental investigations. Due to the different values of the length of the waveguide the injection of a light beam with constant power (in the example 3 mW) leads to different values of the gain at the beginning of the tapered section. As a consequence, the influence of nonlinear properties such as in particular gain saturation varies from structure to structure. In the tapered amplifier with short waveguide (Fig. 7.4(a)) the gain is still rather low in the beginning of the tapered region so that its magnitude at the output facet ($z = L$) is below the value reached in the structure with long waveguide (Fig. 7.4(d)). Furthermore, the different preamplifications obtained in the waveguides of different lengths are a decisive factor which determines the amount of saturation achieved in the tapered section. In the structures with long waveguides of 750 μm (Fig. 7.4(c)) and 1000 μm (Fig. 7.4(d)) the increase within the waveguide is followed by an approximately linear increase in the tapered section. This indicates transition to the saturation regime. In the geometries with short waveguides (Fig. 7.4(a) and Fig. 7.4(b)) the rise in the first parts of the taper still obeys an exponential relation, i.e. corresponds to the small-signal regime, and the saturation is only reached after further propagation toward the output facet. With typical input powers in the mW-regime a length of the waveguide of $\approx$ 750 μm will thus generally be sufficient for efficient saturation in the tapered region of an amplifier. Since the gain at the output facet determines the total output power, the results of the numerical simulation also show that for identical total length of the device and identical injection current densities a tapered amplifier with a waveguide of several 100 μm can provide much higher output powers than structures with a shorter waveguide.

Transverse light field dynamics

The beam quality of a semiconductor laser amplifier is an important factor for various applications such as e.g. nonlinear frequency conversion or the coupling of the emitted radiation into e.g. a resonator or a fiber. The spatial beam profile at the output facet is thereby determined by the dynamic interaction between the light fields and the charge carrier plasma occurring during the propagation of the light signal. As the simulations show, these lead to dynamic optical patterns and beam shaping. Our microscopic theory on the basis of a Maxwell–Bloch description can be used to simulate this spatio-temporal beam propagation and to investigate the influence of the geometry. The intensity distributions at the output facets and within the active layer thereby give good insight into the underlying physical processes and may provide hints for optimization. Furthermore, the temporally averaged distributions at the output facet can be compared to representative experimental measurements [34]. The beam quality of a tapered amplifier is to a large degree determined by the transverse light field dynamics. Thus in the following we will restrict our analysis to the variation of the angle in the tapered section of the amplifier and keep the overall length of the device as

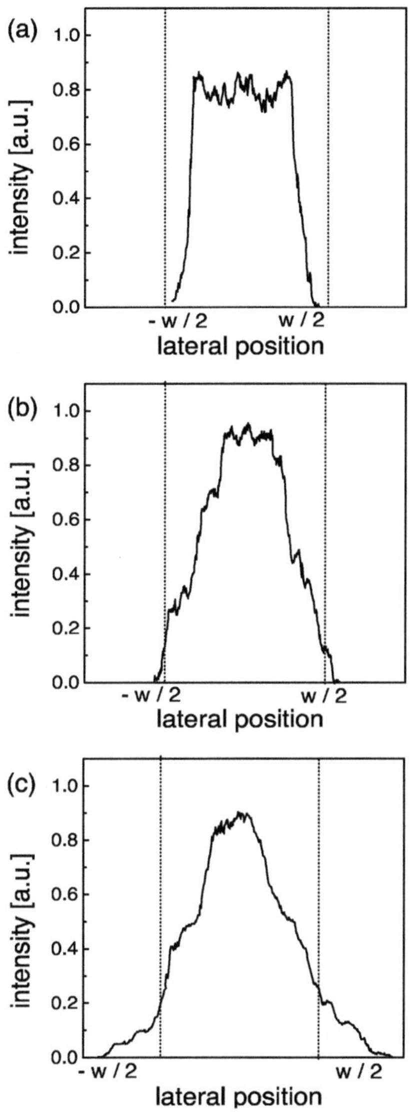

Fig. 7.5. Temporal average of the nearfield intensity distribution at the output facet of tapered amplifiers. The length of the active region is 2 mm (including the waveguide length 500 μm) and the emission wavelength is 915 nm. The angle of the taper is (**a**) 4.7°, (**b**) 6.0°, (**c**) 8.8°. The dotted line denotes the beam width a corresponding Gaussian beam would have

well as the waveguide length at a constant typical value (2 mm and 1 mm respectively). As reference value one usually uses the half divergence angle of a Gaussian beam (beam waist corresponding to the width of the entrance facet w_1, wavelength λ) given by $\alpha = 2\lambda/(n_l \pi w_1)$ (n_l is the refractive index of the active area). For $w_1 = 3.5$ µm, $n_l = 3.35$ and $\lambda = 920$ nm of typical InGaAs semiconductor lasers one gets a value of $\approx 5.7°$ for the Gaussian diffraction angle.

The dependence of the spatial beam properties on the angle in the tapered section (i.e. larger, smaller or identical to the respective Gaussian divergence angle) can be examined with the calculated nearfield intensity distributions at the output facet of tapered amplifiers. Figure 7.5(a) shows as examples calculated nearfield profiles (temporally averaged over ≈ 5 µs) for a tapered amplifier (width of the waveguide 3.5 µm, emission wavelength 920 nm, refractive index 3.35) with an angle of 4.7°, i.e. smaller than the respective Gaussian angle. The modulations that can be seen in Fig. 7.5 originate from the dynamic light–matter interactions within the active area. The modulation depth is thereby not only determined by the resonator-internal dynamics but also by material inhomogeneities and may thus vary from device to device. The simulations use uniform material properties and a uniform carrier injection.

Due to its diffraction the diverging light signal would cover a larger region than the area given by the dimensions of the taper in Fig. 7.5(a) and consequently is laterally "cut" by the absorbing layers leading to the characteristic "table top shape". Figure 7.5(b) shows a nearfield intensity distribution of a tapered amplifier in which the angle of the tapered region has been adjusted to the Gaussian diffraction angle, i.e. 5.7°. In this geometry the diffraction of the light field is not impeded by the lateral confinement of the taper so that the width of the beam approaching the output facet is mostly determined by its diffraction. Thus the shape of the nearfield is given by a Gaussian profile. If one further increases the angle of the tapered region, i.e. the angle in the active area is significantly larger than the diffraction angle of the respective Gaussian beam, the transverse degree of freedom gets more important and a multitude of additional physical processes may arise. In this case it is in particular the amount of amplified spontaneous emission (ASE) by spontaneous recombination within the active area and the longitudinal boundary conditions given by the facet reflectivity of the resonator that affect the spatio-temporal dynamics (e.g. transverse light field dynamics and counterpropagation effects) and thus determine the spatial beam properties of the emitted radiation. The example shown in Fig. 7.5(c) refers to a structure with a good antireflection coating ($\approx 10^{-4}$). This geometry leads to a rather uniform lateral enlargement of the width of the signal beam. In contrast, the spatio-temporal dynamics in a tapered amplifier with a poor quality coating ($> 10^{-2}$) is determined by a strong influence of the spontaneous emission leading to complex wave mixing processes and filamentation which will be analyzed in Sect. 7.3.2. The spatial nearfield profiles displayed in Fig. 7.5 are the result of dynamic light–matter interactions occurring during forward and

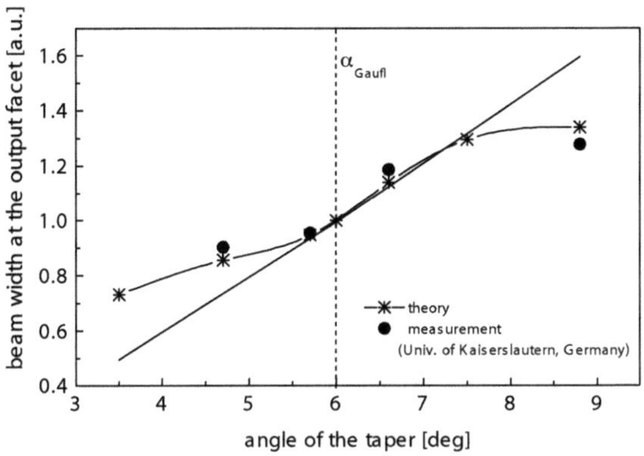

Fig. 7.6. Beam width at the output facet of InGaAs tapered amplifiers (length of the active area: 2 mm, length of the waveguide: 1 mm, width of the waveguide: 3.5 μm) against the tapering angle. Shown are theoretical values of the beam widths as calculated with the microscopic Maxwell–Bloch equations and the results of an experimental measurement [62]. The theoretical values are connected with a spline-function to guide the eye. The line shows the idealized case of a beam width that increases linearly with the tapering angle

backward propagation of the injected light signal and the ASE. Due to different characteristic time scales of intensity and carrier density, these typical shapes are not established after one single passage of the signal in the amplifier but dynamically form on time scales of many resonator round trips. In order to quantify the deviation of the geometry-dependent beam properties one can extract the width of the calculated intensity profiles by temporally averaging the distributions at the output facet. Figure 7.6 shows the results of the modelling and the results of an experimental measurement [34]. The values are scaled to the width of a respective Gaussian beam. If the tapering angle is identical with the Gaussian divergence angle (in this case $\approx 6°$), the good spatial overlap between the active area and the propagating light field leads to almost unaffected propagation and uniform amplification of the light signal. If on the other hand the angle of the tapered section deviates significantly from the Gaussian divergence beam the width of the beam waist is given by both the diffraction of light and the spatial distribution of the carrier density in the active area. The actual value of the beam width then lies in between the Gaussian width (i.e. the scaled value '1' in Fig. 7.6) and a value assuming a uniform covering of the entire width of the active area (shown as line in Fig. 7.6). In tapered geometries with smaller angle the lateral confinement of the active area and the beam guiding via the induced refractive index profile counteracts the lateral diffraction-induced enlargement leading to a value < 1 in Fig. 7.6. However this value is larger than the width of a uniform Gaussian distribution at the output facet (i.e. above the line) since

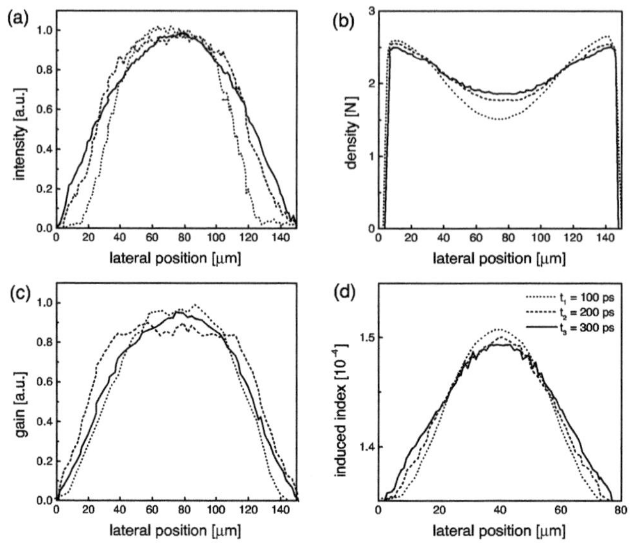

Fig. 7.7. Snapshots of the intensity (**a**), carrier density (**b**), gain (**c**) and refractive index (**d**) at the output facet of a tapered amplifier (InGaAs, length of the active area: 2 mm, width of the facet: 150 μm, waveguide length: 1 mm, tapering angle: 8.8°), 100 ps, 200 ps und 300 ps after the light injection. The injection current density is $1.5 \cdot 10^7$ A/m^2, and the input power is 3 mW

the light diffraction produces a reshaping with high intensity values near the lateral confining layers. If on the other hand the tapering angle is larger than the Gaussian angle the high inversion on both sides of the signal beam leads to lateral enlargement of the beam (Fig. 7.7), i.e. to a value > 1 but below the line assuming a uniform enlargement of the beam with increasing facet width.

Spatio-temporal beam shaping

The dynamic beam shaping occurring in the amplifier may be visualized by characteristic snapshots of lateral intensity distributions at the output facet. As an example we consider results for a tapered amplifier (total length of the active area 2 mm, length of the waveguide 1 mm, width at the output facet 150 μm) where the angle of the taper was set to 8.8°, i.e. significantly larger than the corresponding Gaussian angle. Figure 7.7 shows snapshots of the intensity (a), the carrier density (b), the gain (c) and the induced refractive index (d) taken at the output facet ($z = L$), 100 ps, 200 ps and 300 ps after the injection of the signal light field.

Within the first 100 ps after the injection of the light beam (corresponding to approximately five resonator round trips) the propagation in the tapered region is mostly determined by the diffraction of the light. The lateral beam width of the amplified signal is thus to a good approximation identical to the width given by the "unperturbed" propagation of a Gaussian beam which obeys the relation $w(z) = w_1 \left[1 + (L_2/z_0)\right]^{1/2} \approx L_2 \tan \alpha$ where L_2 is the length of the tapered section. For the considered geometry this yields a width $w(z)$ of ≈ 100 µm at the output facet. In contrast the angle of 8.8° of the tapered section of the amplifier described in Fig. 7.7 implies a width of the output facet of ≈ 150 µm. Thus at $t = 100$ ps after the injection of the signal the lateral width of the beam waist is significantly smaller than the width of the output facet (Fig. 7.7(a)). At the same time, due to the spatially selective induced recombination the distribution of the carrier density (Fig. 7.7(b)) shows significant spatial hole burning in the lateral center of the amplifier. Since the reduction of the carrier density is accompanied by an increase of the refractive index, the induced refractive index profile shows similar behavior (Fig. 7.7(d)). The spatial gain depends on both the intensity (via the optical fields) and on the carrier inversion (via the microscopic distributions $f^{e,h}$ and p). As a consequence, the spatial gain at the output facet (Fig. 7.7(c)) is dynamically shaped by both distributions. In the lateral center the gain is mostly determined by the high intensity of the signal, whereas it is the high carrier density in particular which is responsible for the gain values in the lateral regions on both sides of the light beam. The deviation between the Gaussian beam waist and the width of the output facet thus leads to a corresponding deviation between the profiles of the intensity (Fig. 7.7(a)) and gain (Fig. 7.7(c)). On the other hand, this gain profile corresponds to the spatially dependent induced emission and thus is a measure for the intensity distribution in the next time step. Thus the shape and in particular the width of the gain profile $t = 100$ ps after the light injection are the origin of the lateral enlargement of the intensity profile which can be seen at $t = 200$ ps in Fig. 7.7(a).

The dynamic mutual influences of the intensity, carrier density, gain and refractive index thus are the origin of deviations from the unperturbed Gaussian beam propagation. Due to the different time scales of the carrier density and light field dynamics a characteristic temporal delay of a few ps exists between a change in one of these quantities and the re-emergence of the respective "optical pattern" in the other quantity. After a few resonator round trips (typically a few hundred ps) these counteracting processes, i.e. the lateral enlargement of the beam produced by the high inversion on both sides of the light signal on the one hand and the self focussing via induced refractive index on the other hand, result in a quasistationary state. In this regime the variations of the intensity and density have asymptotically decreased compared to their initial dynamics immediately after the injection but due to the various characteristic time scales of the underlying physical processes still vary in space and time. A comparison of the snapshots in Fig. 7.7 confirms that the changes in the distributions at the output facet occurring between 200 and 300 ps are much smaller than in the time regime between 100 and 200 ps.

In order to further quantify the influence of the geometry on relevant output quantities we have additionally calculated the fraction of the emitted power contained in the central lobe of the farfield for the different geometries. We thereby found a slight reduction of ≈ 1 to 4 % (depending on current density and input power) of the power when the tapering angle differed from the Gaussian angle.

The results of our modelling show that the tapered amplifier with appropriate geometry provides good spatial beam quality. The variation of the geometry shows a slight dependence of the beam quality on the angle of the tapered section. The dependence of the beam quality on the geometry gets stronger if additionally the facet reflectivities of the structures are high (10^{-2}). This is the case since the increased counterpropagation leads to a strong influence of the amplified spontaneous emission which may result in complex wave mixing and gain competition. As an example, in the following we will present numerical results of the spatio-temporal behavior in a tapered amplifier (InGaAs, length of the active area 2 mm) with an angle of 8.8° and a facet reflectivity of 10^{-2} and compare it to the dynamics in an 'optimized' structure (angle $\theta = 6.0°$ of the tapered section, facet reflectivity 10^{-4}).

Filamentation

The spatio-temporal dynamics leading to characteristic beam properties of tapered amplifiers with different geometries can be visualized with spatial intensity distributions in the active area. Figure 7.8 shows snapshots of a tapered amplifier with an angle of 8.8° in the tapered section, Fig. 7.8(a–d)), and a tapered amplifier which is adjusted to the corresponding Gaussian angle (6.0°, Fig. 7.8(e–h)). The facet reflectivity of the geometry with the large tapering angle is 10^{-2} while for the geometrically adjusted structure a value of 10^{-4} has been used in the simulations. The first snapshot is taken at 500 ps after the optical injection and the time between successive plots is 5 ps. The injection current density is $2.5 \cdot 10^7$ A/m^2. In the situation visualized in Fig. 7.8(a–d)) dynamic transversal intensity filaments can be observed. They are a result of the spatio-temporal interaction between the injected light field, the spontaneous emission and the active semiconductor medium. The physical effects responsible for the mutual influence of the light fields are the fast carrier dynamics in the frequency domain on the one hand (fs ... ps) and the slow dynamics (ps and ns-regime) of the light propagation and of the spatial density on the other hand. In combination they lead to the formation of characteristic spatial structures in the μm-regime.

The spontaneous emission emerging on both sides of the signal beam (see e.g. the area near the output facet in Fig. 7.8(a)) can spatially migrate in

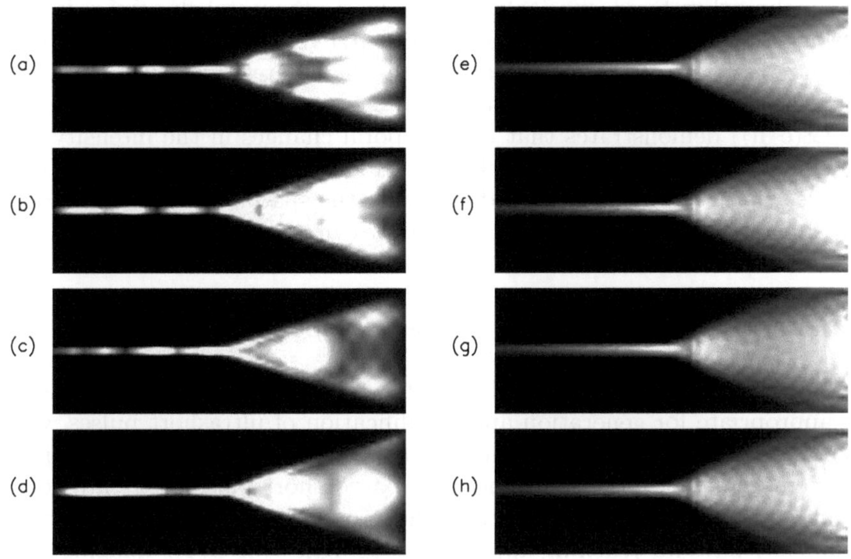

Fig. 7.8. Comparison of a tapered amplifier with increased laser-internal dynamics ((**a**–**d**), angle of the tapered section: 8.8°, facet reflectivity: 10^{-2}) and the spatio-temporal dynamics in an optimized geometry ((**e**–**h**), tapering angle: 6.0°, facet reflectivity: 10^{-4}). The time between successive plots is 5 ps

the propagation direction and in the transverse direction via interaction with the spatially inhomogeneous carrier density distributions. Thereby they may significantly affect the propagation and amplification of the injected signal. In particular, a dynamic refractive index profile builds up favoring the formation of longitudinally and transversally confined optical filaments (see e.g. Fig. 7.8(c)). These areas of increased intensity cause spatial hole burning in the carrier density. Consequently the filaments are surrounded by regions of high gain so that a spatial migration (in the lateral and the longitudinal directions) of optical filaments initiated by the nonlinear coupling between the light fields and the carrier system arises. The transition between the waveguide and the tapered section thereby acts as a source of filaments (Fig. 7.8(a)) which then are amplified and migrate towards the output facet (Fig. 7.8(b)) where a superposition with the reflected wave takes place. In the following time step (Fig. 7.8(c)) the intensity near the output facet is comparatively low and a new filament is born in the transition section restarting the processes of spatial migration and amplification (Fig. 7.8(d)). The snapshots visualize the complex light field dynamics that evolves in a tapered amplifier with improper geometry and high facet reflectivity. It is this spatio-temporal dynamics which then leads to transverse structuring in the temporally averaged nearfield intensity distributions.

If on the other hand one looks at the spatio-temporal dynamics inside the geometrical aligned structure (Fig. 7.8(e–h)) very different behavior can be observed. There the intensity uniformly distributes over the entire tapered section. Furthermore, comparison of the snapshots at different times (Fig. 7.8(e–h)) demonstrates that the temporal changes in the intensity are much smaller than in the case displayed in Fig. 7.8(a–d). However, one can clearly see that even in the adjusted structure the spatial and temporal variations subside with increasing time but never completely disappear.

7.4 Laser Amplifiers with Pulsed Injection

Due to its large amplification bandwidth the semiconductor laser is particularily appropriate for generation and amplification of ultrashort pulses with peak powers exceeding the constant output power of corresponding continuous systems. Furthermore, semiconductor based short pulse lasers are compact and efficient light sources with a high potential for integrated system design. They have far-reaching applications as light sources for imaging systems in medicine and biology [35], high-speed data transfer [36], data storage [35], optical switching [37; 38], optical measurements [39], temporally resolved detection of physical processes [40] as well as nonlinear frequency conversion [41]. To achieve high power levels that are required for many applications, a frequently used method is to amplify the pulsed radiation of e.g. a mode-locked diode oscillators (usually consisting of a single-stripe semiconductor laser in an external cavity configuration) in a large-area semiconductor laser amplifier. In recent years the impressive progress in technology and the large application bandwidth has lead to many experimental and theoretical investigations [36; 42; 43; 44; 45; 46; 47; 48; 49; 50; 51; 52; 53; 54; 55; 56; 57]. In order to obtain optimum amplifier performance in terms of the power and shape of the amplified pulses it is essential to understand the microscopic light–matter interactions (see Chap. 3) that affect the pulse during its propagation through the amplifier, determining characteristic parameters such as peak power as well as spectral and temporal pulse shape. In particular it is the geometry of the laser active area and the interaction of the propagating pulse with the spontaneous emission, which both have a strong influence not only on spatial beam properties but also on the spectral and temporal pulse properties. Here we use the Maxwell–Bloch–Langevin description to predict the laser performance and to explore the effects of the laser-internal physical processes usually concealed in experiments.

Generally, the power as well as the spectral characteristics of the amplified pulses are strongly dependent on the material properties of the amplifying medium and, in particular, on the operating conditions, such as amplifier current, pulse input power and duration. The underlying physical processes are spatio-spectral gain saturation and induced nonlinearities in the carrier system. These processes originate from macroscopic interactions in the active area as well as from microscopic spatio-spectral interactions in the carrier distributions in the valence and conduction bands and in the nonlinear interband polarization (Chap. 3).

7.4.1 Dynamic Wave Mixing and Pulse Shaping

The temporally varying amplitude of an injected light pulse leads to additional degrees of complexity when compared to the situation of continuous-wave amplifiers. Various experimental investigations [58; 59; 60; 61] have demonstrated that the total power emitted by an amplifier system is significantly influenced by spontaneous recombination. In the following we will present numerical results on a laser amplifier that has been optically injected with a coherent light pulse. In accordance with typical experimental investigations the time between successive pulses has been chosen to correspond to a repetition rate of 2.4 GHz of the pulsed laser source (i.e. 440 ps). Figure 7.9 shows typical results of calculated pulse shapes (duration 7 ps, double-sided exponentially decaying pulse shape at the entrance facet) that have passed a tapered amplifier of length 2750 μm. Note that we have set the time origin to the center of the unaffected amplified pulse (i.e. without saturation). The figures show the dependence of the intensity on the amplifier injection current for peak input powers of 80, 160 and 240 mW. At low input power levels (Fig 7.9(a)) the pulse more or less maintains its symmetric shape. An increase in injection current leads to a uniform rise in the pulse amplitude. The temporal intensity fluctuations directly reflect the dynamically varying light–matter interactions in the space and frequency domains. Comparison of the individual pulse shapes in Fig. 7.9 shows that the depth and periodicity of the modulations depend on the current and the propagating pulse. In addition, the large extension of the active area leads to characteristic spatially varying carrier diffusion and light diffraction which produce additional temporal modulations in the emitted pulse via the light field propagation. If the power of the injected signal is sufficiently high to balance the competing process of amplified spontaneous emission an efficient amplification of the injected light pulse is achieved (Fig 7.9(b)). For low injection current the resulting pulse shapes are approximately symmetric. With increasing current a temporally asymmetric pulse shape forms that marks the beginning of saturation. The leading part of the pulse ($t < 0$ in Fig. 7.9) reduces the gain available for the trailing part. Consequently, the trailing part of the pulse experiences less amplification and shows a steep decline. If the input power is further increased (Fig 7.9(c)) the pulse center and the trailing part of the pulse are hardly amplified, even if the injection current is increased. The rise in amplitude then concentrates mainly on the leading pulse part and leads to strong pulse distortion and to an increase in pulse duration. It is important to note that gain saturation is not only achieved by the pulse itself but also by the ASE which gains influence at high currents. Since the ASE reduces the density between the pulses it represents a relevant limiting factor for the

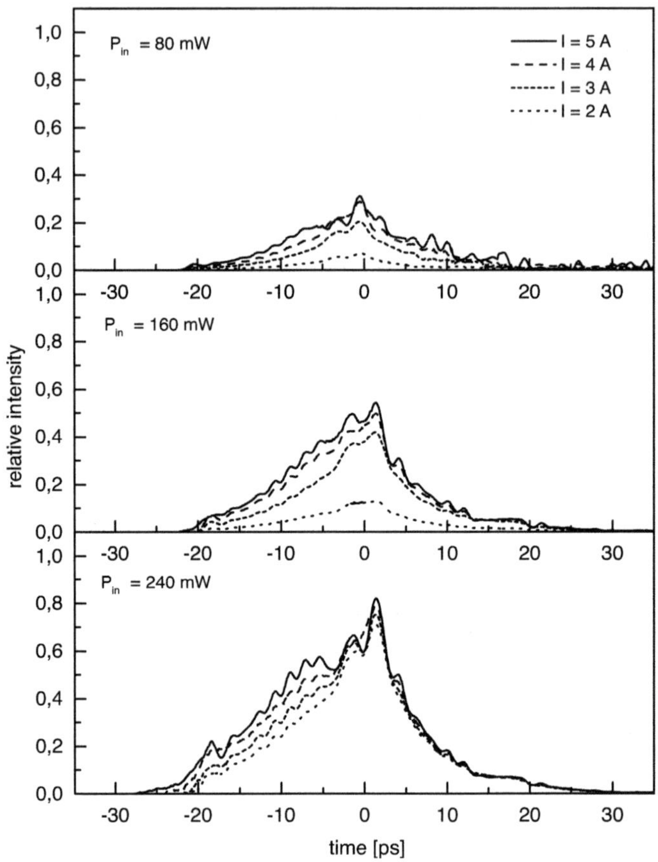

Fig. 7.9. Temporal shape of the amplified pulse after propagation in the active area of the tapered amplifier. The input pulse had a two sided exponentially decaying shape with a pulse duration of 7 ps. The peak input power is (**a**) 80 mW, (**b**) 160 mW and (**c**) 240 mW

gain available for pulse amplification. Both light fields, the field of the ASE and the field of the pulse, thus contribute to saturation.

The modulations and sub-structures that appear in the shape of the pulses are determined by microscopic carrier scattering processes (Chap. 3). The microscopic interaction processes are in general stronger in the leading part of the pulse since the carriers are still near their quasi-equilibrium when their interaction with the light field of the pulse starts. On the other hand, the relative perturbations induced by the trailing part of the pulse are significantly smaller since the carrier system is already excited. For low input power the modulations in the trailing part of the pulse have their origin in the dynamic wave mixing between the light pulse and the spontaneous emission that are of similar magnitude.

7.4.2 Dynamics of Induced and Spontaneous Emission

During the propagation of the light pulse in the active area of a semiconductor laser amplifier the spatial and spectral interactions of the injected (narrow-band) light field with the light field arising from spontaneous recombination of the electron–hole pairs within the resonator lead to an amplification of the injected radiation for sufficiently high input power levels. With the limited number of electron–hole pairs available this will then be at the expense of the gain available for amplified spontaneous emission (ASE). However, between successive pulses the entire inversion is available for the spontaneously emitted light. Consequently, it is the interplay of a propagating coherent light signal with the ASE that affects the emission characteristics of semiconductor laser amplifiers.

For a given injection current the power of the propagating pulse determines the reduction of inversion and thus the gain available for the amplification of spontaneous emission after the pulse has left the laser. In the time interval between successive pulses the injection current leads to a re-establishment of the carrier inversion, i.e. to a refilling of the spatio-spectral hole burnt by the optical pulse. This characteristic gain dynamics then determines the rise of the ASE between two pulses. We note that depending on the power of the signal and ASE a (usually small) amount of spontaneous emission also occurs during the pulse emission. Figure 7.10 shows the dynamics of a picosecond pulse (7.5 ps full width at half maximum) propagating in a single-stripe laser in double-pass amplification. After the emission of the amplified pulse at $t = 0$ ps the emitted total power first stays at a low level and then rises before the next pulse is emitted (at $t = 440$ ps). The re-establishment of inversion can be divided into two characteristic time regimes:

- During the first few 10 ps, the fast inter and intraband carrier dynamics determine the emission properties. In particular, spectral hole burning and carrier heating lead to fast carrier relaxation (see Chap. 3). The spectral inversion re-establishes on a picosecond time scale. The strong nonequilibrium distribution, however, keeps the gain available for amplification of spontaneous emission at a low level. The fast relaxation of the carriers in the frequency domain towards their quasi-equilibrium thus determines the emitted optical radiation.
- On time scales of a few 100 ps the carrier transport dynamics of the carrier density in the space domain determines the dynamics of the emission. The carrier density and consequently the gain that are locally reduced by the induced recombination are partially built up by the injection current.

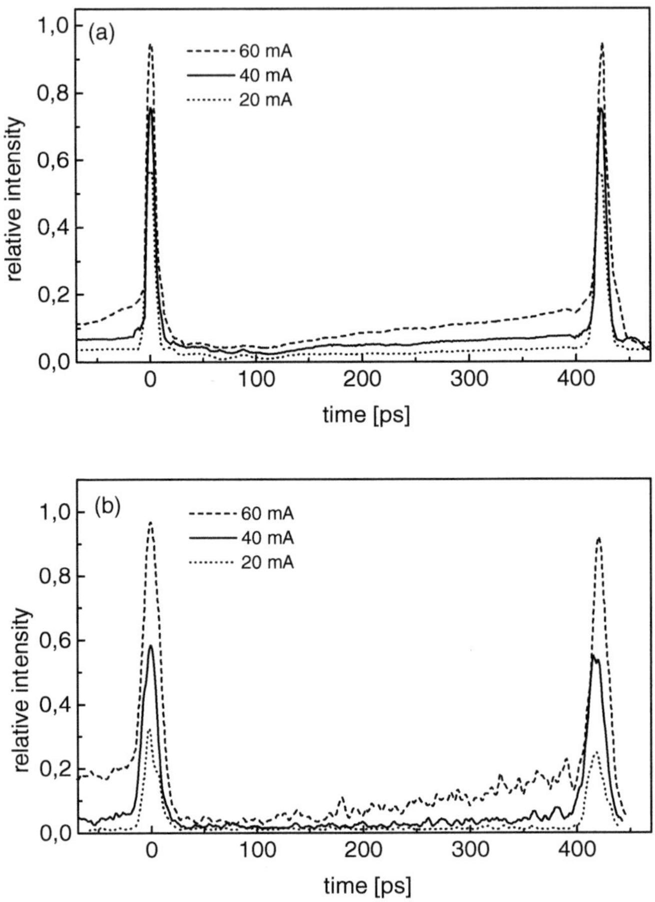

Fig. 7.10. Dynamics of a double-pass amplification of a picosecond pulse (7.5 ps full width at half maximum) propagating in a single-stripe laser (repetition rate 2.4 GHz). (**a**) Results of the numerical modelling and (**b**) experimental measurement [63] at different amplifier currents

In addition to this spectral interplay spatial effects (dimensions/geometry, propagation, diffraction, carrier diffusion, etc.) determine the amplification processes. These are particularly important in high-power (tapered) amplifiers. As an example, Fig 7.11(a–d) illustrates the dynamics of the total emitted radiation, the amplified pulse signal, the ASE and the carrier density at the output facet of a high-power tapered amplifier. The intensity arising from the spectrally narrow-band field contributions shown in Fig. 7.11(b) yields the amplified pulse signal. The corresponding intensity associated with the

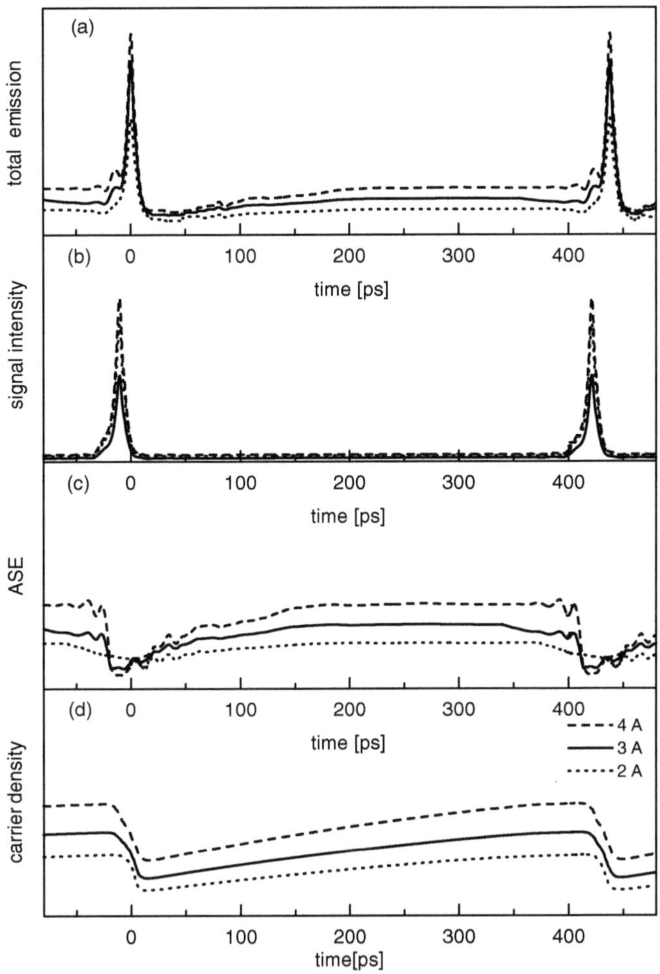

Fig. 7.11. Dynamics of (**a**) the total output power, (**b**) the amplified pulsed radiation, (**c**) the amplified spontaneous emission, (**d**) the carrier density for optical injection of the amplified pulsed radiation of the single-stripe amplifier

broad-band contributions of the optical fields (ASE) is shown in Fig. 7.11(c). In combination both light field contributions constitute the total light field (Fig 7.11(a)). Both reduce the inversion (Fig 7.11(d)). Due to continuous carrier injection through the pump current the ASE rises between successive pulses and decreases during pulse emission. As a result the part of the gain available for amplification of the individual radiation contributions is defined by the amplitudes of both field contributions at every time instance and at each location within the active area.

The spatio-spectral hole created by the pulse is spectrally located at the maximum of the gain available for the ASE so that the spontaneous light field right after the pulse emission experiences a significantly reduced inversion. Thus the ASE remains at a low level for a few ps after the pulse emission. The temporal asymmetries in the emitted power contributions of the ASE thereby indirectly reflect the dynamic changes in the gain induced by the propagating light pulse. As a result the depression in the ASE increases with increasing injection current and the minimum shifts towards earlier time steps according to the increasing temporal asymmetry in the microscopic carrier distributions. The modulations that can be seen in the ASE during the pulse emission indicate partially coherent coupling of both light fields via the inversion of the medium. The dynamics of the total carrier density (Fig 7.11(d)) depends on the injection current and on the reduction of the carrier density induced by the pulse propagating in the active area.

We emphasize that the actual field that exits the amplifier (i.e. the field that would be measured in an experiment) depends on the joint processes of spatio-temporally dependent wave mixing, scattering and transport processes linked with light field propagation and diffraction of the ASE and the coherent signal light field.

7.4.3 Self Phase Modulation

An ultrashort optical pulse propagating in an inverted semiconductor medium dynamically induces temporally varying gain and index distributions. The resulting inhomogeneous amplification and wave-guiding causes a dynamic 're-shaping' of the pulse (Chap. 6). The underlying physical interactions are, in particular, spatio-spectral hole burning, carrier heating and (carrier–carrier and carrier–phonon) relaxation. For high input power values a (spatio-spectral) saturation of the inversion is obtained. Via the nonlinear interband polarization this leads to dynamic changes in both the amplitude and phase of a propagating light signal.

In the following we will discuss the physical origins of these dynamic phase changes and analyze the influence of the spatio-spectral gain saturation on important output parameters such as power and optical spectra. The underlying processes may be discussed on the basis of Fig. 7.12 by considering the variations of the emitted light intensity (Fig. 7.12(a)), the relative nonlinear phase induced by the dynamic changes in refractive index and gain (Fig. 7.12(b)) and the instantaneous relative frequency $\delta\omega = (\omega - \omega_0)$ (ω_0 is the unperturbed frequency of the injected pulse) of the amplified radiation during the emission of the pulse (Fig. 7.12(c)). The interplay of these quantities determines the emission spectra (Fig. 7.12(d)). When the propagating light pulse saturates the amplifier the carrier density is significantly reduced by the leading part of the pulse. This gain reduction causes a temporally asymmetric amplification of the signal field resulting in an asymmetric

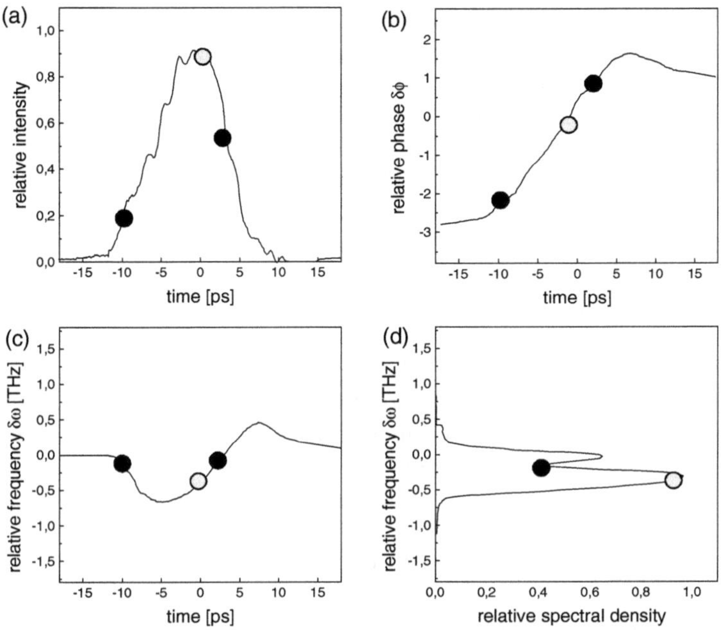

Fig. 7.12. Physical interactions and perpetual influencing of temporal variations and spectral properties: (**a**) temporal variation of the intensity, (**b**) temporal variation of the nonlinear phase, (**c**) induced frequency variation, (**d**) spectral power density

pulse shape (Fig. 7.12(a)). The variations in the carrier density and the induced changes in the gain as well as in the refractive index produce temporal changes in the phase (Fig. 7.12(b)). Thereby the specific temporal variation in phase is determined by slow and fast changes in the carrier system. First, the propagating pulse produces spatio-spectral hole burning and carrier heating (mostly due to stimulated emission) in the microscopic carrier distributions. The excited carriers then relax via carrier–carrier and carrier–phonon interaction towards the lattice temperature on a picosecond time scale. Due to the interband polarization, the microscopic carrier distributions and the propagating power affect the dynamic spectral refractive index and gain via the optical fields. These processes lead to fast time-dependent changes in phase and amplitude of the pulse during propagation. Secondly, the injected pulse power saturates the gain of the amplifier, thereby significantly reducing the spatially dependent total carrier density. This effect leads to slow, long-lived changes in the local gain and refractive index on a time scale of a few hundred ps. Since the pulse duration is short compared to the build-up of the gain (typically a few hundred ps, depending on the current density and on the amount of carrier density reduction as a consequence of the amplification of the passing light pulse), the trailing part of the pulse experiences a gain which has been reduced by the leading part of the pulse, thus causing temporal and spectral asymmetric amplitude and phase changes.

The temporal variations of the nonlinear phase ϕ created by the slow and fast changes in the carrier system is via $\delta\omega \propto -\dot{\phi}$ attributed to corresponding frequency variations (Fig. 7.12(c)). Thus a frequency may appear twice within the pulse envelope. The interference of the optical field contributions at the individual frequencies with their respective relative phase (Fig. 7.12(b)) determines the intensity which can be observed in the optical emission spectrum. The solid and open dots in Fig. 7.12 represent examples of the frequency contribution in the pulse center (open dots) and a phase difference of π between the contributing optical fields of equal frequency leading to a minimum in the spectrum (solid dots). Since the various frequency contributions originate from different parts of the pulse the asymmetric shape of the spectrum is a direct consequence of the asymmetric pulse shape due to (spatio-spectral) gain saturation. The number of minima is given by the maximum relative phase between the field contributions of equal frequency and is typically in the order of one.

Figure 7.14 shows the measured (a) and calculated (b) peak output power P_{peak} of the amplified pulses from a single-stripe laser amplifier as a function of current for input pulses of 15 mW peak power. The linear dependence indicates that the laser operates far from saturation in a regime where nonlinear effects are negligible. The calculation of a typical spectrum of this amplifier (Fig. 7.14(c)) reveals that the temporal and spectral properties of injected pulses are maintained during propagation. In comparison to this single-stripe laser amplifier, a high-power laser amplifier may exhibit more complex behavior. A calculation of the dependence of the output pulse power on the injection current demonstrates that with increasing injection current a saturation effect can be observed (Fig. 7.14). This characteristic behavior can be attributed to the nonlinear effects explained above (Fig. 7.12) which now play a more important role. The strong gain and the high peak power of the pulses in the tapered amplifier lead to various physical processes with different time scales. The fast microscopic interactions affecting the carrier distributions are spatio-spectral hole burning, carrier heating as well as the dynamic spatio-spectral wave mixing between the propagating light pulse and the ASE. The slow macroscopic changes are the long-lived reduction of the carrier density in the active area leading to spatial gain and index profiles. The combination and interplay of the microscopic and macroscopic physical processes causes self phase modulation which leads to spectral broadening and the generation of a second peak on the short wavelength side of the main peak shown in Fig. 7.15. These spectral distortions and modulations, being strongly subject to the time-dependent changes in amplitude, gain, phase, and refractive index are thus a consequence of the saturation behavior and self phase modulation of the picosecond pulses in the high-power tapered am-

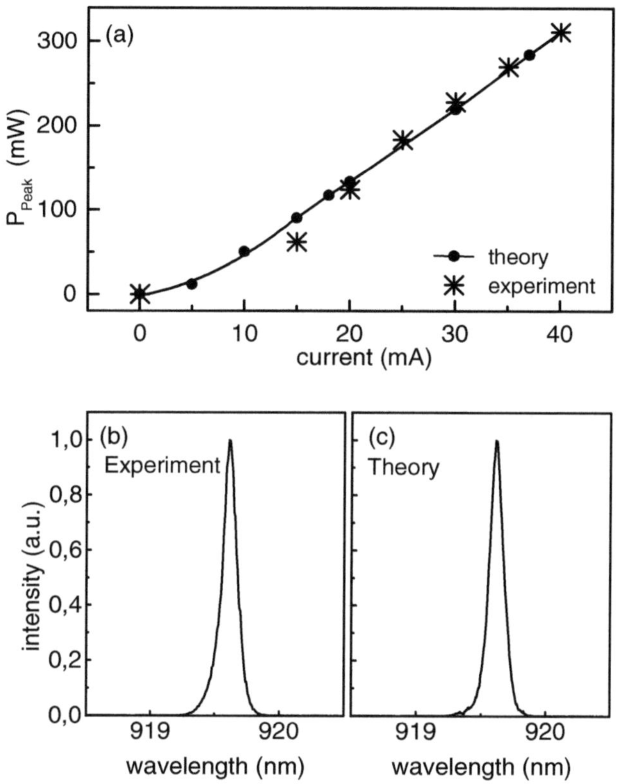

Fig. 7.13. The measured [62] and calculated peak output powers of a single-stripe amplifier as a function of the amplifier current are shown in (**a**). The optical peak input power is 15 mW. Under these operation conditions the spectrum of the output pulse from the single-stripe amplifier is shown as measured (**b**) [62] and calculated (**c**)

plifier. In particular, the gain depletion induced by ASE significantly reduces the saturation power level of high-power amplifiers. Thus, not to include the ASE, would lead to a systematic underestimation of the self phase modulation for the given operation conditions.

7.5 Conclusion

In this chapter we have given an overview of simulations of the spatio-temporal dynamics of various semiconductor laser amplifier configurations. Simulations on the basis of Maxwell–Bloch–Langevin equations include in particular the dynamics of the carrier distributions and the interband po-

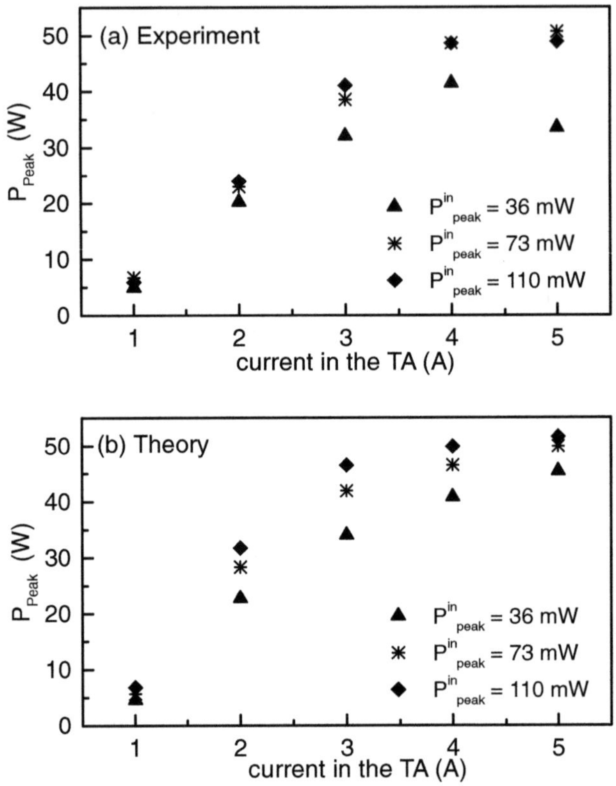

Fig. 7.14. Measured (**a**) [62] and calculated (**b**) peak power of a high-power tapered amplifier as a function of the amplifier current and peak power of the optical input pulse from the amplifier

larization which are self-consistently coupled to the macroscopic constraints imposed by the geometry, electronic contacts, epitaxial structure and facet reflectivities of the semiconductor laser. Taking into account the spectral components of the injected signal light field together with the light field originating from spontaneous electron–hole recombination allows inclusion of the spatio-spectral gain competition on a microscopic level. The microscopic modelling on the basis of the Maxwell–Bloch–Langevin description allows us to determine and analyze on a microscopical level the relevant physical processes such as carrier recombination by induced emission, amplified spontaneous emission and gain saturation as well as dynamic self-focusing. In concert, these effects determine on a fundamental level the performance of high-power semiconductor lasers. In particular, the simulations of typical amplifier configurations allows an analysis of the temporal behavior of the optical

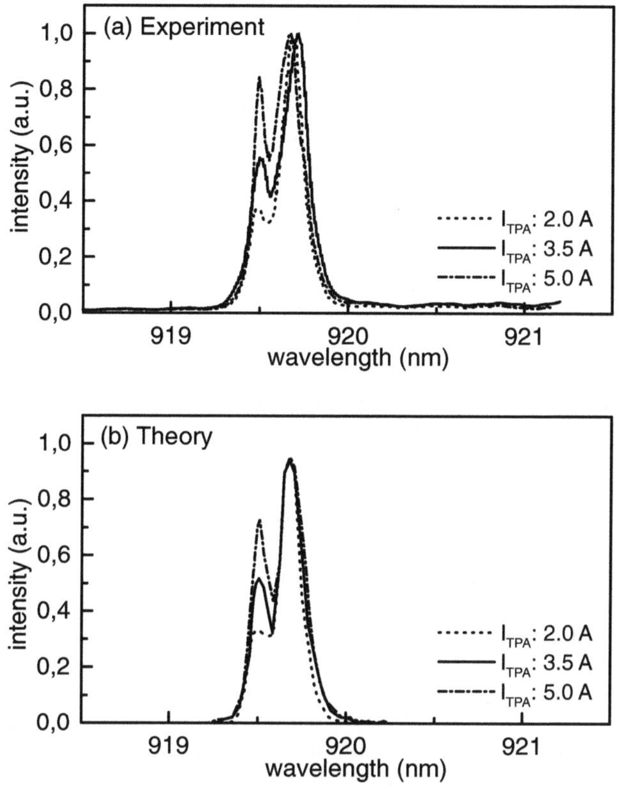

Fig. 7.15. Measured (**a**) [62] and calculated (**b**) spectrum of the output pulse from the tapered amplifier as a function of the amplifier current

fields and the carriers as well as calculation of spatial beam characteristics (farfield, nearfield) and spectra.

References

1. O. Hess, S.W. Koch, and J.V. Moloney. Filamentation and beam propagation in broad-area semiconductor lasers. *IEEE J. Quant. Electr.*, 31:35–43, 1995.
2. E. Gehrig and O. Hess. *High-Power Diode Lasers*, chapter Dynamics of High-Power Diode Lasers. Springer-Verlag, Berlin, 2000.
3. I. Fischer, O. Hess, W. Elsäßer, and E. Göbel. Complex spatio-temporal dynamics in the nearfield of a broad-area semiconductor laser. *Europhys. Lett.*, 35:579–584, 1996.
4. R. Lang, D. Mehuys, A. Hardy, K. M. Dzurko, and D. F. Welch. Spatial evolution of filaments in broad area diode laser amplifiers. *Appl. Phys. Lett.*, 62:1209–1211, 1993.

5. R. Lang, A. Hardy, R. Parke, D. Mehuys, and S. O'Brien. Spontaneous filamentation in broad-area diode laser amplifiers. *IEEE J. Quant. Electr.*, 30:658–693, 1994.
6. J. R. Marciante and G. P. Agrawal. Nonlinear mechanisms of filamentation in broad-area semiconductor lasers. *IEEE J. Quant. Electr.*, 32:590–596, 1996.
7. O. Hess. Spatio-temporal complexity in multi-stripe and broad-area semiconductor lasers. *Chaos, Solitons & Fractals*, 4:1597–1618, 1994.
8. H. Adachihara, O. Hess, E. Abraham, P. Ru, and J. V. Moloney. Spatiotemporal chaos in broad-area semiconductor lasers. *J. Opt. Soc. Am. B*, 10:658–665, 1993.
9. W. E. Plano D. Mehuys, R. S. Gees and D. F. Welch. 1.5 W visible, near-diffraction-limited, tapered-stripe optical amplifier. *Electron. Lett.*, 30:961–962, 1995.
10. L. Goldberg, D. Mehuys, M.R. Surette, and D.C. Hall. High-power, near-diffraction-limited large-area traveling-wave semiconductor amplifiers. *IEEE J. Quant. Electr.*, 29:2028–2043, 1993.
11. L. Goldberg and F. Weller. Broad area high power semiconductor optical amplifer. *Appl. Phys. Lett.*, 58:1357–1359, 1991.
12. D. Mehuys, D. F. Welch, and L. Goldberg. 2.0-W cw diffraction-limited tapered amplifier with diode injection. *Electron. Lett.*, 28:1944–1946, 1994.
13. L. Goldberg, D. Mehuys, and D. C. Hall. 3.3 W cw diffraction limited broad area semiconductor amplifier. *Electron. Lett.*, 20:1082–1084, 1992.
14. D. Mehuys, L. Goldberg, and D.F. Welch. 5.25-W cw near-diffraction-limited tapered-stripe semiconductor optical amplifier. *IEEE Photon. Technol. Lett.*, 5:1179–1182, 1993.
15. S. O'Brien, R. Lang, R. Parker, D. F. Welch, and D. Mehuys. 2.2-W continuous-wave diffraction-limited monolithically integrated master oscillator power-amplifier at 854 nm. *IEEE Photon. Technol. Lett.*, 9:440–442, 1997.
16. T. Mukai, Y. Yamamoto, T. Kimura, and W. T. Tsang, editors. *Semiconductors and Semimetals, Vol.22*, chapter Opitcal Amplification by Semiconductor Lasers. Lighwave Communications Tech., Part E, Integrated Optoelectronics, 1985.
17. T. Saitoh and T. Mukai. *Coherence, Amplification and Quantum Effect in Semiconductor Lasers*, chapter Traveling-wave semiconductor laser amplifiers. Wiley, New York, 1991.
18. M. Sagawa, K. Hiramoto, T. Toyonaka, T. Kikiawa, S. Fujisan, and K. Uomi. High-power, highly-reliable operation of InGaAs/InGaAsP 0.98 μm lasers with an exponential flared stripe. *Electron. Lett.*, 32:2277–2279, 1996.
19. G. Bendelli, K. Komori, S. Arai, and Y. Suematsu. A new structure for high-power TW-SLA. *IEEE Photon. Technol. Lett.*, 3:42–44, 1991.
20. R. Parke et al. 2.0 W CW diffraction-limited operation of a monolithically integrated master oscillator power amplifier. *IEEE Photon. Technol. Lett.*, 5:297–300, 1993.
21. E. Gehrig, O. Hess, and R. Wallenstein. Modelling of the performance of high-power diode amplifier systems with an opto-thermal microscopic spatio-temporal theory. *IEEE J. Quant. Electr.*, 35:320–331, 1999.

22. E. Gehrig and O. Hess. Nonequilibrium spatio-temporal dynamics of the Wigner-distributions in broad-area semiconductor lasers. *Phys. Rev. A*, 57:2150–2163, 1998.
23. O. Hess and T. Kuhn. Spatio-temporal dynamics of semiconductor lasers: Theory, modeling and analysis. *Prog. Quant. Electr.*, 20:85–179, 1996.
24. C. Z. Ning, J. V. Moloney, A. Ega, and R. A. Indik. A first-principles fully space-time resolved model of a semiconductor laser. *Quantum Semiclass. Opt.*, 9:1–11, 1997.
25. A. Egan, C. Z. Ning, J. V. Moloney, R. A. Indik, M. W. Wright, D. J. Bossert, and J. G. McInerney. Dynamic instabilities in master oscillator power amplifier semiconductor lasers. *IEEE J. Quant. Electr.*, 34:166–169, 1998.
26. J. V. Moloney, R. A. Indik, and C. Z. Ning. Full space-time simulation for high-brightness semiconductor lasers. *IEEE Photon. Technol. Lett.*, 9:731–733, 1997.
27. O. Hess and T. Kuhn. Maxwell-Bloch equations for spatially inhomogeneous semiconductor lasers II: Spatio-temporal dynamics. *Phys. Rev. A*, 54:3360–3368, 1996.
28. C. Z. Ning, R. A. Indik, and J. V. Moloney. Effective Bloch equations for semiconductor lasers and amplifiers. *IEEE J. Quant. Elect.*, 33:1543–1550, 1997.
29. Z. Dai, R. Michalzik, P. Unger, and K. J. Ebeling. Numerical simulation of broad-area high-power semiconductor laser amplifiers. *IEEE J. Quant. Electr.*, 343:2340–2254, 1997.
30. P. M. W. Skovgaard, J. G. McInerney, J. V. Moloney, R. A. Indik, and C. Z.Ning. Enhanced stability of MFA-MOPA semiconductor lasers using a nonlinear trumpet shaped flared. *IEEE Photon. Technol. Lett.*, 9:1220–1222, 1997.
31. L. Goldberg, M. R. Surette, and D. Mehuys. Filamentation formation in tapered GaAlAs optical amplifier. *Appl. Phys. Lett.*, 62:2304–2306, 1993.
32. R. J. Lang, A. Hardy, R. Parke, D. Mehuys, S. O'Brien, J. Major, and D. Welch. Numerical analysis of flared semiconductor laser amplifiers. *IEEE J. Quant. Electr.*, 29:2044–2051, 1993.
33. J. N. Walpole. Semiconductor amplifiers and lasers with tapered gain regions. *Opt. Quant. Electron.*, 28:623–645, 1996.
34. C. Seibert and R. Wallenstein. University of Kaiserslautern, Germany, 1999.
35. G. P. Agrawal. *Semiconductor lasers: past, present, and future.* AIP Press, Woodbury, 1995.
36. P. J. Delfyett, L. T. Florez, N. Stoffel, T. Gmitter, N. C. Andreadakis, Y. Silberberg, J. P. Heritage, and G. A. Alphonse. High-power ultrafast laser diodes. *IEEE J. Quant. Electr.*, 28:2203–2219, 1992.
37. P. W. Smith. Applications of all-optical switching and logic. *Phil Trans. R. Soc. Land.*, 313:159–165, 1984.
38. S. Scholz, O. Hess, and R. Rühle. Dynamic cross-waveguide optical switching with a nonlinear photonic band-gap structure. *Opt. Expr.*, 3:28–34, 1998.
39. P. Vasil'ev. *Ultrafast diode lasers: fundamentals and applications.* Artech House, Boston, 1995.
40. S. L. Shapiro. *Ultrashort Light Pulses.* Springer-Verlag, Berlin, 1977.

41. L. Goldberg and D. A. V. Kliner. Tunable UV generation at 286 nm by frequency tripling of a high-power mode-locked semiconductor laser. *Opt. Lett.*, 20:1640–1642, 1995.
42. G. P. Agrawal and N. A. Olsson. Self-phase modulation and spectral broadening of optical pulses in semiconductor laser amplifiers. *IEEE J. Quant. Electr.*, 25:2297–2305, 1989.
43. T. Saitoh and T. Mukai. Gain saturation characteristics of traveling-wave semiconductor laser amplifiers in short optical pulse ampification. *IEEE J. Quant. Electr.*, 26:2086–2094, 1990.
44. Y. Lai, K. L. Hall, E. P. Ippen, and G. Eisenstein. Short pulse gain saturation in InGaAsP diode laser amplifiers. *IEEE Photon. Technol. Lett.*, 2:711–713, 1990.
45. M. Trippenbach and Y. B. Band. Effects of self steepening and self-frequency shifting on short-pulse splitting in dispersive nonlinear media. *Phys. Rev. A*, 57:4791–4803, 1998.
46. C. Headley III and G. P. Agrawal. Simultanous amplification and compression of picosecond optical pulses during raman amplification in optical fibers. *J. Opt. Soc. Am. B*, 10:2383–2389, 1993.
47. V. I. Tolstikhin, M. Willander, and A. N. Mamaev. Picosecond pulse shaping using dynamic carrier heating in a gain-switched semiconductor laser. *J. Appl. Phys.*, 78:2955–2958, 1995.
48. M. Artoni and R. Loudon. Quantum theory of optical-pulse propagation through an amplifying slab. *Phys. Rev. A*, 57:622–627, 1998.
49. K. Sutkus, K. Shum, R. R. Alfano, and P. J. Delfyett. Effect of carrier heating on the wavelength chirp of ultrashort laser pulses in semiconductor traveling wave amplifiers. *IEEE Photon. Technol. Lett.*, 6:372–374, 1994.
50. A. Mecozzi and J. Mørk. Saturation induced by picosecond pulses in semiconductor optical amplifiers. *J. Opt. Soc. Am. B*, 14:761–770, 1997.
51. G. P. Agrawal and N. A. Olsson. Amplification and compression of weak picosecond optical pulses by using semiconductor-laser amplifiers. *Opt. Lett.*, 14:500–502, 1989.
52. K. L. Hall, J. Mark, E. P. Ippen, and G. Eisenstein. Femtosecond gain dynamics in InGaAsP optical amplifiers. *Appl. Phys. Lett.*, 56:1740–1742, 1990.
53. A. Dienes, J. P. Heritage, M. Y. Hong, and Y. H. Chang. Time- and spectral-domain evoluation of subpicosecond pulses in semiconductor optical amplifiers. *Opt. Lett.*, 17:1602–1604, 1992.
54. C. T. Hultgren and E. P. Ippen. Ultrafast refractive index dynamics in AlGaAs diode laser amplifiers. *Appl. Phys. Lett.*, 59:635–637, 1991.
55. A. V. Uskov, J. R. Karin, J. E. Bowers, J. G. McInerney, and J. L. Bihan. Effects of carrier cooling and carrier heating in saturation dynamics and pulse propagation through bulk semiconductor absorbers. *IEEE J. Quant. Electr.*, 34:2162–2171, 1998.
56. R. A. Indik, R. Binder, M. Mlejnek, J. V. Moloney, S. Hughes, A. Knorr, and S. W. Koch. Role of plasma cooling, heating, and memory effects in subpicosecond pulse propagation in semiconductor amplifiers. *Phys. Rev. A*, 53:3614–3620, 1996.

57. N. A. Olsson and G. P. Agrawal. Spectral shift and distortion due to self-phase modulation of picosecond pulses in 1.5 μm optical amplifiers. *Appl. Phys. Lett.*, 55:13–15, 1989.
58. L. Goldberg, D. Mehuys, and D. Welch. High-power mode-locked compound laser using a tapered semiconductor amplifier. *IEEE Photon. Technol. Lett.*, 6:1070, 1994.
59. A. Mar, R. Helkey, J. Bowers, D. Nehuys, and D. Welch. Mode-locked operation of a master oscillator power amplifier. *IEEE Photon. Technol. Lett.*, 6:1067–1069, 1994.
60. S. Gee, G. Alphonse, J. Connolly, and P. J. Delfyett. High-power mode-locked external cavity semiconductor laser using inverse bow-tie semiconductor optical amplifiers. *IEEE J. of select. topics in Quant. Electron.*, 4:209, 1998.
61. D. Woll. University of Kaiserslautern, 1999. private communication.
62. D. Woll and R. Wallenstein, University of Kaiserslautern, Germany, 1999.
63. J. Schumacher and R. Wallenstein. University of Kaiserslautern, Germany, 1999.

8 Mesoscopic Theory for Quantum Dot Lasers

In this chapter we present a mesoscopic theory for the spatio-temporal carrier and light field dynamics in quantum dot lasers. Quantum dot Maxwell–Bloch equations (QDMBEs) are set up that mesoscopically describe the spatio-temporal light field and inter/intra-level carrier dynamics in each quantum dot (QD) of a typical QD ensemble in quantum dot lasers. In particular, this includes the influence of disorder, spontaneous luminescence, counterpropagation of amplified spontaneous emission and induced recombination as well as carrier diffusion in the wetting layer (quantum well media) of the quantum dot laser. Intradot scattering via emission and absorption of phonons, as well as the scattering with the carriers and phonons of the surrounding wetting layer are dynamically included on a mesoscopic level. The spatio-temporal light field dynamics reveals a characteristic interplay of spontaneous and stimulated emission in quantum dot lasers that depends on disorder in the quantum dot ensemble, typical spatial fluctuations in size and energy levels of the quantum dots and irregularities in the spatial distribution of the quantum dots in the active layer. Those effects are simulated via statistical methods. They are shown to directly affect the propagation of an ultrashort pulse in a quantum dot waveguide. The strong influence of the localized carrier dynamics is seen in the selective depletion and re-filling of quantum dot energy levels. Results of the calculated time dependent emission characteristics of an excited quantum dot structure reveal the dynamic interplay between carrier capture and intra-dot carrier dynamics leading to characteristic relaxation dynamics and spectral mode competition.

8.1 Introduction

The quantum dot laser (QDL) is a complex nonlinear system in which the spatio-temporal dynamics of the propagating light fields is fundamentally linked with the physical properties of an ensemble of QDs. To unravel the complexity of this interplay a profound theoretical analysis of this novel laser type is necessary, in particular for interpretation of recent experimental results. Even more, clear insight is needed for technological design of quantum dot lasers with improved physical properties. In recent years the impressive technological progress in the field of quantum dot lasers has lead to the development of various new theories that specifically focus on the physical properties of QDs and QDLs. While the investigation of the electronic and optical properties of QDs represents a formidable task all by itself, for the QDL we have to set up a theoretical basis that combines the particular quantum optics of an ensemble of QDs with the particular material properties of QDs. Our mesoscopic theory bridges theoretical descriptions of microscopic material properties of QDs with macroscopic phenomenological laser theories. In Sect. 8.2 we derive and explain our theory which is based on spatially resolved Maxwell–Bloch equations for QDs. Section 8.3 shows results of our numerical modeling.

8.2 Theoretical Description: Quantum Dot Maxwell–Bloch Equations

In this chapter we will derive the quantum dot Maxwell–Bloch equations (QDMBEs). The QDMBEs consist of semiconductor Bloch equations for an ensemble of QDs (QDSBEs). Thereby, in particular, we will include various carrier relaxation processes (Sect. 8.2.2). To represent the dynamic subwavelength variations in the light field dynamics we will derive a wave equation tailored for quantum dot lasers from Maxwell's equations. The coupled system of partial differential equations then constitutes the quantum dot Maxwell–Bloch equations that model on a mesoscopic basis the dynamic light–matter interactions occurring within a QD laser.

Figure 8.1 schematically shows the typical geometry of a QDL. The active layer contains an ensemble of spatially distributed QDs that are embedded in the quantum well wetting layer (WL). Depending on the epitaxial growth

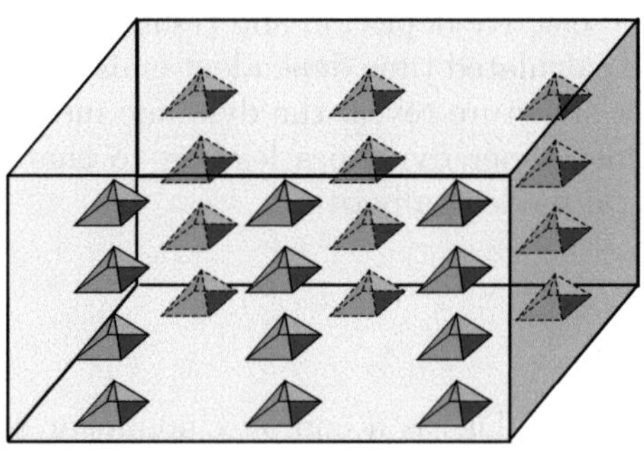

Fig. 8.1. Schematic of an idealized active layer of a quantum dot laser: columns of identical pyramidal quantum dots are aligned on a perfect grid

process the laser may consist of several layers defining vertical "QD stacks" (columns). Light propagates within the active layer in the resonator predominately along the longitudinal (z) direction. This dynamics of the light fields is described by Maxwell's wave equations ("MWEs") considering the spatio-temporal changes of the light fields propagating in the forward ('+') and backward ('−') directions in the laser (see Fig. 8.2). The layered vertical

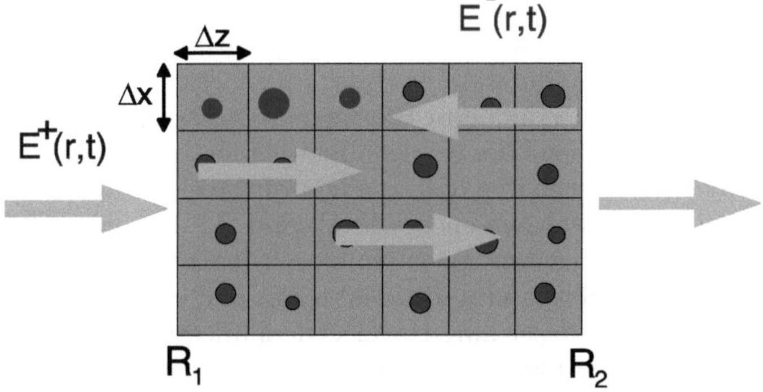

Fig. 8.2. Representation of the quantum dot laser model geometry. The counter-propagating light fields ($E^{\pm}$) spatio-temporally couple with carriers in the ensemble of quantum dots. Characteristic fluctuations in size and location of the quantum dots are effectively represented on a numerical grid with equally spaced grid points in the lateral (x) and propagation (z) direction (details see text)

(y) structure is considered via effective material and device parameters. In particular, these are the effective refractive index and the guiding properties of the layer as well as the physical properties of the QD stack (vertically averaged energy levels, damping rates or QD size). The vertically averaged physical properties characterize an 'effective' QD. The properties that enter the QDSBEs in a self-consistent way are the energy levels, the initial occupation of the levels (established e.g. via optical or electrical pumping) and the size of the QDs.

Via the polarization of the active QD medium, the light fields are locally coupled to the dynamics of the carriers and to the inter-level dipole dynamics (described on the basis of the QDSBEs). Thereby the individual time scales of the underlying optical and electronic processes spanning a temporal regime from femtoseconds up to nanoseconds are mesoscopically represented. In particular, their spatial and spectral characteristics are fully taken into account and include, e.g. the localization of the dots in the medium, fluctuations in size and shape of the QDs, the spatially dependent light field propagation and diffraction as well as spatially dependent scattering processes and carrier transport.

The time-dependent calculation of the carrier distributions and the light field dynamics allow for explicit consideration of the individual time scales of the various interaction processes. The relevant time scales range from the femtosecond regime (for the fast carrier scattering processes) up to the picosecond and nanosecond regime (for the dynamics of the propagating light fields and of the spatial carrier density).

8.2.1 Carrier Dynamics Within a Quantum Dot

Starting from the single particle density matrices for the electrons, $n^{\mathrm{e}} = \langle c^{\dagger} c \rangle$, and holes, $n^{\mathrm{h}} = \langle d^{\dagger} d \rangle$, and for the inter-level polarization, $p = \langle d^{\dagger} c \rangle$, where c and d are the local annihilation operators for electrons and holes respectively, one can derive semiconductor Bloch equations specifically for quantum dots. The resulting quantum dot semiconductor Bloch equations ('QDSBEs') mesoscopically describe the dynamic changes of the electron and hole distributions inside the dot (for each energy level) and the dynamics of the (inter-level) dipoles. If one considers an ensemble of quantum dots as the active medium in a quantum dot laser additional terms and effects are of relevance. These are contributions describing the electrical injection of carriers (pumping) Λ^{e} (including Pauli-blocking), induced recombination (with generation rate $g^{\mathrm{e,h}}$), spontaneous recombination of the carriers (Γ_{sp}), carrier–carrier and carrier–phonon scattering for the intra-dot relaxation ($\partial n^{\mathrm{e,h}}/\partial t|_{\mathrm{QD}}^{\mathrm{c-ph}}$) and the interaction with the wetting layer ($\partial n^{\mathrm{e,h}}/\partial t|_{\mathrm{QD-WL}}$). The dynamics of the occupation of electrons (e, level index 'i') and holes (h, level index 'j'), $n^{\mathrm{e,h}}$, and the dynamics of the inter-level polarizations $p^{\pm}$ (coupled to the forward (+) and backward (−) propagating optical fields) within a QD are then governed by the equations of motion

$$
\begin{aligned}
\frac{\partial n^{\mathrm{e}}(i)}{\partial t} &= \Lambda^{\mathrm{e}}(i)\left(D^{\mathrm{e}}(i) - n^{\mathrm{e}}(i)\right) + g^{\mathrm{e}}(i) - \gamma_{\mathrm{nr}} n^{\mathrm{e}}(i) \\
&\quad - \sum_j \Gamma_{\mathrm{sp}} n^{\mathrm{e}}(i)\cdot n^{\mathrm{h}}(j) + \left.\frac{\partial n^{\mathrm{e}}(i)}{\partial t}\right|_{\mathrm{QD}}^{\mathrm{c-ph}} + \left.\frac{\partial n^{\mathrm{e}}(i)}{\partial t}\right|_{\mathrm{QD-WL}} \\
\frac{\partial n^{\mathrm{h}}(j)}{\partial t} &= \Lambda^{\mathrm{h}}(j)\left(D^{\mathrm{h}}(j) - n^{\mathrm{h}}(j)\right) + g^{\mathrm{h}}(j) - \gamma_{\mathrm{nr}} n^{\mathrm{e}}(i) \\
&\quad - \sum_i \Gamma_{\mathrm{sp}} n^{\mathrm{h}}(j)\cdot n^{\mathrm{e}}(i) + \left.\frac{\partial n^{\mathrm{e}}(i)}{\partial t}\right|_{\mathrm{QD}}^{\mathrm{c-ph}} + \left.\frac{\partial n^{\mathrm{e}}(i)}{\partial t}\right|_{\mathrm{QD-WL}} \\
\frac{\partial p^{\pm}(j,i)}{\partial t} &= -(\mathrm{i}\bar{\omega}(j,i) + \gamma_{\mathrm{p}})p^{\pm}(j,i) - \frac{\mathrm{i}}{\hbar}\left[n^{\mathrm{e}}(i) + n^{\mathrm{h}}(j)\right]\mathcal{U}^{\pm} \\
&\quad - \frac{\mathrm{i}}{\hbar}\delta\mathcal{U}_{\mathrm{nl}}^{\pm} + F_{\mathrm{p}} q^{\mathrm{P}} + \left.\frac{\partial p^{\pm}(j,i)}{\partial t}\right|_{\mathrm{QD}}^{\mathrm{p-ph}}
\end{aligned}
\tag{8.1}
$$

where γ_{nr} represents the rate due to nonradiative recombination and τ_{p} denotes the dephasing time of the inter-level dipole. The pump term

$$
\Lambda^{\mathrm{c}}(l) = \Gamma_{\mathrm{QDS}} \frac{I\eta}{eh} \frac{n_{\mathrm{eq}}^{\mathrm{c}}(l)}{\sum_l n_{\mathrm{eq}}^{\mathrm{c}}(l)\left(D^{\mathrm{e,h}}(l) - n^{\mathrm{c}}(l)\right)}
\tag{8.2}
$$

mesoscopically represents the carrier injection and includes the pump-blocking effect ($\mathrm{c} = \mathrm{e,h}$ and $l = i, j$ for electrons and holes respectively). It depends on the absolute injection current, I, pump efficiency η, and the thickness of the active area, h. $D^{c}(l)$ denotes the degeneracy of an end energy level (i.e. the maximum occupation with carriers). Γ_{QDS} describes the reduction of the pump efficiency resulting from the vertically arranged QDs, i.e. the "spatial overlap" between carrier injection and a vertical stack of QDs in the medium.

The generation rates given by

$$\begin{aligned} g^{\mathrm{e}}(i) &= \mathrm{Re}\Bigg[\frac{\mathrm{i}}{\hbar}\sum_{j}[(\mathcal{U}^{+}p_{\mathrm{nl}}^{+*}(j,i)+\mathcal{U}^{-}p_{\mathrm{nl}}^{-*}(j,i)) \\ &\quad - (\mathcal{U}^{+*}p_{\mathrm{nl}}^{+}(j,i)+\mathcal{U}^{-*}p_{\mathrm{nl}}^{-}(j,i))]\Bigg] \\ g^{\mathrm{h}}(j) &= \mathrm{Re}\Bigg[\frac{\mathrm{i}}{\hbar}\sum_{i}[(\mathcal{U}^{+}p_{\mathrm{nl}}^{+*}(j,i)+\mathcal{U}^{-}p_{\mathrm{nl}}^{-*}(j,i)) \\ &\quad - (\mathcal{U}^{+*}p_{\mathrm{nl}}^{+}(j,i)+\mathcal{U}^{-*}p_{\mathrm{nl}}^{-}(j,i))]\Bigg] \end{aligned} \tag{8.3}$$

depend on the inter-level polarization p and on the optical field contributions of spontaneous and induced emission constituting the local field $\mathcal{U}^{\pm}$. The Langevin noise term $F_{\mathrm{p}}q^{\mathrm{p}}$ describes dipole fluctuations [1] with amplitude $F_{\mathrm{p}} = \frac{\Gamma\sqrt{2\hbar\epsilon_{\mathrm{r}}}}{n_{1}^{2}L\sqrt{\epsilon_{0}\omega_{0}}}$. The local fields $\mathcal{U}^{\pm} = \boldsymbol{d}(j,i)\boldsymbol{E}^{\pm} + \delta\mathcal{U}^{\pm}$ are composed of the optical light field contributions $E^{\pm}$ as well as those induced by Coulomb screening in each quantum dot and by the Coulomb interactions between the carriers in the QD and the carriers in the wetting layer, $\delta\mathcal{U}$. $\boldsymbol{d}(j,i)$ is the inter-level dipole matrix element. The inter-level polarization depends via $\bar{\omega}(j,i) = \hbar^{-1}\left(\mathcal{E}^{\mathrm{e}} + \mathcal{E}^{\mathrm{h}}\right) - \omega$ (ω is the frequency of the propagating light fields) on the carrier energies $\mathcal{E}^{\mathrm{e,h}}$ that are given by

$$\mathcal{E}^{c}(l) = \epsilon^{c}(l) + \delta\mathcal{E}^{c}(l), \tag{8.4}$$

with the unperturbed level energies $\epsilon^{\mathrm{e,h}}$ (i.e. neglecting the carrier dynamics). The characteristic level energies $\epsilon^{c}(l)$ of the unperturbed QD are taken from microscopic material calculations [2] and are self-consistently included in the theory. The Coulomb-induced screening that leads to a renormalization of these energy levels and also results in additional local field contributions strongly depends on the specific QD design (size, shape). These respective corrections have been determined in detailed calculations (e.g. [3; 4; 5]) and are represented in the QDSBEs (8.1) in the form of spatially dependent energies ($\hbar\bar{\omega}$ and local field contributions $\delta\mathcal{U}^{\pm}$).

8.2.2 Carrier Relaxation Dynamics

In each quantum dot the relaxation of the electrons and holes is determined by a variety of physical mechanisms. These are the intra-dot relaxation ($\partial n^{\mathrm{e,h}}(l)/\partial t|_{\mathrm{QD}}$) via acoustic and optical phonons or multi-phonon processes and scattering ($\partial n^{\mathrm{e,h}}(i)/\partial t|_{\mathrm{QD-WL}}$) between the carriers in the QD and the carriers and phonons of the wetting layer. The physical properties of an individual QD (size, shape, energy levels) and the phonon distribution thereby determine the relevance of the various relaxation processes. Here we will use dynamic scattering rates for carrier–phonon relaxation processes on the basis of microscopically determined matrix elements for the respective interaction. The elastic scattering between the QD carriers and the carriers of the wetting layer will be considered on the basis of perturbation theory.

Intra-dot Relaxation

The scattering rates for carrier–phonon intra-dot relaxation generally include emission ('−') and absorption ('+') of longitudinal acoustical (LA) phonons, longitudinal optical (LO) phonons, and, in particular, multiphonon processes (± 2 LO, ± 2 LA, and ± LO ± LA). They are determined on the basis of microscopic calculations, allowing a self-consistent mesoscopic inclusion of all scattering processes that are relevant in QD lasers via a dependence of the rates on the spatially and temporally varying carrier and phonon distributions within the QD and the surrounding layers. The quantum kinetic equations of motion of the single particle density matrices with respect to the carrier–phonon Hamiltonian are given by

$$
\begin{aligned}
\left.\frac{\partial n^{\mathrm{e}}}{\partial t}\right|_{\mathrm{QD}}^{\mathrm{e-ph}}(i) &= 2\sum_{\boldsymbol{q}}\sum_{i_1,i_2}\Big[\,\mathrm{Re}[s^{\mathrm{e}}_{\boldsymbol{q}}(i_1,i)] - \mathrm{Re}[s^{\mathrm{e}}_{\boldsymbol{q}}(i,i_2)]\Big] \\
\left.\frac{\partial n^{\mathrm{h}}}{\partial t}\right|_{\mathrm{QD}}^{\mathrm{h-ph}}(j) &= 2\sum_{\boldsymbol{q}}\sum_{j_1,j_2}\Big[\,\mathrm{Re}[s^{\mathrm{h}}_{\boldsymbol{q}}(j_1,j)] - \mathrm{Re}[s^{\mathrm{h}}_{\boldsymbol{q}}(j,j_2)]\Big] \\
\left.\frac{\partial p}{\partial t}\right|_{\mathrm{QD}}^{\mathrm{p-ph}}(j,i) &= \sum_{\boldsymbol{q}}\Big[\sum_{j_1}[t^{(+)}(j_1,i) - t^{(-)*}(j_1,i)] \\
&\quad - \sum_{i_1}[t^{(+)}(j,i_1) - t^{(-)*}(j,i_1)]\Big]
\end{aligned}
\tag{8.5}
$$

where i and j are the initial (in the case of phonon emission) or final (in the case of phonon absorption) levels of the electrons (i) and holes (j) and i_1, i_2, j_1, j_2 denote intermediate levels from which in and out scattering occurs via the emission and absorption of phonons. The intraband density matrices

$$
\begin{aligned}
s^{\mathrm{e}}_{\boldsymbol{q}}(i_1,i_2) &= \frac{\mathrm{i}}{\hbar}\left\langle g^{\mathrm{e}}_{\boldsymbol{q}} c^{\dagger}_{i_1} b_{\boldsymbol{q}} c_{i_2}\right\rangle \\
s^{\mathrm{h}}_{\boldsymbol{q}}(j_1,j_2) &= \frac{\mathrm{i}}{\hbar}\left\langle g^{\mathrm{h}}_{\boldsymbol{q}} d^{\dagger}_{j_1} b_{\boldsymbol{q}} d_{j_2}\right\rangle
\end{aligned}
\tag{8.6}
$$

depend on the carrier (c_i, d_j) and phonon ($b_{\boldsymbol{q}}$) creation and annihilation operators ($\boldsymbol{q}$ denotes the phonon wavevector). They describe energy relaxation. $g^{\mathrm{e,h}}_{\boldsymbol{q}}$ is the coupling constant of the respective carrier–phonon interaction [6]. The inter-level density matrices

$$
\begin{aligned}
t_{\boldsymbol{q}}^{(+)}(j,i) &= \frac{\mathrm{i}}{\hbar}\left\langle g_{\boldsymbol{q}}^{\mathrm{p}} d_j b_{\boldsymbol{q}} c_i \right\rangle \\
t_{\boldsymbol{q}}^{(-)*}(j,i) &= \frac{\mathrm{i}}{\hbar}\left\langle g_{\boldsymbol{q}}^{\mathrm{p}} d_j b_{\boldsymbol{q}}^{\dagger} c_i \right\rangle
\end{aligned}
\tag{8.7}
$$

describe phase relaxation processes.

We start from the Heisenberg picture to derive a quantum kinetic equation of motion for the phonon assisted density matrices. To include the emission and absorption of phonons contributions up to second order have to be considered. Factorization of the two-particle matrices into two single-particle density matrices, e.g. $\left\langle c^{\dagger} c^{\dagger} b_{\boldsymbol{q}'}^{\dagger} b_{\boldsymbol{q}} \right\rangle = n^{\mathrm{e}} n_{\boldsymbol{q}} \delta_{\boldsymbol{q},\boldsymbol{q}'}$, and using the Markov approximation (i.e. assuming slowly varying distributions) leads to a semiclassical description where only the expectation values of the electrons, holes and the inter-level polarization appear and cross correlation effects are neglected. The reduced equations of motion then read

$$
\begin{aligned}
\frac{\partial s^{\mathrm{e}}(i_1,i_2)}{\partial t} =& -\left(\frac{\mathrm{i}}{\hbar}(\mathcal{E}_{i_1} - \mathcal{E}_{i_2} - \mathcal{E}_{\mathrm{ph}}) + \gamma_q^{\mathrm{e}}\right) s^{\mathrm{e}}(i_1,i_2) \\
& + \frac{1}{\hbar^2}\left|g_q^{\mathrm{e}}\right|^2 \Big[(n_q+1)\, n^{\mathrm{e}}(i_1)\left(D^{\mathrm{e}}(i_2) - n^{\mathrm{e}}(i_2)\right) \\
& - n_q n^{\mathrm{e}}(i_2)\left(D^{\mathrm{e}}(i_1) - n^{\mathrm{e}}(i_1)\right)\Big] \\
& - \sum_j \left(\mathrm{i}\omega(j,i_1) + \gamma_q^{\mathrm{h}}\right)^* t^{(+)}(j,i_2) \\
& - \sum_j \left(\mathrm{i}\omega(j,i_2) + \gamma_q^{\mathrm{h}}\right) t^{(-)}(j,i_1) \\
& - \sum_j \frac{1}{\hbar^2} g_q^{\mathrm{e}} g_q^{\mathrm{h}*} p^*(j,i_1) p(j,i_2) \\
\frac{\partial s^{\mathrm{h}}(j_1,j_2)}{\partial t} =& -\left(\frac{\mathrm{i}}{\hbar}(\mathcal{E}_{j_1} - \mathcal{E}_{j_2} - \mathcal{E}_{\mathrm{ph}}) + \gamma_q^{\mathrm{h}}\right) s^{\mathrm{h}}(j_1,j_2) \\
& + \frac{1}{\hbar^2}\left|g_q^{\mathrm{h}}\right|^2 \Big[(n_q+1)\, n^{\mathrm{h}}(j_1)\left(D^{\mathrm{h}}(j_2) - n^{\mathrm{h}}(j_2)\right) \\
& - n_q n^{\mathrm{h}}(j_2)\left(D^{\mathrm{h}}(j_1) - n^{\mathrm{h}}(j_1)\right)\Big] \\
& - \sum_i \left(\mathrm{i}\omega(j_1,i) + \gamma_q^{\mathrm{e}}\right)^* t^{(+)}(j_2,i) \\
& - \sum_i \left(\mathrm{i}\omega(j_2,i) + \gamma_q^{\mathrm{e}}\right) t^{(-)}(j_1,i) \\
& - \sum_j \frac{1}{\hbar^2} g_q^{\mathrm{h}} g_q^{\mathrm{e}*} p^*(j_1,i) p(j_2,i)
\end{aligned}
$$

$$\begin{aligned}\frac{\partial t^{(+)}(j,i)}{\partial t} =& -\left(\frac{\mathrm{i}}{\hbar}(\mathcal{E}_j+\mathcal{E}_i-\mathcal{E}_{\mathrm{ph}})+\gamma_q^{\mathrm{p}}\right)t^{(+)}(j,i)\\ &-\sum_j\frac{1}{\hbar^2}\left|g_q^{\mathrm{p}}\right|^2\left[n_q n^{\mathrm{h}}(j)+(n_q+1)\left(D^{\mathrm{h}}(j)-n^{\mathrm{h}}(j)\right)\right]p(j,i)\\ &-\sum_i\frac{1}{\hbar^2}\left|g_q^{\mathrm{p}}\right|^2\left[n_q n^{\mathrm{e}}(i)+(n_q+1)\left(D^{\mathrm{e}}(i)-n^{\mathrm{e}}(i)\right)\right]p(j,i)\\ &+\sum_{i_1}\left(\mathrm{i}\omega(j,i_1)+\gamma_q^{\mathrm{e}}\right)^* s^{\mathrm{e}}(i_1,i)\\ &-\sum_{j_1}\left(\mathrm{i}\omega(j_1,i)+\gamma_q^{\mathrm{h}}\right)s^{\mathrm{h}}(j_1,j)\end{aligned} \tag{8.8}$$

where the damping arising from higher order terms is included in $\gamma_q^{\mathrm{c}} = \gamma_q^{\mathrm{c}}(l_1)+\gamma_q^{\mathrm{c}}(l_2)$ and γ_q^{p}.

Typical time scales of the phonon dynamics are slow in comparison with the carrier dynamics. Moreover, due to conservation of energy only phonons with energies close to resonance with the level energies of the QDs contribute significantly to the carrier–phonon interactions. Thus one can adiabatically eliminate the dynamics of the density matrices. Furthermore, the phonon distributions are approximated by their quasi-equilibrium distribution given by the respective Bose statistics, $n_q = 1/(\exp(\hbar\omega_q/(kT))-1)$ with phonon frequency ω_q. Insertion of the resulting expression for the matrices $s^{\mathrm{e,h}}$, $t^{(\pm)}$ yields the following rates describing the change in level occupation by in and outscattering processes

$$\begin{aligned}\left.\frac{\partial n^{\mathrm{e}}(i)}{\partial t}\right|_{\mathrm{QD}}^{\mathrm{e-ph}} =& \sum_{i_1>i}2\frac{|g^{\mathrm{e}}|^2}{\hbar}\mathcal{L}^{\mathrm{e}}(i_1,i)\\ &\Big[(n_q+1)n^{\mathrm{e}}(i_1)\left(D^{\mathrm{e}}(i)-n^{\mathrm{e}}(i)\right)\\ &-n_q n^{\mathrm{e}}(i)\left(D^{\mathrm{e}}(i_1)-n^{\mathrm{e}}(i_1)\right)\Big]\\ &-\sum_{i_1<i}2\frac{|g^{\mathrm{e}}|^2}{\hbar}\mathcal{L}(i,i_1)\\ &\Big[(n_q+1)n^{\mathrm{e}}(i)\left(D^{\mathrm{e}}(i_1)-n^{\mathrm{e}}(i_1)\right)\\ &-n_q n^{\mathrm{e}}(i1)\left(D^{\mathrm{e}}(i)-n^{\mathrm{e}}(i)\right)\Big]\\ &-\sum_j\Big[\sum_{j_1<j}[2\frac{g^{\mathrm{h}}g^{\mathrm{e}*}}{\hbar}\mathcal{L}(j,j_1)p(i,j_1)p^*(i,j)\\ &-2\frac{g^{\mathrm{h}*}g^{\mathrm{e}}}{\hbar}\mathcal{L}(j,j_1)p^*(i,j_1)p(i,j)]\end{aligned}$$

$$
\begin{aligned}
&-\sum_{j_1>j}[2\frac{g^{\mathrm{h}}g^{\mathrm{e}*}}{\hbar}\mathcal{L}(j_1,j)p(i,j_1)p^*(i,j)\\
&-2\frac{g^{\mathrm{h}*}g^{\mathrm{e}}}{\hbar}\mathcal{L}(j_1,j)p^*(i,j_1)p(i,j)]\Bigg]
\end{aligned}
$$

$$
\begin{aligned}
\left.\frac{\partial n^{\mathrm{h}}(j)}{\partial t}\right|_{\mathrm{QD}}^{\mathrm{h-ph}} = &\sum_{j_1>j} 2\frac{\left|g^{\mathrm{h}}\right|^2}{\hbar^2}\mathcal{L}^{\mathrm{h}}(j_1,j)\\
&\Big[(n_q+1)n^{\mathrm{h}}(j_1)\left(D^{\mathrm{h}}(j)-n^{\mathrm{h}}(j)\right)\\
&-n_q n^{\mathrm{h}}(j)\left(D^{\mathrm{h}}(j_1)-n^{\mathrm{h}}(j_1)\right)\Big]\\
&-\sum_{j_1<j} 2\frac{\left|g^{\mathrm{h}}\right|^2}{\hbar^2}\mathcal{L}^{\mathrm{h}}(j,j_1)\\
&\Big[(n_q+1)n^{\mathrm{h}}(j)\left(D^{\mathrm{h}}(j_1)-n^{\mathrm{h}}(j_1)\right)\\
&-n_q n^{\mathrm{h}}(j_1)\left(D^{\mathrm{h}}(j)-n^{\mathrm{h}}(j)\right)\Big]\\
&-\sum_i\Bigg[\sum_{i_1<i}[2\frac{g^{\mathrm{e}}g^{\mathrm{h}*}}{\hbar}\mathcal{L}(i,i_1)p(i_1,j)p^*(i,j)\\
&-2\frac{g^{\mathrm{e}*}g^{\mathrm{h}}}{\hbar}\mathcal{L}(i,i_1)p^*(i_1,j)p(i,j)]\\
&-\sum_{i1>i}[2\frac{g^{\mathrm{e}}g^{\mathrm{h}*}}{\hbar}\mathcal{L}(i_1,i)p(i_1,j)p^*(i,j)\\
&-2\frac{g^{\mathrm{e}*}g^{\mathrm{h}}}{\hbar}\mathcal{L}(i_1,i)p^*(i_1,j)p(i,j)]\Bigg]
\end{aligned}
\tag{8.9}
$$

The function $\mathcal{L}$ describes the dependence of the carrier–phonon interaction on the contributing QD level energies and the energy of the respective phonon. It also contains the damping resulting from higher order contributions. For carriers interacting with an optical (LO) phonon, $\mathcal{L}$ can be expressed by a Lorentzian line shape

$$
\mathcal{L}(l_1,l_2) = \frac{\tau_{\mathrm{LO}}^{-1}}{\hbar^{-2}(\mathcal{E}_{l_1}-\mathcal{E}_{l_2}-\mathcal{E}_{\mathrm{LO}})^2+\tau_{\mathrm{LO}}^{-2}} \tag{8.10}
$$

with $l=i$ for electrons and $l=j$ for holes. The lifetime τ_{LO} with

$$
\tau_{\mathrm{LO}}^{-1} = \frac{2\pi}{\hbar^2}\left|V^{\mathrm{Anh}}\right|^2\sum_q \delta\left(2\omega_{\mathrm{LA}}(q)-\omega_{\mathrm{LO}}\right)[2n_q+1] \tag{8.11}
$$

includes the decay of optical phonons into two acoustical phonons via the anharmonic interaction potential V^{Anh}. The lifetime of acoustical phonons is usually much longer than the lifetime of optical phonons. As a result, the function $\mathcal{L}$ can be approximated by a δ - function, $\mathcal{L}^{c}(l_1, l_2) = \delta(\mathcal{E}_{l_1} - \mathcal{E}_{l_2} - \mathcal{E}_{\mathrm{ph}})$ in the case of direct interaction of carriers with acoustical phonons.

In addition to the emission and absorption of one single phonon the influence of multi-phonon relaxation has to be considered. Among all multi-phonon processes the most relevant ones are the emission/absorption of an LO phonon accompanied by the absorption/emission of an acoustical phonon ($\pm$ LO $\pm$ LA) and the emission/absorption of two acoustical phonons ($\pm$ LA $\pm$ LA). The respective relaxation terms derived in analogy to the above read

$$
\begin{aligned}
\left.\frac{\partial n^{c}}{\partial t}\right|_{\mathrm{c-2ph}}(l) = 2\sum_{l_1>l}\sum_{l_1>l>l_2} \frac{1}{\hbar}\Big| & g_1^{c}(q_1) g_2^{c}(q_2) \mathcal{L}_{q_1}(l_1, l_2) \\
& + g^{e}(q_1) g^{e}(q_2) \mathcal{L}_{q_2}(l, l_2)\Big|^2 \\
& \Big[(n_{q_2} n_{q_1} + 1) n^{c}(l_1)\left(D^{c}(l) - n^{c}(l)\right) \\
- \quad & (n_{q_2} + 1) n_{q_1} n^{c}(i)\left(D^{c}(l_1) - n^{c}(l_1)\right)\Big] \\
& + 2\sum_{l_1>l}\sum_{l_1>l_2>l} \frac{1}{\hbar}\Big| g_1^{c}(q_1) g_2^{c}(q_2) \mathcal{L}_{q_1}(l_2, l) \\
& + g_1^{c}(q_1) g_2^{c}(q_2) \mathcal{L}_{q_2}(l_1, l_2)\Big|^2 \\
& \Big[(n_{q_2} + 1)(n_{q_1} + 1) n^{c}(l_1)\left(D^{c}(l) - n^{c}(l)\right) \\
- \quad & n_{q_2} n_{q_1} n^{c}(i)\left(D^{c}(l_1) - n^{c}(l_1)\right)\Big] \\
& + 2\sum_{l_1>l}\sum_{l_2>l_1>l} \frac{1}{\hbar}\Big| g_1^{c}(q_1) g_2^{c}(q_2) \mathcal{L}_{q_1}(l_2, l) \\
& + g_1^{c}(q_1) g_2^{c}(q_2) \mathcal{L}_{q_2}(l_2, l_1)\Big|^2 \\
& \Big[n_{q_2}(n_{q_1} + 1) n^{c}(l_1)\left(D^{c}(l) - n^{c}(l)\right) \\
- \quad & (n_{q_2} + 1) n_{q_1} n^{c}(i)\left(D^{c}(l_1) - n^{c}(l_1)\right)\Big] \\
& + 2\sum_{l>l_1}\sum_{l>l_1>l_2} \frac{1}{\hbar}\Big| g_1^{c}(q_1) g_2^{c}(q_2) \mathcal{L}_{q_1}(l, l_2) \\
& + g^{e}(q_1) g^{e}(q_2) \mathcal{L}_{q_2}(l_1, l_2)\Big|^2
\end{aligned}
$$

$$
\begin{aligned}
& \Big[(n_{q_2}+1)n_{q_1}n^{\mathrm{c}}(l_1)\left(D^{\mathrm{c}}(l)-n^{\mathrm{c}}(l)\right) \\
- \quad & (n_{q_2}+1)n_{q_1}n^{\mathrm{c}}(i)\left(D^{\mathrm{c}}(l_1)-n^{\mathrm{c}}(l_1)\right)\Big] \\
& +2\sum_{l>l_1}\sum_{l>l_2>l_1}\frac{1}{\hbar}\Big|g_1^{\mathrm{c}}(q_1)g_2^{\mathrm{c}}(q_2)\mathcal{L}_{q_1}(l,l_2) \\
& +g_1^{\mathrm{c}}(q_1)g_2^{\mathrm{c}}(q_2)\mathcal{L}_{q_2}(l_2,l_1)\Big|^2 \\
& \Big[n_{q_2}n_{q_1}n^{\mathrm{c}}(l_1)\left(D^{\mathrm{c}}(l)-n^{\mathrm{c}}(l)\right) \\
- \quad & (n_{q_2}+1)(n_{q_1}+1)n^{\mathrm{c}}(i)\left(D^{\mathrm{c}}(l_1)-n^{\mathrm{c}}(l_1)\right)\Big] \\
& +2\sum_{l_1>l}\sum_{l_2>l>l_1}\frac{1}{\hbar}\Big|g_1^{\mathrm{c}}(q_1)g_2^{\mathrm{c}}(q_2)\mathcal{L}_{q_1}(l_2,l_1) \\
& +g_1^{\mathrm{c}}(q_1)g_2^{\mathrm{c}}(q_2)\mathcal{L}_{q_2}(l_2,l)\Big|^2 \\
& \Big[(n_{q_2}+1)n_{q_1}n^{\mathrm{c}}(l_1)\left(D^{\mathrm{c}}(l)-n^{\mathrm{c}}(l)\right) \\
- \quad & n_{q_2}(n_{q_1}+1)n^{\mathrm{c}}(i)\left(D^{\mathrm{c}}(l_1)-n^{\mathrm{c}}(l_1)\right)\Big]
\end{aligned}
\tag{8.12}
$$

where $\mathcal{L}_{q_1}$ and $\mathcal{L}_{q_1}$ depend on the energy and damping rate of the respective phonon.

Scattering Processes Between Quantum Dots and the Wetting Layer

In addition to the intra-dot relaxation, the dynamics of the quantum dot laser depends on the carrier–carrier and carrier–phonon scattering processes that occur between the QDs and the wetting layer in which they are embedded. Those are both inelastic emission and absorption of phonons as well as elastic collision processes. For the inelastic scattering processes we will consider the inelastic Coulomb interaction between QD carriers and the 2D carrier plasma of the wetting layer via Auger recombination, the ionization of a QD via excitation of carriers by absorption of a phonon as well as carrier capture from the wetting layer in a (up to then unoccupied) state of the QD by emission of a phonon, i.e.

$$
\left.\frac{\partial n^{\mathrm{c}}(l)}{\partial t}\right|_{\mathrm{QD-WL}}=\left.\frac{\partial n^{\mathrm{c}}(l)}{\partial t}\right|_{\mathrm{QD-WL}}^{\mathrm{Aug}}+\left.\frac{\partial n^{\mathrm{c}}(l)}{\partial t}\right|_{\mathrm{QD-WL}}^{\mathrm{c-ph}}. \tag{8.13}
$$

The relaxation rates

$$\left.\frac{\partial n^{\mathrm{c}}}{\partial t}\right|_{\mathrm{QD-WL}}^{\mathrm{c-ph}}(l) = \left[f^{\mathrm{e}}_{\mathrm{QD-WL}} N^{\mathrm{c}}_{\mathrm{WL}} \left(D^{\mathrm{c}}(l) - n^{\mathrm{c}}(l)\right)(n_q + 1) \right.$$
$$\left. - n^{\mathrm{c}} n_q \left(\Delta\mathcal{E}(l)\right) \right] \mathcal{L}(l, \infty) \qquad (8.14)$$

describe the interactions between the discrete energy levels of the QDs and the states of the surrounding quantum well layer via emission and absorption of optical and acoustical phonons, where the level description ∞ in $\mathcal{L}(l,\infty)$ refers to the respective energy of the valence and conduction bands of the wetting layer. In (8.14) the scaling factor $f^{\mathrm{e}}_{\mathrm{QD-WL}}$ represents the fraction of wetting layer states to which a single (effective) QD couples. It is determined by the dot density and by the epitaxial structure defining e.g. the potential barrier between the QD and the surrounding layers. $n_q(\Delta\mathcal{E}(l))$ denotes the phonon number available at energy values higher than the energy given by the potential step between the respective QD level and the wetting layer. The Auger carrier capture kinetics may be attributed to the following processes: (1) A QD electron or hole in the wetting layer collides with a 2D electron and is captured by the QD. The final state of the second 2D electron is then a wetting layer state of higher energy. (2) A 2D hole is captured via Coulomb scattering with a QD electron by the dot while the electron is excited into a wetting layer state. We will represent the two types of processes by the rates [7]

$$\left.\frac{\partial n^{\mathrm{e}}}{\partial t}\right|_{\mathrm{QD-WL}}^{\mathrm{Aug}}(i) = -\sum_j B_{\mathrm{he}} N^{\mathrm{h}}_{\mathrm{WL}} n^{\mathrm{e}}(i)(D^{\mathrm{h}}(j) - n^{\mathrm{h}}(j))$$
$$+ C_{\mathrm{ee}} {N^{\mathrm{e}}_{\mathrm{WL}}}^2 (D^{\mathrm{e}}(i) - n^{\mathrm{e}}(i))$$
$$+ C_{\mathrm{eh}} N^{\mathrm{e}}_{\mathrm{WL}} N^{\mathrm{h}}_{\mathrm{WL}} (D^{\mathrm{e}}(i) - n^{\mathrm{e}}(i))$$
$$\left.\frac{\partial n^{\mathrm{e}}}{\partial t}\right|_{\mathrm{QD-WL}}^{\mathrm{Aug}}(j) = B_{\mathrm{he}} N^{\mathrm{h}}_{\mathrm{wl}} \sum_i n^{\mathrm{e}}(i)(D^{\mathrm{h}}(j) - n^{\mathrm{h}}(j))$$
$$+ C_{\mathrm{hh}} {N^{\mathrm{h}}_{\mathrm{WL}}}^2 (D^{\mathrm{h}}(j) - n^{\mathrm{h}}(j))$$
$$+ C_{\mathrm{he}} N^{\mathrm{h}}_{\mathrm{WL}} N^{\mathrm{e}}_{\mathrm{WL}} (D^{\mathrm{h}}(j) - n^{\mathrm{h}}(j)) \qquad (8.15)$$

In the wetting layer the carriers are not as strongly localized as in the quantum dot islands and may therefore diffuse within the layer. With $N^{\mathrm{e,h}}_{\mathrm{WL}}$ denoting the local density of electrons (c = e) and holes (c = h), the dynamics of wetting layer carriers is represented by the diffusion equation

$$\frac{\partial N^{\mathrm{c}}_{\mathrm{WL}}}{\partial t} = \frac{J}{ed} + \left.\frac{\partial N^{\mathrm{c}}}{\partial t}\right|_{\mathrm{QD-WL}}$$
$$- \gamma_{\mathrm{sp}} N^{\mathrm{e}}_{\mathrm{WL}} N^{\mathrm{h}}_{\mathrm{WL}} - \gamma^{\mathrm{nr}}_{\mathrm{WL}} N^{\mathrm{c}}_{\mathrm{WL}}, \qquad (8.16)$$

with a pump term describing carrier injection and a rate for nonradiative emission processes. In (8.16) the change in carrier density due to Auger relaxation can be expressed as [7]

$$\begin{aligned}
\left.\frac{\partial N^{\mathrm{e}}}{\partial t}\right|_{\mathrm{QD-WL}} &= B_{\mathrm{he}} N_{\mathrm{WL}}^{\mathrm{h}} n^{\mathrm{e}}(i)(D^{\mathrm{h}}(j) - n^{\mathrm{h}}(j)) n_{\mathrm{QD}} \\
&\quad - C_{\mathrm{ee}} {N_{\mathrm{WL}}^{\mathrm{e}}}^{2} (D^{\mathrm{e}}(i) - n^{\mathrm{e}}(i)) n_{\mathrm{QD}} \\
&\quad - C_{\mathrm{eh}} N_{\mathrm{WL}}^{\mathrm{e}} N_{\mathrm{WL}}^{\mathrm{h}} (D^{\mathrm{e}}(i) - n^{\mathrm{e}}(i)) n_{\mathrm{QD}} \\
\left.\frac{\partial N^{\mathrm{e}}}{\partial t}\right|_{\mathrm{QD-WL}} &= -B_{\mathrm{he}} N_{\mathrm{WL}}^{\mathrm{h}} f^{\mathrm{e}}(i)(D^{\mathrm{h}}(j) - n^{\mathrm{h}}(j)) n_{\mathrm{QD}} \\
&\quad - C_{\mathrm{hh}} {N_{\mathrm{WL}}^{\mathrm{h}}}^{2} (D^{\mathrm{h}}(i) - n^{\mathrm{h}}(i)) n_{\mathrm{QD}} \\
&\quad - C_{\mathrm{he}} N_{\mathrm{WL}}^{\mathrm{h}} N_{\mathrm{WL}}^{\mathrm{e}} (D^{\mathrm{h}}(j) - n^{\mathrm{h}}(i)) n_{\mathrm{QD}}
\end{aligned} \tag{8.17}$$

In (8.15) and (8.17) B_{he}, C_{ee}, C_{eh} and C_{he} are the respective Auger capture coefficients which we take from the detailed calculation in [7]. n_{QD} is the dot density.

The elastic Coulomb scattering processes between the QDs and the wetting layer are treated on the basis of perturbation theory [8; 9]. Elastic collisions do not change the occupation of the levels. However, they may lead to significant changes in electronic energies and damping that result in a spectral shift and spectral broadening represented by an energy correction term $\delta\mathcal{E}^{\mathrm{e,h}}$ and the dipole damping τ_{p}. These spatio-temporally varying quantities are self-consistently included in the QDSBEs. They lead to spatially dependent line shapes and frequency differences $\bar{\omega}$ between the frequency of the propagating light field and the eigen frequencies of the spatially localized QDs in the laser structure. The shift of the emission frequency and the carrier damping rate resulting from the elastic Coulomb scattering between QD and the surrounding layer are

$$\begin{aligned}
\Delta\omega_{\mathrm{QD-WL}}^{\mathrm{e,h}} &= \sqrt{2kT/m^{\mathrm{e,h}}} {\sigma^{\mathrm{e,h}}}_{\omega} N_{\mathrm{WL}}^{\mathrm{e,h}} \\
\Delta\gamma_{\mathrm{QD-WL}}^{\mathrm{e,h}} &= \sqrt{2kT/m^{\mathrm{e,h}}} {\sigma^{\mathrm{e,h}}}_{\gamma} N_{\mathrm{WL}}^{\mathrm{e,h}}
\end{aligned} \tag{8.18}$$

where $\sigma_{\gamma}^{\mathrm{e,h}}$ and $\sigma_{\omega}^{\mathrm{e,h}}$ denote the intersection areas of the scattering processes given by [9]

$$\begin{aligned}
\sigma_{\gamma}^{\mathrm{e,h}} &= \int_0^{\infty} 2\mathrm{d}b_{\mathrm{QD}} \left(1 - \cos\left(\int_{-\infty}^{\infty} \mathrm{d}t \Delta\omega_{\mathrm{QD}}^{\mathrm{e,h}}(t)\right)\right) \\
\sigma_{\omega}^{\mathrm{e,h}} &= \int_0^{\infty} 2\mathrm{d}b_{\mathrm{QD}} \sin\left(\int_{-\infty}^{\infty} \mathrm{d}t \Delta\omega_{\mathrm{QD}}^{\mathrm{e,h}}(t)\right)
\end{aligned} \tag{8.19}$$

with

$$\begin{aligned}
\Delta\omega_{\mathrm{QD}}^{\mathrm{e,h}} &= -\frac{C_3^{\mathrm{e,h}}}{r_{\mathrm{QD}}^3} - \frac{C_4^{\mathrm{e,h}}}{r_{\mathrm{QD}}^4} \\
C_3^{\mathrm{e,h}} &= \frac{\pm e^2}{4\pi\varepsilon_0 n_{\mathrm{eff}}^2 \hbar} \frac{\beta h_{\mathrm{QD}}^2}{2}
\end{aligned}$$

$$C_4^{\mathrm{e,h}} = \frac{e^2}{16\pi^2\varepsilon_0^2 n_{\mathrm{eff}}^4 \hbar} \frac{k_{\mathrm{F}}}{\hbar} \tag{8.20}$$

The sign $\pm$ in $C_3^{\mathrm{e,h}}$ refers to the situation where the carriers in the QD and in the wetting layer that participate in the collision process have equal (+) or different (−) sign, respectively. b_{QD} is the spatially dependent collision parameter, h_{QD} and r_{QD} are the height and the radius of the QD. n_{eff} is the effective index of the material. The coefficient β is a measure of the linear Stark effect (resulting from expressing the component of the dipole in the direction of the QD axis as βh_{QD}) and k_{F} is a coefficient describing the quadratic Stark effect which can be estimated from the dipole moment and the eigen energies [10]. The collision-induced correction $\Delta\omega_{\mathrm{QD-WL}}^{\mathrm{e,h}}$ is added to the spatially dependent energy renormalization $\delta\mathcal{E}$ and $\Delta\gamma_{\mathrm{QD-WL}}^{\mathrm{e,h}}$ contributes to the damping rates $\gamma^{\mathrm{e,h}}$ and γ^{p} in the density matrices. We note that the spatial dependence of b_{QD}, r_{QD} and h_{QD} modeled in the form ($p = b_{\mathrm{QD}}, h_{\mathrm{QD}}, r_{\mathrm{QD}}$)

$$p_{\mathrm{QD}} = p_{\mathrm{QD}}^{\mathrm{av}} \left(1 + p_{\mathrm{QD}}^{\mathrm{fluc}}\right), \tag{8.21}$$

where $p_{\mathrm{QD}}^{\mathrm{av}}$ denotes the average value and $p_{\mathrm{QD}}^{\mathrm{fluc}}$ is the spatially dependent fluctuation, represents an arbitrarily distributed ensemble of quantum dots of varying size and shape. $b_{\mathrm{QD}}^{\mathrm{fluc}}$ considers for example the spatial fluctuation of the collision parameter resulting from a spatial localization of the QDs in the laser. The higher the amplitude of the fluctuations the higher is the degree of disorder in the spatial distribution of the effective QDs. r_{QD} and h_{QD} are the average radius and height of the QDs, respectively. This leads to spatially dependent energy corrections and damping rates.

The QDSBEs including the dynamic intra-dot scattering and the interactions with the wetting layer constitute a fundamental basis for a microscopic analysis of the relevant physical processes such as the influence of many-body interactions, spontaneous recombination, carrier relaxation and carrier injection. Via the generation rate and the dipole dynamics at each location in the laser structure the carrier dynamics within the QD (8.1) and the wetting layer (8.16) fundamentally linked to the light field dynamics that, in turn, is described by a suitable wave equation.

8.2.3 Optical Field Dynamics: Counterpropagation and Diffraction

Spatio-temporal light field dynamics plays a major role for relevant physical quantities such as the spatio-spectral gain and induced index of the system that, in combination with the complex carrier dynamics, determine output quantities of the laser system, i.e. emission wavelength, spectral bandwidth, saturation properties and temporal emission characteristics. A realistic theoretical treatment consequently requires full consideration of the spatially and temporally varying optical fields (associated with spontaneous and induced emission processes) that are mesoscopically coupled to the dynamics of the electrons and holes in the QDs. The QD ensemble represents a strongly inhomogeneous gain medium with spatially distributed QDs with individual material properties (dielectric constant, refractive index, etc.). This spatial inhomogeneity is even more intensified by the spatio-temporal dynamics of the carrier distributions in the QDs and in the wetting layer as well as by the nonlinear interaction of both carrier systems with each other. One may immediately sense that these space and time-dependent variations lead to strong phase changes during the propagation of the light fields in the laser resonator. Consequently the calculation of the light field dynamics has to include the temporal and spatial changes of the field amplitudes in an appropriate manner.

We start from Maxwell's equations for the optical field $\boldsymbol{E}$ and the polarization $\boldsymbol{P}$ and the material equations and derive the wave equation

$$\frac{1}{\varepsilon_0}\nabla\nabla\cdot\boldsymbol{P}+\nabla^2\boldsymbol{E}-\frac{1}{c^2}\frac{\partial^2}{\partial t^2}\boldsymbol{E}=\mu_0\frac{\partial^2}{\partial t^2}\boldsymbol{P}\,, \tag{8.22}$$

where ε_0 and μ_0 are the permittivity and the permeability in vacuum respectively and c is the velocity of light. Insertion of the ansatz

$$\begin{aligned}\boldsymbol{E}&=e^{\mathrm{i}\beta z-\mathrm{i}\omega t}(\boldsymbol{E}_\mathrm{T}+e_zE_z)\\ \boldsymbol{P}&=e^{\mathrm{i}\beta z-\mathrm{i}_\omega t}(\boldsymbol{P}_\mathrm{T}+e_zP_z)\end{aligned} \tag{8.23}$$

for the optical fields and the polarization leads to

$$\begin{aligned}&\left(-\beta^2+2\mathrm{i}\beta\frac{\partial}{\partial z}+\frac{\partial^2}{\partial z^2}\right)(\boldsymbol{E}_\mathrm{T}+e_zE_z)\\ &+\nabla_\mathrm{T}^2(\boldsymbol{E}_\mathrm{T}+e_zE_z)-\nabla_\mathrm{T}\nabla_\mathrm{T}-\nabla_\mathrm{T}\boldsymbol{E}_\mathrm{T}\\ &-\nabla_\mathrm{T}\left(\mathrm{i}\beta+\frac{\partial}{\partial z}\right)e_zE_z-\left(\mathrm{i}\beta+\frac{\partial}{\partial z}\right)\nabla_\mathrm{T}\boldsymbol{E}_\mathrm{T}\\ &+\frac{1}{c^2}\left(\omega^2+2\mathrm{i}\omega\frac{\partial}{\partial t}-\frac{\partial^2}{\partial t^2}\right)(\boldsymbol{E}_\mathrm{T}+e_zE_z)\\ =&-\mu_0\left(\omega^2+2\mathrm{i}\omega\frac{\partial}{\partial t}-\frac{\partial^2}{\partial t^2}\right)(\boldsymbol{P}_\mathrm{T}+e_zP_z)\end{aligned} \tag{8.24}$$

with the propagation constant β and frequency ω. With the main field propagation in parallel and anti-parallel to the resonator axis in (8.24) we may neglect the mixed derivatives, $\nabla_\mathrm{T}\frac{\partial}{\partial z}$, $\frac{\partial}{\partial z}\nabla_\mathrm{T}$. Similarly, $\nabla_\mathrm{T}\boldsymbol{E}_\mathrm{T}\approx-\mathrm{i}\beta E_z$ such that the deviates $\nabla_\mathrm{T}\nabla_\mathrm{T}E$ and $\nabla_\mathrm{T}\mathrm{i}\beta E_z$ can also be safely omitted. Disregarding the second order derivate $\frac{\partial^2}{\partial t^2}\boldsymbol{P}$ of the polarization (in analogy to the microscopic Bloch equations where one implicitly assumes a linear response function), we finally obtain the following effective wave equation for the counterpropagating (+,−) optical fields in a QD laser:

$$\nabla_\mathrm{T}^2\boldsymbol{E}^\pm\pm2\mathrm{i}\beta\frac{\partial}{\partial z}\boldsymbol{E}^\pm+\frac{\partial^2}{\partial z^2}\boldsymbol{E}^\pm$$

$$+\frac{2i\omega}{c^2}\frac{\partial}{\partial t}\boldsymbol{E}^{\pm} - \frac{1}{c^2}\frac{\partial^2}{\partial t^2}\boldsymbol{E}^{\pm}$$
$$= -\mu_0\omega^2\boldsymbol{P}^{\pm} - 2i\omega\frac{\partial}{\partial t}\boldsymbol{P}^{\pm} + F_{\mathrm{E}}q^{\mathrm{E}} \tag{8.25}$$

The Langevin noise term $F_{\mathrm{E}}q^{\mathrm{E}}$ that has been added to (8.25) has been derived from Quantum Maxwell–Bloch equations [1]. It considers spontaneous light field fluctuations that depend via $F_{\mathrm{E}} = (\sqrt{2\hbar\omega_0})/(\sqrt{\epsilon_{\mathrm{r}}\epsilon_0})$ on the specific material parameters and on the emission wavelength of the device. $q^{\mathrm{E}}(\boldsymbol{r},t)$ obeys the correlation relation

$$\langle q^{\mathrm{E}}(\boldsymbol{r},t)q^{\mathrm{E}}(\boldsymbol{r}',t')\rangle = \kappa\delta(\boldsymbol{r}-\boldsymbol{r}',t-t') \tag{8.26}$$

where $\kappa = 1/(2L)\ln[R_1R_2]$ corresponds to the damping rate of the resonator. The polarization of the active semiconductor medium

$$\boldsymbol{P}^{\pm} = V^{-1}\sum_{i,j}\boldsymbol{d}(i,j)\,p^{\pm}(i,j) \tag{8.27}$$

is the source of the optical fields (V denotes the normalization volume of the crystal).

Our derivation of the ‘quantum dot Maxwell–Bloch equations’ (‘QDMBEs’) reflects the spirit of describing the (spatio-temporal) dynamics of (spatially inhomogeneous) semiconductor lasers on the basis of Maxwell–Bloch equations [11]. The QDMBEs mesoscopically consider the dynamics of the carrier distributions in the dots and the inter-level dipoles together with the spatio-temporal dynamics of the optical fields (including spontaneous light fields, amplified spontaneous emission (ASE) and induced recombination).

Specific laser configurations of an actual device that is characterized by its geometry, mirror reflectivity, current injection, etc. are fundamentally included in our description. They enter the theory in the form of boundary conditions for the dynamically varying optical fields and the carriers, like the pump term of the QDMBEs. The laser cavity induces additional counterpropagation and wave-guiding effects which superimpose the carrier–field dynamics. The resulting complex dynamic spatio-spectral interactions between the QDs, the optical fields and the surrounding layers influence the emission properties (e.g. the temporal behavior of the optical fields, emission spectra). In the following Sect. 8.3 we will discuss selective results of numerical simulations that illustrate the interplay of light field and carrier dynamics in quantum dot lasers.

8.3 Coupled Spatio-Temporal Light Field and Inter/Intra-Level Carrier Dynamics in Quantum Dot Lasers

In the following, we will present selective results of numerical simulations based on QDMBEs. Specifically, we will consider spontaneous and induced light emission in the active dot medium and analyze the influence of spatial inhomogeneities (in quantum dot parameters such as dot size, level energies, dipole matrix elements) on the spatio-temporal light field and carrier dynamics. For specificity, the QDL structure is assumed to consist of three dot layers (InAs/GaAs [2]) with a dot density of 10^{10} cm^{-2}. The dots are assumed to be of pyramidal shape with base length 12 nm and with three energy and five hole levels. The length of the laser is 1 mm, its width (of the active zone) 10 µm. At every location in the medium the QDSBEs are coupled to a diffusion equation describing the spatial distribution of the carriers in the surrounding layers via dynamic scattering terms. The general representation of physical properties and components of a QDL by the QDMBEs is sketched in Fig. 8.2. In simulation of the spatio-temporal dynamics the spatial dependence of the carriers in the wetting layer (WL) and the propagating light fields ($\boldsymbol{E}^{\pm}$) are considered via a numerical grid with equally spaced grid points in the lateral (x) and propagation (z) directions. The local distribution of QDs is defined by spatial coordinates with respect to this grid. Thereby the spatial distance between the position of each QD and the center of the respective cell (with length Δz and width Δx) is saved in a spatially dependent variable that is used for the collision rates between the QD and the wetting layer. Each mesh (size $\Delta x \cdot \Delta z$) contains the following information: number of (effective) QDs in the mesh, N_{QD} (note that a "hole" in the spatially distributed QDs, i.e. $N_{\text{QD}} = 0$, is also possible), position of the QDs (i.e. their distance from the center of the area $(\Delta x \cdot \Delta z)$) and the individual material properties of the QDs. The specific laser configuration is defined by the size of the medium and the reflectivities of the facets (R_1, R_2) and enters the theoretical description as boundary conditions for the optical light fields.

A convenient way to visualize the spatio-temporal light dynamics of spontaneous and induced emission processes is in the form of temporal snapshots of the spatial light fields and of carrier distributions. To additionally grasp the complex microscopic carrier relaxation dynamics we will focus on the dynamics of the level occupations. On route, we will analyze the influence of spatially varying quantum dot properties (e.g. dot size, level energies, dipole matrix elements) on the spatio-temporal light field and carrier dynamics.

8.3.1 Spatio-Temporal Light Field Dynamics: Interplay of Spontaneous and Induced Emission

One of the very characteristic properties of a laser is the build-up of coherence in the light field from the initial spontaneous emission. In the quantum dot laser whose active medium is fundamentally characterized by the spatially inhomogeneous ensemble distribution of active quantum dot sources, one would expect that this transition is determined by this very feature together with the dependence on the (electrical or optical) excitation of the system. For low electrical injection current Fig. 8.3 shows typical snapshots of the spatial light field distribution in our model quantum dot laser structure. The time interval between the snapshots – showing the speckle distribution that is a characteristic of the quantum dot medium – is 3 ps. In this example we have assumed the very ideal case where the distribution of the dots in the structure is uniform (i.e. the dots are positioned with constant dot-to-dot distance). Furthermore for each dot within the structure we have used an identical set of parameters for dot size, level energies and dipole matrix element. The injection of carriers has been chosen such that the occupation of the energy levels of the dots is near transparency. In spite of the "ideal" conditions assumed for the laser structure, spatial fluctuations in light and carrier distribution arise. They are the result of spontaneous light fluctuations, microscopic carrier relaxation dynamics and the nonlinear coupling between the light fields and the charge carrier plasma. The carrier dynamics within each dot is determined by processes such as carrier injection, spectral hole burning, intra-dot carrier relaxation via phonon emission and absorption and carrier–carrier and carrier–phonon interaction with the wetting layers as well as screening. We will later focus on them in more detail. For now we can see that for the light field dynamics the underlying physical processes consist of both coherent (in the case of e.g. induced recombination) and incoherent contributions (e.g. spontaneous emission, carrier relaxation). Consequently they vary from dot to dot even when identical dot parameters and an ideal uniformity of the dot distribution in the layers are assumed.

The interplay of incoherent and coherent interactions yields a spatially varying number of electrons and holes in the energy levels of the quantum dots. Together, the spontaneous and induced light emitted by a quantum dot then contributes to the forward and backward propagating light fields and is thus transfered to the neighboring dots leading to complex spatio-temporally varying light–matter interactions. The propagating light fields, on the other hand, experience a spatially dependent modification via the interaction with the quantum dot ensemble. In combination with the diffraction of the light field this leads to a spatially varying light field dynamics (Fig. 8.3(a–c)). The nonlinear and inhomogeneous light–matter interaction and the carrier dynamics affect the spatial charge carrier density at the same time. For the time frame of Fig. 8.3(c), Fig. 8.4 shows the distribution of electrons as an example. The spatially varying level occupation and the formation of char-

(a)

(b)

(c)

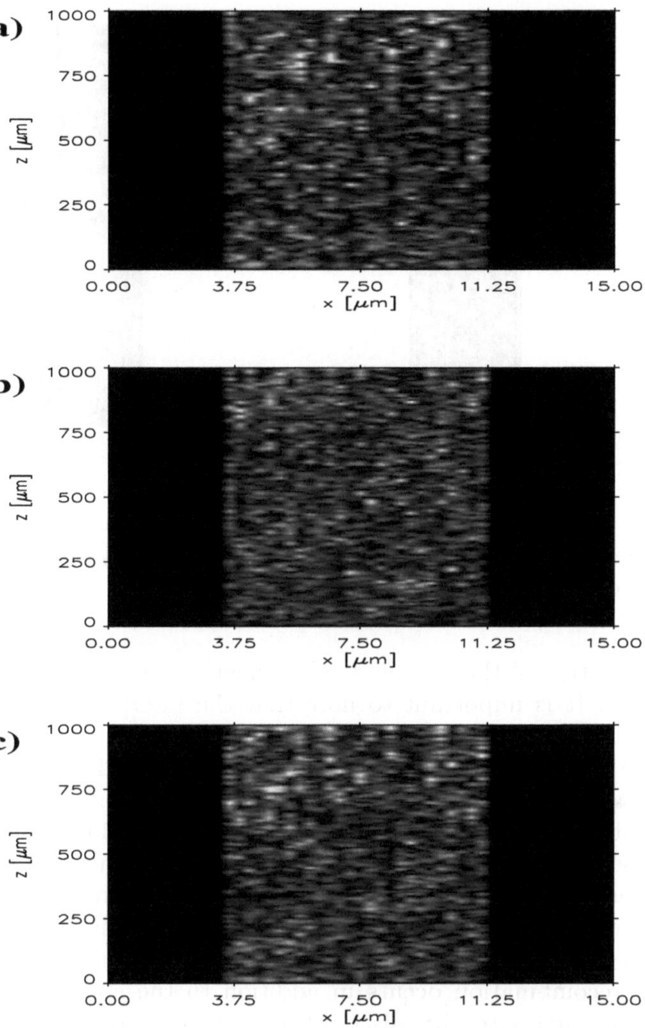

Fig. 8.3. Dynamics of luminescence pattern of spontaneous emission of an idealized quantum dot laser with perfect and uniform dot arrangement and identical parameters for dot size, level energies and dipole matrix elements. Bright colors indicate high levels of intensity. The time between successive snapshots is 3 ps

acteristic optical patterns are a direct consequence of spontaneous light fluctuations and scattering. The microscopic intra-dot scattering of the carriers within the dots via emission and absorption of phonons, the interaction of the 'dot carriers' with the carriers and the phonons of the wetting layer and the nonlinear coupling to the propagating light fields leads to a spatially

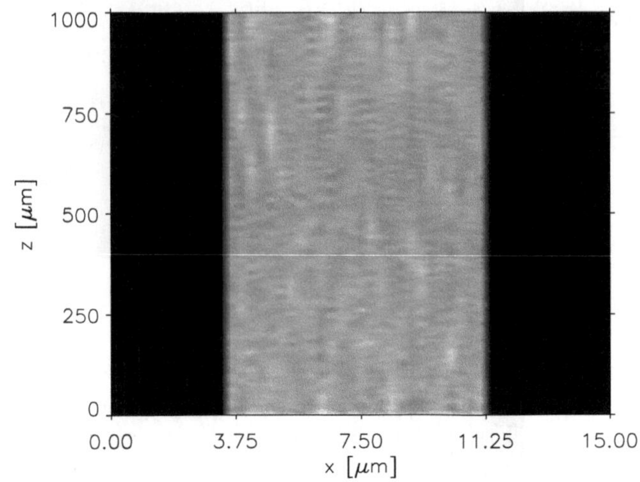

Fig. 8.4. Snapshot of the spatial electron distribution corresponding to the luminescence pattern of Fig. 8.3(c)

varying occupation of the dots and subsequently to complex transverse carrier dynamics. It is important to note that the interplay of light with the carriers results in a spatio-temporally varying occupation although we have assumed the "ideal" case of uniform carrier injection and regular matrix-like positioning of quantum dots that each have identical properties (size, level energies, matrix elements). The spatio-temporal light field dynamics changes if we increase the excitation level (carrier injection) by increasing the respective pump term in the Bloch equations so that the dots are almost completely filled with carriers. In this case the snapshots (again with time steps of 3 ps) of Fig. 8.5(a–c) show the result of a significant inversion: light amplification by induced recombination occurs in addition to the spontaneous emission processes. The first intensity distribution is taken 100 ps after the initial excitation of the dots. In the longitudinal (z) direction one can observe dynamically varying intensity modulations. These longitudinal structures are typical for the onset of laser oscillations of a device immediately after start-up. They are a measure of the characteristic internal coherence length scales that typically lie in the μm-regime. In time the structures lead to intensity spiking and relaxation oscillations in the light emission. In the lateral direction, i.e. parallel to the output facet, the intensity is rather uniform when compared to Fig. 8.3. This uniformity originates from induced emission processes which now play a major role in the over-all behavior of the device: The initial filling of the dots establishes a carrier inversion and thus a high gain. Due to the increased influence of induced emission processes a spatio-temporal coherence builds up that via the propagating light fields is

(a)

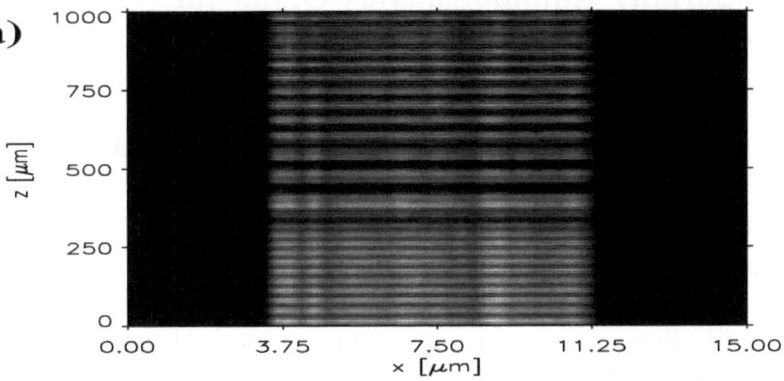

(b)

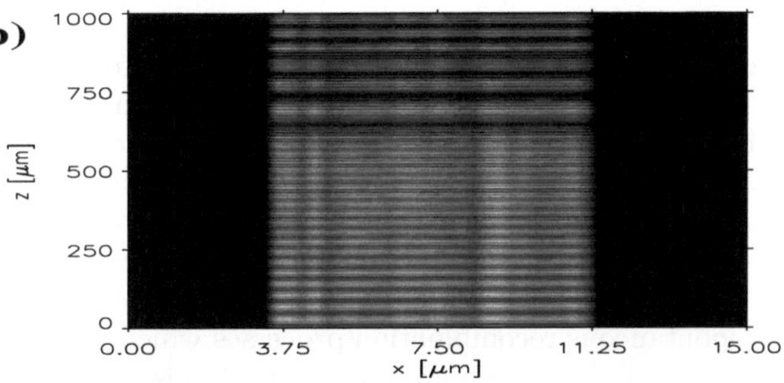

(c)

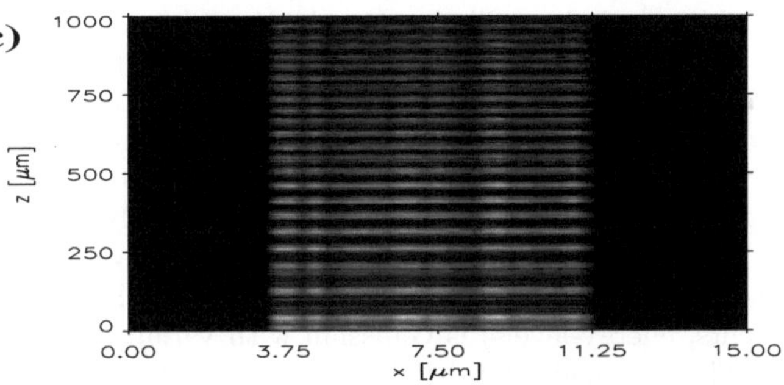

Fig. 8.5. Spatio-temporal dynamics of stimulated emission in quantum dot lasers pumped above threshold. The time interval between successive snapshots is 3 ps

transferred in both time and spatial dimensions. The coupling of the carriers in the dots with the propagating light fields in combination with the high gain characterizing the dot medium may then lead to a narrow-band stable laser output.

8.3.2 The Spatially Inhomogeneous Quantum Dot Ensemble: Influence of Disorder

Contrary to the ideal situation assumed so far, slight dot-to-dot variations in size, energy levels and material parameters exist in real quantum dot laser systems. In addition, the dots are not equally positioned on a grid within the layers. The respective variations in quantum dot parameters and dot-to-dot distance depends on the material system and the epitaxial growth process of the particular quantum dot system. In the numerical simulations in the following we will discuss some first steps in the analysis of disorder in quantum dot lasers and its influence on the spatio-temporal light fields and carrier dynamics and on emission spectra. In particular, we will focus on spatial fluctuations in the position of the quantum dots, their size, energy levels, dipole matrix elements and scattering rates that vary from dot to dot in a real laser device. Thereby we can systematically change the spatial fluctuation of the individual parameters. In this section we will focus on the influence of the variation of dot size in the inhomogeneously distributed quantum dot ensemble that constitutes the active laser material. A consequence of this will be a variation of energy levels and scattering matrix elements.

The dynamics of the light fields in quantum dot lasers is determined by induced and spontaneous recombination processes which, in turn, depend on the spatio-spectral carrier distribution in the dot levels. The mutual influence of light and matter is particular strong during the start-up regime of the laser and leads to characteristic oscillations in the time domain. Figure 8.6 shows the first 10 ns (after start-up) of the dynamics of the optical nearfields (left) and carrier distribution (right) at the output facet of an InGaAs quantum dot laser. Its width and cavity length are 50 μm and 500 μm, respectively, and the density of quantum dots (self-organized growth) is 10^{10} cm^{-2}. In the top frames of Fig. 8.6 the QD parameters deviate only slightly from their average values (1 % variance) while in the middle and lower frames the variance of the parameter values of the spatially distributed QDs (i.e. their size, dipole matrix elements, energy-levels) is Gaussian with variances of 4 % and 8 %, respectively. The nearfield intensity distributions show modulations on a picosecond time scale. They originate from dynamic interactions between the light fields and the dot carriers ranging from the femtosecond time scale (in the case of microscopic carrier scattering) up to the picosecond and nanosecond time scales (reflecting the resonator round trip time of the propagating light fields and the slow build-up and decay of the spatial carrier density). In combination the light diffraction and the spatially dependent interaction of light with the carriers in the dots and in the wetting layer lead to formation of dynamic optical patterns. The time scales of the carrier dynamics thereby are transformed into characteristic interaction lengths like the coherence length via the propagation of the light fields. In combination with the diffraction of the light fields this leads to transverse modulations in the μm regime.

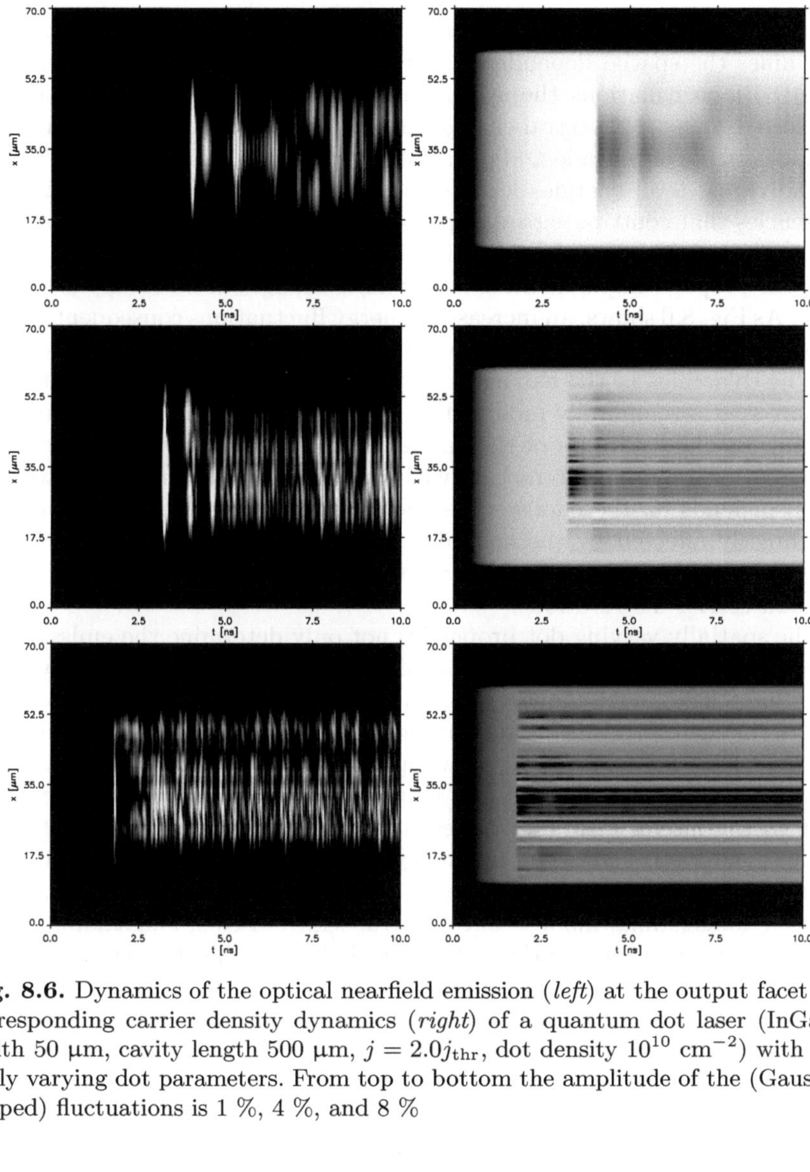

Fig. 8.6. Dynamics of the optical nearfield emission (*left*) at the output facet and corresponding carrier density dynamics (*right*) of a quantum dot laser (InGaAs, width 50 µm, cavity length 500 µm, $j = 2.0 j_{\mathrm{thr}}$, dot density 10^{10} cm^{-2}) with spatially varying dot parameters. From top to bottom the amplitude of the (Gaussian shaped) fluctuations is 1 %, 4 %, and 8 %

Dynamic carrier capture and escape, the complex intra-dot level dynamics and the diffraction of the propagating light fields lead to a transverse coupling of the field distribution arising at the individual dots. Via the light field propagation these fluctuations are then transfered to respective longitudinal (z) and temporal (t) changes. Due to the coupling and interplay of spatial with temporal degrees of freedom these material inhomogenieties thus affect both the transverse nearfield distribution as well as the dynamics of the light fields. With increasing fluctuation amplitude a characteristic filament structure evolves. The spatial variation of dot size and level energies directly determines the spatial dependence of the nearfield. Via the term $\mathrm{i}\bar{\omega} + \gamma_{\mathrm{p}}$ of the QD Bloch equations the spatial fluctuations in the dot properties are transferred to spatio-spectral changes in the inter-level polarization. This, in turn, affects the amplitude and phase of the propagating light fields. Thereby the individual level energies lead to locally varying transition energies and frequencies that contribute to the spatial and spectral properties of the propagating light fields. Via the spatial light field polarization these are passed on to the propagating light fields as dynamic changes in both amplitude and phase. As Fig. 8.6 shows, an increase in energy fluctuations consequently leads to corresponding fluctuations in the light field dynamics. Via the generation rate the dynamics of the carrier system depends on the light fields that in turn are spatio-temporally modified by the spatially varying dot parameters. In addition, the dynamics of the carrier relaxation processes (phonon emission or absorption and interaction with carriers and phonons of the wetting layers) depends on the energy differences of the levels involved and thus is also directly affected by the spatially varying dot parameters. As a consequence, dynamic characteristic filament structures evolve in both the light field and carrier distributions.

The spatially varying dot properties not only determine the emission dynamics, they also induce dynamic changes in the real and imaginary parts of the light fields. Our numerical simulation allows us to analyze the influence of the spatially varying dot properties on the spatio-spectral emission characteristics. Figure 8.7(a–c) shows spatially resolved emission spectra corresponding to Fig. 8.6 with 1 %, 4 % and 8 % fluctuations. The vertical axis refers to the lateral position at the output facet of the quantum dot laser; the horizontal axis shows the frequency dependence in a range of 600 GHz. With a cavity length of the device of 500 μm this corresponds to approximately 7 longitudinal modes. The spectral width of each longitudinal mode is determined by a large variety of physical effects. The characteristic times of induced and spontaneous recombination define a lower limit for a laser linewidth. However, the real spectral width is significantly broadened. First, in large-area lasers the transverse degree of freedom leads to a characteristic transverse migration of the light fields that is determined by the dynamic interplay of light diffraction, dynamic self-focusing as well as carrier scattering and relaxation. As a consequence, a group of transverse modes arises for each longitudinal mode. Second, the transverse dynamics is determined by inhomogeneous broadening resulting e.g. from the spatially varying quantum dot parameters. They affect both the coherent light–matter coupling (via the carrier dependence of the generation rate) and the incoherent processes such as carrier–carrier and carrier–phonon interactions.

Figure 8.7 clearly demonstrates that an increase in fluctuation amplitude affects both the spectral as well as the spatial degree of freedom. The spatially varying transition energies lead to significant broadening of the emission spec-

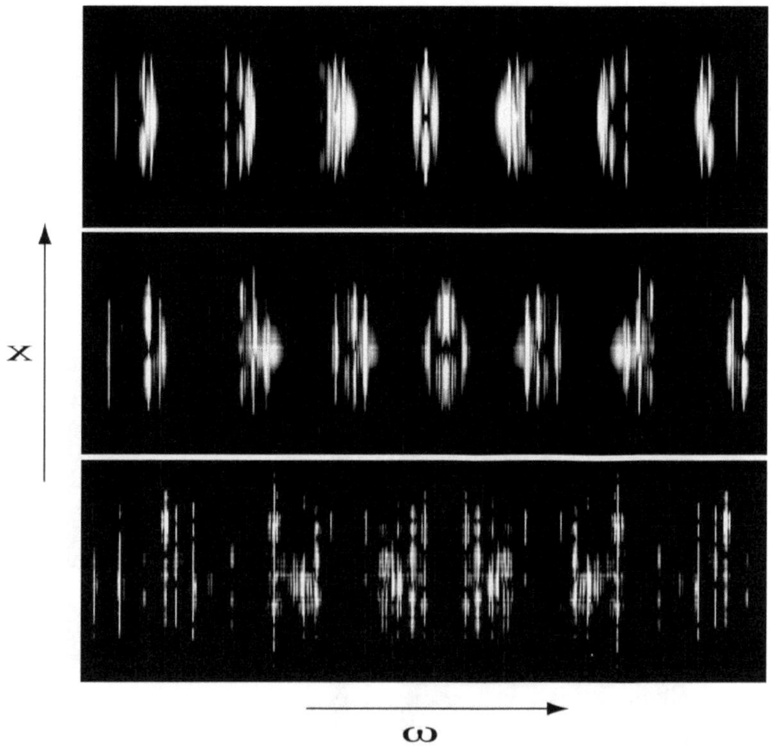

Fig. 8.7. Emission spectra of a quantum dot laser with Gaussian fluctuations with amplitude (from top to bottom) 1 %, 4 % and 8 %. The vertical axis refers to the lateral position at the output facet of the quantum dot laser, the horizontal axis shows the frequency dependence in a range of 600 GHz. With a cavity length of the device of 500 μm this corresponds to approximately 7 longitudinal modes

tra. This originates from the variance in transition energy and (indirectly) the intra-dot and inter-dot scattering dynamics. The light propagation and diffraction in combination with carrier scattering and relaxation not only lead to a coupling of longitudinal and transverse degrees of freedom but also to a coupling of spectral and transverse dynamics. As a consequence, the spatially resolved emission spectra show inhomogenieties in both the transverse dimension (i.e. over the lateral extension of the quantum dot broad-area device) and the spectrum.

8.3.3 Ultrashort Pulse Propagation in a Quantum Dot Waveguide

A typical means of probing the internal dynamics of a quantum dot laser is to inject an optical pulse into the laser (cf. Chap. 3) and analyze the ultrafast dynamics of the output signal. Thus for this configuration one additionally has to consider a light pulse propagating in a laser waveguide (width of the structure 10 μm, length 1 mm) filled with an inverted quantum dot ensemble. The snapshots displayed in Fig. 8.8 show the intensity (a–c) and the carrier density (d–f) in the active area of a QD laser during the propagation of an ultrashort light pulse whose frequency corresponds to the transition energy of the QDs. The injection current density has been chosen such that the population of the dots in the layers of the QD waveguide are significantly above the transparency. For the dot-to-dot fluctuation a variance of 5 % has been assumed. The time between successive plots is 3 ps. Figures 8.8(a) and (d)

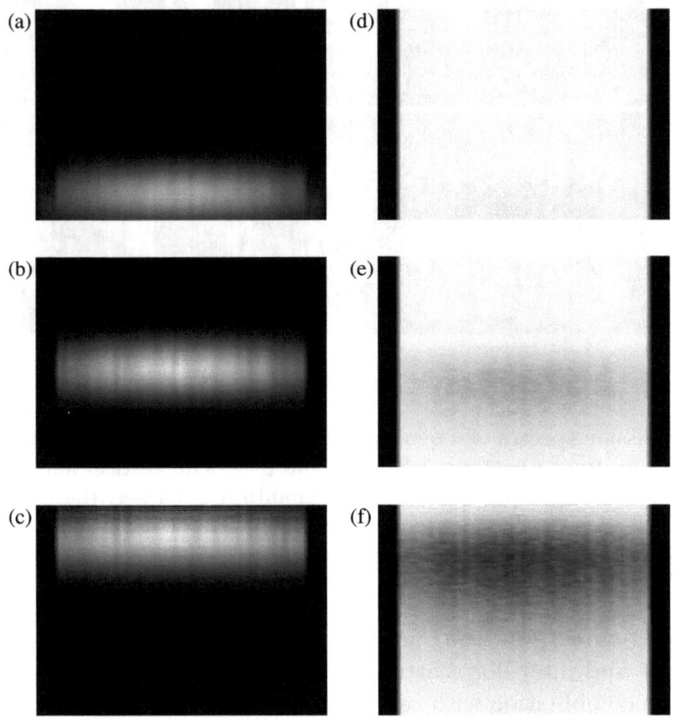

Fig. 8.8. Propagation of a femtosecond pulse (full width at half maximum of 500 fs) tuned to resonance of an inverted quantum dot ensemble. (**a**–**c**): snapshots of the light field and (**d**–**f**): corresponding snapshots of the carrier density. The time between successive snapshots is 3 ps

represent the spatial distributions of the intensity (a) and the carrier density (d) immediately after optical injection. It is important to note that initially the lateral spatial shape of the injected light field is Gaussian shaped with a width (FWHM) of 6 μm and the temporal profile of the pulse is chosen Gaussian as well with a full width at half maximum of 500 fs. Immediately after injection the light pulse starts to interact with the ensemble of populated QDs. During its propagation through the laser the pulse locally reduces the population in the dots established by the injection current by induced recombination. With continuing propagation the light pulse is significantly amplified (Fig. 8.8(b)). Due to the nonlinear light–matter interaction between the pulse and the spatially distributed dots a complex spatio-temporal behavior arises. It is directly reflected in the dynamic spatial structures in both intensity and carrier density. The pulse is laterally structured and temporally distorted via the interaction with the dots (Fig. 8.8(b,c)). At the same time, spatial hole burning effects can be observed in the carrier distribution Fig. 8.8(e,f). The partial refilling of the dots – determined by carrier injection, carrier capture and thermalization via carrier relaxation – defines a finite "response time" of the QD medium. As a consequence, the spatial extension of the hole burnt by the light pulse (Fig. 8.8(f)) significantly exceeds the spatial area covered by the optical pulse (Fig. 8.8(e)). Thus, a light pulse propagating in a QD waveguide is strongly modified by the nonlinear interaction with the QD system. The amount of spatial and temporal distortions the light pulse experiences thereby strongly depends on both spatial effects (such as dot density, uniformity of the dot distribution, spatial fluctuations) and microscopic "spectral" effects (determined by the characteristic relaxation times and microscopic dot properties). The spatial effects displayed in Fig. 8.8 are thus – via the coupling between wave equation and QD-Bloch equations – the result of microscopic physical effects. In order to analyze these dynamics in the following we will focus on the population changes induced by a light pulse.

8.3.4 Inter and Intra-Level Carrier Dynamics of the Quantum Dot Ensemble

While the optical output signal carries information on the carrier dynamics of the quantum dot ensemble, it does not directly reveal the carrier dynamics within each energy level. In our QDMBEs the inter and intra-level dynamics are automatically represented and calculated. The level dynamics allows visualization of the microscopic interactions occurring within the dots. As an example we will concentrate on results calculated for the hole level occupation during the propagation of a light pulse (500 fs). The respective electron level occupations show qualitatively similar behavior. We will focus on three physically different situations: absorbing, transparent and amplifying QD media. Figure 8.9(a) shows the situation where the energy levels are almost empty at the start of the calculation. In this case the pulse leads to optical excitation of the carrier system. If on the other hand the dots are initially significantly filled (Fig. 8.9(c)) an effect similar to the well-known spectral hole burning occurs. Depending on the dipole matrix elements for the individual states and depending on the frequency detuning of the pulse with respect to the frequency of the respective electron and hole states a

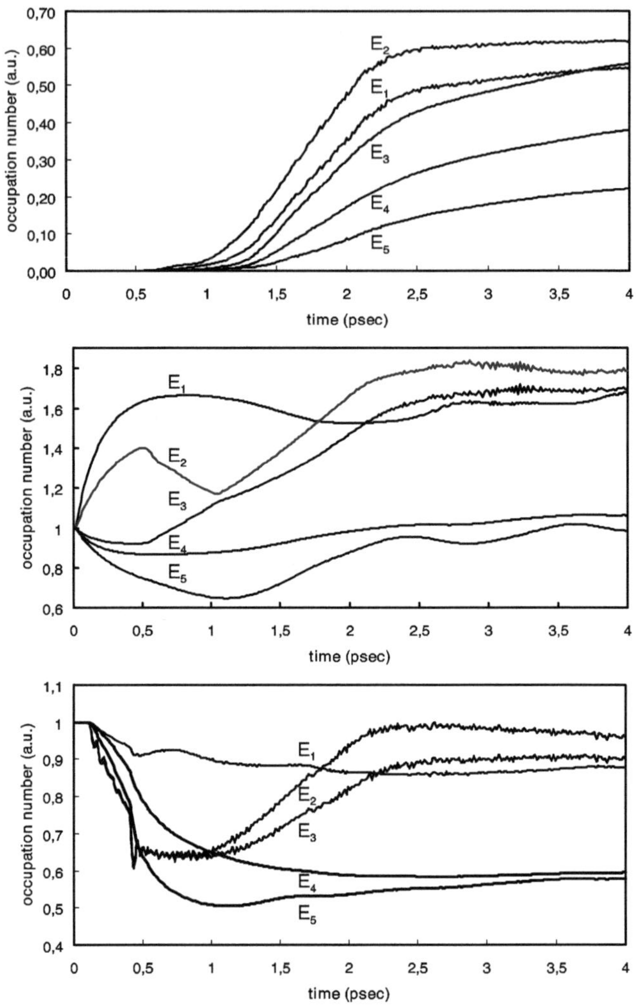

Fig. 8.9. Occupation of hole levels during the propagation of a light pulse: (**a**) absorption, (**b**) transparency, (**c**) amplification

reduction of the individual level occupation and a partial refilling via carrier injection and microscopic scattering processes occurs. The microscopic scattering processes involved in this "level-burning" are determined by emission and absorption of phonons, multi-phonon interactions and the interaction of the carriers in the dots with the carriers and phonons of the wetting layer.

The magnitude of the various "channels" for relaxation mechanisms thereby depend on the QD energy levels, on the energy difference to the surrounding layers and on the coupling of a dot to its nearest neighbors. These dot-to-medium and dot-to-dot interactions are thereby determined by the dot density and the light propagation that mesoscopically couples the QDs. It is consequently both specific material properties and dynamic light–matter interactions that decide whether the dots are isolated from their environment or behave like a thermalized ensemble. In passing we may remark that it is a decisive advantage of the mesoscopic spatially resolving that every intermediate state of the dots is allowed without restriction to the limiting cases of fully isolated dots or a completely thermalized set of dots.

The physical situation is particularly interesting if the dot medium is near transparency (Fig. 8.9(b)). Due to the individual matrix elements the occupation of the various levels may rise or be reduced even though the sum of all contributions stays more or less at a constant value. In particular, the dynamic changes in the level occupation and the dynamic saturation of individual levels may lead to the situation that the laser first saturates a specific level and then changes from one inter-level transition to another one.

8.3.5 Luminescence of Optically Excited Quantum Dot Media

In Sect. 8.3.3 we have considered the propagation of a resonant light pulse in an inverted quantum dot medium, i.e. an electrically pumped QDL. While this certainly represents the preferred mode of operation of quantum dot lasers in most applications, in many current experimental set-ups, however, one investigates the luminescence of optically excited quantum dot media by an (ultra-) short optical pump pulse. In the quantum dot laser this corresponds to an approximately δ-shaped excitation of carriers into one or more high energy carrier reservoirs either in the dots themselves (direct optical pumping) or by carrier capture from the optically pumped wetting layer (indirect pumping). Since the dynamic interplay between the dots and the wetting layer is determined by a large variety of relaxation processes involving the dot-carriers and the carriers in the wetting layer, the excitation of the dots via the wetting layer represents a particularly interesting case. In the following we will thus focus on the case of an ensemble of (initially empty) dots that is dynamically filled from the wetting layer (the high energy carrier reservoir). The dynamic coupling between the dot-carriers and the carriers of the wetting layer is determined by the density of quantum dots, the individual dot properties and the epitaxial growth process. These factors contribute simultaneously and lead to a characteristic response of the quantum dot system. Thus the resulting luminescence is influenced by and directly reflects the multitude of physical quantities that are involved in the characteristic excitation and relaxation processes. These are characteristic (material) properties like the optical matrix elements for the various (intra-dot and dot-wetting layer) carrier and carrier–phonon interactions, the transition matrix elements of the dot levels involved, the energies of the dot levels and the wetting layer states, the dot density and the spatial dot distribution. Since a detailed analysis and variation of all these parameters (some of which are presently not even known in detail) involves extensive simulations in the following we will restrict ourselves to investigation of the influence of the filling degree, the coupling strength and the energy levels. Thereby we implicitly assume the remaining parameters to be (spatially-dependent) constants.

Figures 8.10, 8.11 and 8.12 show the characteristic luminescence, i.e. the laterally averaged intensity at the output facet of an optically pumped QDL structure. The degree of initial filling of the high energy reservoir and the coupling strength between the dots and their environment determines the time constants for carrier capture into the dots and the degree of dot filling. In combination this may lead to very different characteristic emission behavior discussed below.

Influence of Excitation Strength

The three curves displayed in Fig. 8.10 show the dependence of the luminescence on the filling degree of the carrier reservoir. The solid, grey and dotted

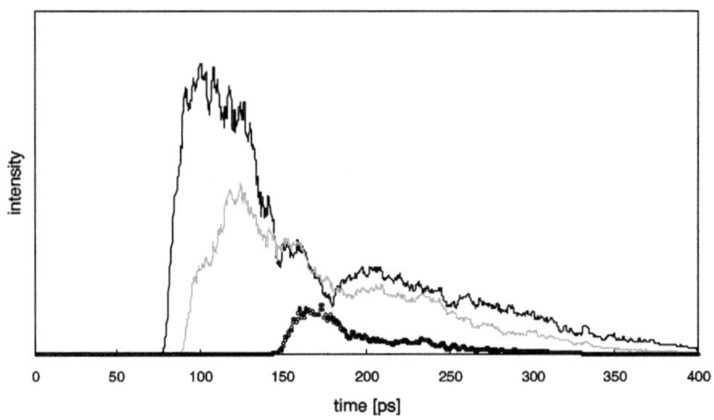

Fig. 8.10. Dependence of the QD luminescence on excitation level. *Solid*, *grey* and *dotted* lines correspond to completely, half and significantly less than half filled wetting layer states, respectively

lines show the situation where all, half and significantly less than half of the available wetting layer states are filled with carriers, respectively. For a moderate initial excitation the loading of the QDs with carriers is comparatively slow leading to delayed onset of light emission. An increase in the initial carrier filling of the reservoir provides a higher inversion in the dots resulting in intense light emission. The variation of the reservoir not only determines the instant and intensity of the light emission, it also affects the shape of the curve: For high excitation a selective saturation of individual transitions may occur. As a consequence, a second transition can be involved leading to mode beating, temporal modulations or a second peak in the emission curve.

Influence of Coupling Strength

In order to analyze the influence of the coupling strength we have calculated the intensity and level occupation in the dots in relation to the fraction of wetting layer states to which the dots couple (normalized to a unit cell $\Delta z \times \Delta x$). In a given laser structure this value is determined by the dot density and by the potential step between the dots and their environment determined by the size and shape of the dots as well as the particular material systems and epitaxial growth processes. The resulting dynamic behavior of the emitted intensity is depicted in Fig. 8.11. First, a high coupling strength

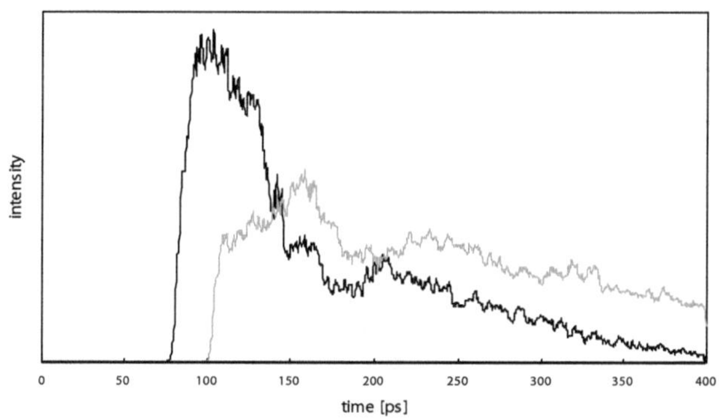

Fig. 8.11. Dependence of the QD luminescence on the coupling strength. High coupling strength (*black*) provides a fast filling and efficient refilling of the dots that are partially depleted via induced emission processes. Weak coupling (*grey*) leads to a gradual depletion of the carrier reservoir

provides a faster filling of the (initially) empty dots. As a consequence, the dot occupation reaches the characteristic threshold value that is apparent in Fig. 8.11 at an earlier time step. Second, it enables efficient refilling of the dots that are partially depleted via induced emission processes. In combination this leads to an intense peak in the emission curve (black). A weak coupling (grey), on the other hand, leads to a slow depletion of the carrier reservoir. As a result it delays the onset of light emission and "stretches" the shape of the emission curve in time.

Influence of Quantum Dot Size and Growth: Variation of Energy Levels

Variation in the size and epitaxial growth of a quantum dot has a direct consequence for its energy levels. These variations in eigen energies enter directly in the QD Bloch equations. In the following we will consider two channels of transitions with the highest transition matrix elements for two different cases: (1) a QD system with close transition energies (i.e. separated by less than the LO phonon energy) and (2) a QD system where the carrier levels belonging to the two most dominant transitions differ by an energy much higher than the LO phonon energy. For these two examples Fig. 8.12

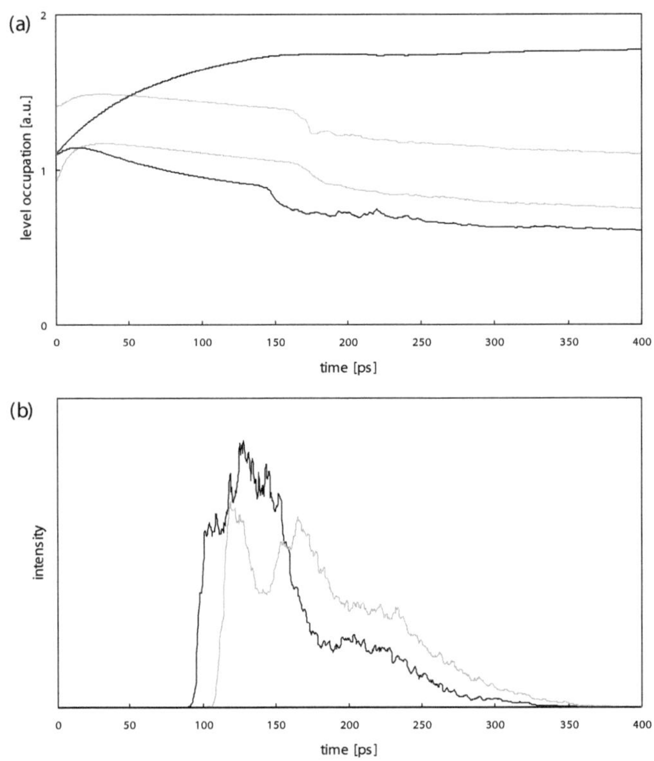

Fig. 8.12. Dependence of the QD luminescence on the transition energies. The black lines pertain to a QDL with the highest matrix elements separated by more than the LO phonon energy. The grey curves illustrate the case where the respective transition energies are very close to each other. The respective emission properties represented by the solid curve in Fig. 8.12(b) shows one intense peak belonging to the main transition. The dashed curves visualize the carrier occupation for the dot system with close transition energies

shows the temporal behavior of the (electron) level occupations (Fig. 8.12(a)) and the resulting emission curve (Fig. 8.12(b)) after the initial excitation of the QDs. The black lines in Fig. 8.12(a) pertain to the QDL where the transitions with the highest matrix elements are separated by more than the LO phonon energy. The grey curves correspond to the situation where the respective transition energies are very close to each other. For the example with higher level separation the carriers populating the two QD levels are mainly decoupled: the carrier recombination is mostly restricted to one level (belonging to the transition with the highest dipole matrix element) whereas the second level absorbs carriers from the reservoir. The respective emission properties represented by the solid curve in Fig. 8.12(b) show one intense peak belonging to the main transition. For the same excitation conditions the dashed curves visualize the carrier occupation for the dot system with close transition energies. In this situation the two main carrier levels interact and "interfere" via dynamic carrier and phonon scattering. The resulting emission curve (dashed line) shows two maxima. The specific shape of the temporal emission characteristics is thus a direct consequence of the dynamic interplay of competing transitions and spectral modes.

8.4 Conclusion

In conclusion, we have set up a mesoscopic theory on the basis of a Maxwell–Bloch description. The resulting QDMBEs consist of coupled spatio-temporally resolved wave equations and QD Bloch equations for the electron and hole levels within each quantum dot of a quantum dot ensemble inside a quantum dot laser.

We have presented results of our numerical simulations that aim to mesoscopically represent realistic QD laser structures. The simulations include, in particular, microscopic QD properties, spatially dependent QD parameters and fluctuations, spatially inhomogeneous light propagation and dynamic scattering. The carrier scattering processes are considered on a mesoscopic level and include both the intra-dot relaxation and the interactions between the QD carriers and the surrounding layers. The specific laser configuration of a model device is considered via the macroscopic boundary conditions and constraints. The QDMBEs allow calculation and visualization of spatial distributions of the light field intensity and carriers. Furthermore, calculation of level occupations provides a detailed analysis of the various relaxation processes. For a specific set of parameters the Quantum Dot Maxwell–Bloch equations allow a microscopically founded interpretation of the QD properties obtained from experimental investigations. The mesoscopic theory discussed in this chapter may thus establish a basis for linking the microscopic analysis of QD material properties with the quantum electronics of modern quantum dot laser systems.

References

1. H. F. Hofmann and O. Hess. Quantum Maxwell-Bloch equations for spatially inhomogenous semiconductor lasers. *Phs. Rev. A*, 59:2342–2358, 1999.
2. O. Stier, M. Grundmann, and D. Bimberg. Electronic and optical properties of strained quantum dots modeled by 8-band k·p theory. *Phys. Rev. B*, 59:5688–5701, 1999.
3. M. Braskén, M. Lindberg, D. Sundholm, and J. Olsen. Full configuration interaction calculations of elctron-hole correlation effects in strain-induced quantum dots. *Phys. Rev. B*, 61:7652–7655, 2000.
4. K. Oshiro, K. Akai, and M. Matsuura. Size dependence of polaronic effects on an exciton in a spherical quantum dot. *Phys. Rev. B*, 59:10850–10855, 1999.
5. P. G. Bolcatto and C. R. Proetto. Shape and dielectric mismatch effects in semiconductor quantum dots. *Phys. Rev. B*, 59:12487–12498, 1999.
6. J. Schilp, T. Kuhn, and G. Mahler. Electron-phonon quantum kinetics in pulse excited semiconductors: Memory and renormalization effects. *Phys. Rev. B.*, 50:5435–5447, 1994.
7. A. V. Uskov, J.+McInnerney, F. Adler, H. Schweizer, and M. H. Pilkuhn. Auger carrier capture kinetics in self-assembled quantum dot structures. *Appl. Phys. Lett.*, 72:58–60, 1998.
8. J. L. Pan and P. L. Hagelstein. Collisional processes involved in the population kinetics of semiconductor quantum-dot lasers. *Phys. Rev. B*, 49:2554–2575, 1994.
9. A. V. Uskov, K. Nishi, and R. Lang. Collisional broadening and shift of spectral lines in quantum dot lasers. *Appl. Phys. Lett.*, 74:3081–3083, 1999.
10. G. Bastard, editor. *Wave Mechanics Applied to Semiconductor Heterostructures.* Halsted, New York, 1988.
11. O. Hess and T. Kuhn. Maxwell-Bloch equations for spatially inhomogeneous semiconductor lasers I: Theoretical description. *Phys. Rev. A*, 54:3347–3359, 1996.

9 Delayed Optical Feedback and Control of Spatio-Temporal Dynamics

9.1 Introduction: Spatio-Temporally Chaotic Semiconductor Lasers

In spite of the tremendous improvements in laser processing technologies there are still two major situations which continue to be serious sources of instabilities. First, due to its very high gain and outcoupling rate, the semiconductor laser is very sensitive to delayed optical feedback (DOF) caused by distant reflecting surfaces such as an optical fiber. Second, in high-power in-plane edge-emitting lasers and vertical-cavity surface-emitting semiconductor lasers the nonlinear interaction of spatial with temporal degrees of freedom leads to chaotic spatio-temporal instabilities. Clearly, for practical reasons it is highly desirable to understand the nature of these complex temporal and spatio-temporal processes and, building on this knowledge, to develop schemes that may allow us to control and suppress them.

Up to now the destabilizing mechanism in broad-area and vertical-cavity surface-emitting lasers is closely related to the nonlinear interaction of multiple transverse modes. Similar to the situation for broad-area lasers, both temporally and spatially irregular behavior corresponding to deterministic spatio-temporal chaos is observed in coupled multi-stripe laser arrays [1]. As this constitutes a severe problem when good beam quality at high intensities is desired, there have recently been initiatives to apply concepts from the field of chaos control for stabilization. To stabilize spatio-temporal instabilities and chaos a number of schemes using delayed optical feedback [2] and stabilization of travelling waves by means of delayed optical feedback and spatial filtering in a broad-area laser model [3; 4] were demonstrated in recent theoretical work. While theoretical investigations of delay-induced temporal instabilities in single-stripe lasers using the Lang–Kobayashi rate equations have been a subject of intense research until today, the investigation of spatio-temporal phenomena occurring in multi-stripe laser arrays [5] and broad-area lasers [6] has evolved to an issue of great interest in the last few years. In recent theoretical work [7] both destabilizing mechanisms were combined in investigations on the influence of delayed optical feedback on spatio-temporal dynamics.

In the following we apply delayed optical feedback to achieve suppression of spatio-temporal instabilities. Thus we can stabilize stationary operation in the fundamental transverse mode from an originally spatio-temporally chaotic state. We first give a brief survey of the model equations and next discuss a complex eigenmode analysis via the Karhunen–Loève algorithm, which yields a decomposition of the complex optical field into a set of orthonormal transverse modes and the individual oscillation frequencies of these modes. We then show that this permits setting up optical feedback conditions which lead to steady state operation in the fundamental transverse mode by suppression of higher order transverse modes.

9.2 Transverse Lang–Kobayashi Model for Coupled Lasers

To represent the transverse light field diffraction and carrier diffusion effects responsible for the coupling of lasers in arrays we extend the plane–wave Lang-Kobayashi (LK) model [8] to include transverse effects (TLK) [7]. Due to the structural similarity of these model equations with the LK-model, while still being comparatively simple in structure (in fact, it is the most simple transverse extension of the LK equations), we will explicitly discuss this model here. Generally, the transverse dependence is caused, on the one hand, by transverse coupling via optical diffraction (diffraction coefficient D_p) and charge carrier diffusion (D_f) and, on the other hand, by the transverse inhomogeneity of stripe-geometry lasers. We thus obtain the following set of nonlinear coupled partial differential equations for the complex optical field $\boldsymbol{E}(x,t)$ and the charge carrier density $N(x,t)$:

$$\begin{aligned}\frac{n_\mathrm{l}}{c}\partial_t \boldsymbol{E} &= \mathrm{i}D_\mathrm{p}\partial_x^2 \boldsymbol{E} - \left[\gamma_\mathrm{E} + \mathrm{i}\eta(x)\right]\boldsymbol{E} + \Gamma(x)\left[g(N) + \mathrm{i}k_0\delta n(N)\right]\boldsymbol{E} \\ &\quad + \frac{1}{2L}\gamma_\mathrm{R}\mathrm{e}^{\mathrm{i}\Phi}\boldsymbol{E}(x,t-\tau)\,,\end{aligned} \tag{9.1}$$

$$\partial_t N = \Lambda(x) + D_\mathrm{f}\partial_x^2 N - \gamma_\mathrm{nr}N - \frac{2\epsilon_0 c}{\hbar\omega_0 n_\mathrm{l}}g(N)|\boldsymbol{E}|^2\,. \tag{9.2}$$

The transversely varying parameters $\Lambda(x)$, $\eta(x)$ and $\Gamma(x)$ describe current injection via stripe electrodes of a given width w, index guiding through transverse index steps located below the electrodes and the transversely varying confinement factor, respectively [7]. Further parameters are the nonradiative decay rate γ_nr of the carrier density, the refractive index of the active layer n_l, and the carrier frequency and vacuum wavenumber, ω_0 and k_0, respectively. The variation of the optical gain and the refractive index of the active medium with the carrier density is approximated by the phenomenological linear gain function $g(N) = a(N - N_0)$ where a is the linear gain coefficient, N_0 is the carrier density at transparency and $\delta n = -\alpha a N/k_0$. The linewidth enhancement factor α is fixed at $\alpha = 2$.

The distributed mirror loss is represented by the damping constant $\gamma_{\mathrm{E}} = -\frac{\log\sqrt{R_1R_2}}{2L}$, where R_1 and R_2 are the power reflectivities of the front and rear facets, respectively. Delayed optical feedback is represented by the feedback parameters γ_{R} (feedback strength), τ (delay time) and Φ (feedback phase).

9.3 Eigenmode Analysis

In order to characterize the spatio-temporal complexity in the laser output, it is highly desirable to find out how many optical modes are involved and how they are spatially structured. To this end, eigenmode analysis via Karhunen–Loève decomposition (KLD) has been successfully applied [9; 10]. Given a time series of a spatially extended system (from experiment or numerical simulation), this method provides a decomposition into an orthonormal set of eigenmodes. In its original form, this algorithm computes eigenmodes for real input data. In optics, one therefore generally uses data on the output intensity. On this basis, however, the structure of the complex field modes and their respective frequencies are not directly available. As it turns out, for efficient stabilization schemes information on both spatial modes and frequency distribution is crucial. For convenience we briefly add the analysis presented in [11] where the traditional eigenmode analysis has been extended to complex input data pertaining to the spatially resolved optical field.

The complex eigenmodes p_n corresponding to transverse modes $p_n(x)$ of the optical field are obtained by solving the eigenvalue problem of the Hermitian covariance matrix:

$$\boldsymbol{C}p_n = \lambda_n p_n,$$

$$\text{where} \quad C_{jk} = \frac{1}{T}\int_0^T E^*(x_j,t)E(x_k,t)\,\mathrm{d}t \quad j,k = 1\ldots N_x$$

and T and N_x denote the length of the time series and the number of transverse grid points, respectively. The real eigenvalues λ_n yield the relative importance of the complex eigenmodes p_n, which form an orthonormal set, meaning that

$$\sum_{k=1}^{N_x} p_m^*(x_k)p_n(x_k) = \delta_{mn}. \tag{9.3}$$

The time varying modal amplitudes are also complex quantities. They govern the dynamics of the individual eigenmodes and are obtained according to

$$a_n(t) = \sum_{k=1}^{N_x} p_n^*(x_k)E(x_k,t)\,. \tag{9.4}$$

It is the frequency spectra of these modal time series $a_n(t)$ that yield the oscillation frequencies of the complex eigenmodes and will later help us in establishing suitable conditions for achieving stabilizing feedback. From the eigenmodes and expansion coefficients the original complex optical field can be reconstructed with

$$E(x_k, t) = \sum_n a_n(t) p_n(x_k). \tag{9.5}$$

The same procedure can be carried out for the carrier density $N(x,t)$. Since $N(x,t)$ is real, its eigenmodes q_n and modal amplitudes $b_n(t)$ are real quantities. The reconstruction of the carrier density is obtained with

$$N(x_k, t) = \bar{N}(x_k) + \sum_n b_n(t) q_n(x_k) \tag{9.6}$$

where $\bar{N}$ is the time averaged carrier density profile. The eigenmodes obtained from complex KLD can be used as an orthonormal basis for mode projection, in analogy to the Galerkin procedure. In this way, the set of partial differential equations in (9.1)–(9.2) can be reduced to a set of ordinary differential equations [11].

9.4 Stabilization of Chaotic Laser Arrays

Depending on the separation between the laser stripes, semiconductor laser arrays are strongly, moderately or weakly coupled [1; 12; 13]. In the regime of strong coupling (small separation between the lasers) the dynamical behavior is strongly dependent on the amount of external pumping via the electrical current. With increasing pump current, spatio-temporal instabilities arise in multi-stripe lasers due to the nonlinear interaction of multiple transverse modes. Not surprisingly, the higher the number of transversely coupled laser stripes, the larger the spatio-temporal complexity. The basic effects, however, can already be demonstrated with the simplest laser array, the twin-stripe laser. Therefore, we start our discussion with this simple array.

For low values of the pump current, the twin-stripe laser operates in a steady state of coherent light emission with one single transverse mode. However, when we increase the pump current above a critical threshold value J_c, an instability leading to spontaneous periodic intensity pulsations sets in. Several dynamical regimes with alternating pulsations in the two stripes are observed [11]. Further increase of the current drives the laser into a both spatially and temporally irregular regime (Fig. 9.1). It is found that the value of J_c varies strongly with the amount of charge carrier diffusion, which is quite significant in semiconductor lasers.

Eigenmode analysis reveals that the spatio-temporal dynamics is governed by the two eigenmodes shown in Fig. 9.2. The first antisymmetric mode, which will be referred to as the fundamental mode, represents a mode where

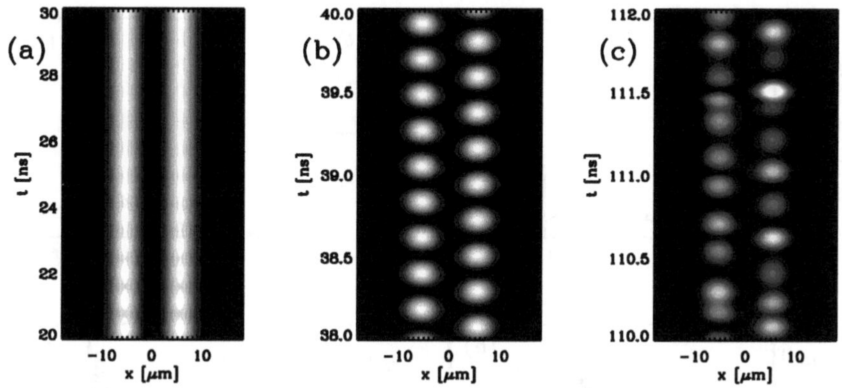

Fig. 9.1. Spatio-temporal dynamics of the intensity $I(x,t) = (1 - R_1)\epsilon_0 c/n_1|E(x,t)|^2$ for the twin-stripe laser in the regime of coherent continuous wave light emission ($J = 30\,\text{mA}$) (**a**), in the periodic regime ($J = 45\,\text{mA}$) (**b**) and in the chaotic regime ($J = 70\,\text{mA}$) (**c**) for an intermediate diffusion of $D_\text{f} = 4{\cdot}10^{-4}\,\frac{\text{m}^2}{\text{s}}$. Light shading corresponds to high intensity values

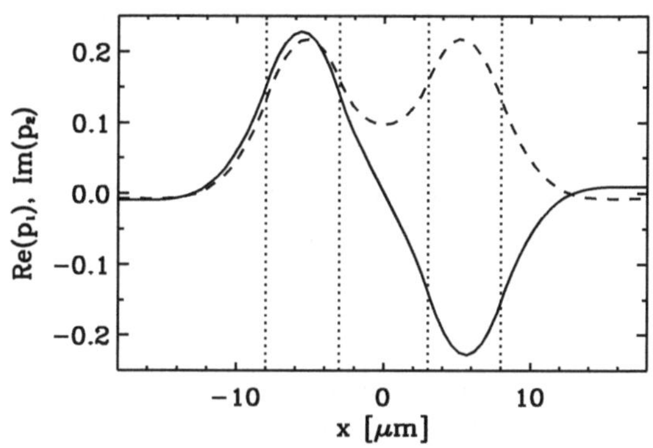

Fig. 9.2. Eigenmodes $p_1(x)$ and $p_2(x)$ for the twin-stripe laser. Shown are the real part of the first eigenmode (*solid line*) and the imaginary part of the second eigenmode (*dashed line*). The vertical lines indicate the locations of the laser stripes

the two adjacent laser stripes oscillate with opposite phase. This mode represents the "natural" operation condition of stable transversely coupled laser arrays. Due to the antisymmetric shape of the fundamental mode, its modulus vanishes in the center of the gap separating the laser stripes. The second eigenmode is symmetric with respect to the two lasers. The time-dependent expansion coefficients associated with it represent in-phase oscillation of the stripes. In contrast to the anti-symmetric mode, its modulus is clearly non-vanishing in the center of the gap. Therefore, the symmetric eigenmode can

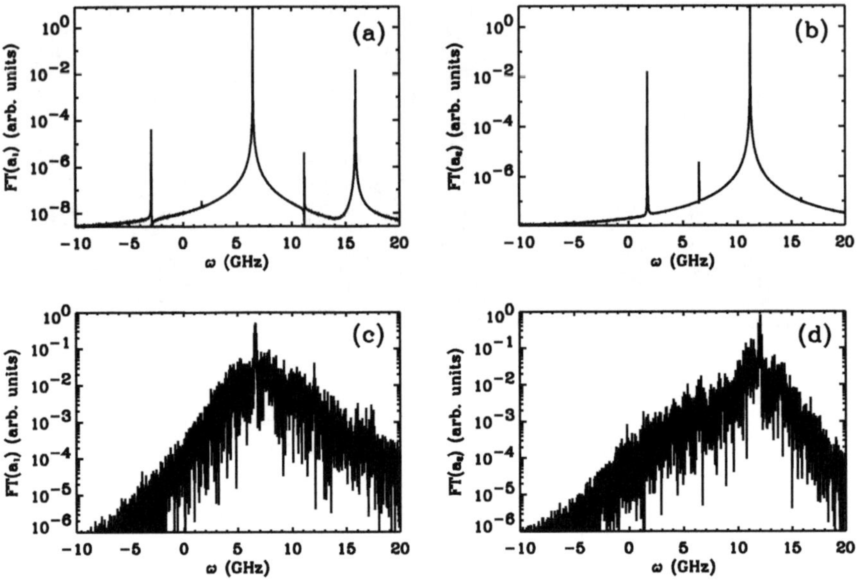

Fig. 9.3. Frequency spectra of the time-varying modal amplitudes $a_1(t)$ and $a_2(t)$ of the two eigenmodes in the periodic regime ($J = 45\,\mathrm{mA}$), (**a**, **b**), and in the chaotic regime ($J = 70\,\mathrm{mA}$) (**c**, **d**)

profit from the charge carriers diffusing towards the gap more than the fundamental mode. This effect becomes more and more important as the pump current increases until, at the critical current value, the symmetric mode becomes involved in the dynamics. In the above-mentioned periodic regime it coexists with the fundamental mode, leading to transverse mode-beating. This beating generates the periodic intensity pulsations, whose frequency equals the difference between the mode frequencies. For high currents there is mode competition rather than coexistence, which leads to chaotic behavior. The frequencies $\Omega_{1,2}$ pertaining to the first and second eigenmodes are obtained via the corresponding frequency spectra of the time series $a_{1,2}(t)$ (Fig. 9.3). In the periodic regime, the frequencies emerge as sharp lines, while in the chaotic regime the peaks are considerably broadened.

Using the frequencies $\Omega_{1,2}$ of the transverse optical eigenmodes, we are able to establish a destructive interference condition in the symmetric mode, in order to re-obtain steady state operation in the fundamental transverse mode. In a steady state condition we can easily realize the effects of the delay term. Imagine that the complex equation (9.1) for $E = R \cdot \mathrm{e}^{\mathrm{i}\theta}$ is written as two real equations for amplitude R and phase $\theta = \omega t$. Then the delay term produces a contribution $\gamma_{\mathrm{R}} R(x) \cos(\Phi - \omega\tau)$ in the amplitude equation while a term $\gamma_{\mathrm{R}} \sin(\Phi - \omega\tau)$ arises in the equation for the phase. This means that

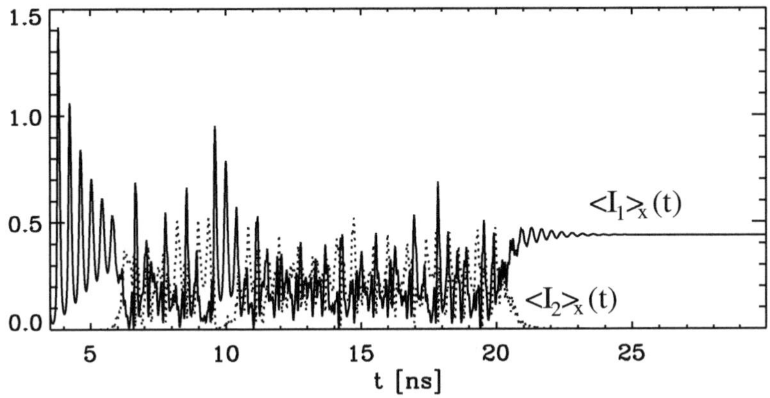

Fig. 9.4. Time series of the transversely averaged intensities $\langle I_1 \rangle_x(t)$, $\langle I_2 \rangle_x(t)$ pertaining to the two eigenmodes (*solid line*: fundamental mode, *dashed line*: symmetric mode). Stabilization of the antisymmetric fundamental transverse mode is achieved after switching on time delay at $t_0 = 20$ ns. For $t < t_0$, $\gamma_R = 4 \cdot 10^{-2}$, $\tau = \Phi = 0$ and for $t > t_0$, $\gamma_R = 4 \cdot 10^{-2}$, $\tau = 0.0862$ ns, $\Phi = 0.543$

there is a contribution to the gain of strength γ_R times a cosine dependence on the overall phase $\Psi = \Phi - \omega\tau$ which governs the interference condition. The contribution in the phase equation amounts to a frequency shift depending on γ_R and Ψ.

The contribution of delayed optical feedback to the gain can be graphically interpreted as the presence of a cosine shaped filter in the frequency domain which, for appropriately chosen feedback parameters τ and Φ, can cause destructive interference for light of a particular frequency. This effect can be used to provide damping in a transverse mode oscillating at that frequency.

Let ω_1 be the frequency of the fundamental mode and ω_2 that of the mode that we aim to suppress. The two parameters τ and Φ permit setting up the two interference conditions:

$$\Phi - \omega_2\tau = (n + \frac{1}{2}), \tag{9.7}$$

$$\Phi - \omega_1\tau = m \qquad n, m = 0, \pm1, \pm2 \ldots \tag{9.8}$$

(here and in the following, all phases are written in units of 2π). Equation (9.7) represents destructive interference in the transverse mode to be suppressed, while (9.8) describes constructive interference in the fundamental transverse mode, which seems to be a natural choice. Then τ and Φ are determined by the following equations as

$$\tau = \frac{1}{2(\omega_2 - \omega_1)} \tag{9.9}$$

$$\Phi = \omega_1 \tau \tag{9.10}$$

where the integers n and m were chosen such that the two frequencies for which the interference conditions are provided correspond to adjacent extrema of the cosine filter.

For a chaotic twin-stripe laser with $J = 80\,\mathrm{mA}$, $D_f = 4.0 \cdot 10^{-4} \frac{\mathrm{m}^2}{\mathrm{s}}$ (which is well within the chaotic range) we obtain the modal frequencies $\Omega_1 = 6.3\,\mathrm{GHz}$ and $\Omega_2 = 12.1\,\mathrm{GHz}$. Identifying $\Omega_{1,2}$ with $\omega_{1,2}$ from (9.9) and (9.10) then yields the values $\tau = 0.0862\,\mathrm{ns}$ and $\Phi = 0.543$. Using these parameters we indeed obtain stable continuous-wave operation in the fundamental antisymmetric mode. The stabilizing effect of delayed optical feedback is demonstrated in Fig. 9.4. For $t < t_0 = 20\,\mathrm{ns}$, $\gamma_\mathrm{R} = 4 \cdot 10^{-2}$, $\tau = \phi = 0$, and for $t > t_0$, τ and Φ are set to the above values. After the chaotic interval the symmetric eigenmode is quickly damped to zero while the antisymmetric eigenmode approaches continuous-wave emission via damped relaxation oscillations. When stabilization is achieved and there is exactly constructive interference, the original system of equations, i.e. without delayed feedback, is effectively recovered. In this case only one mode, for which $\Psi = 0$, exists. Thus, the delay term in the phase equation is zero and the delay term in the amplitude equation reduces to a slightly enhanced reflectivity (γ_R). In this sense, our procedure bears some analogy to genuine schemes of controlling chaos where, upon successful control, the controlling force vanishes [14; 15].

Eigenmode analysis of multi-stripe lasers shows that the maximum number of relevant eigenmodes in the complex spatio-temporal regimes is equal to the number of stripes. Thus, in principle, several transverse modes have to be suppressed in order to stabilize laser arrays with more than two stripes. In the case of the five-stripe array, where four relevant transverse modes exist besides the fundamental mode, stabilization of the fundamental mode can still be achieved. An example where the five-stripe laser is brought from a state of spatio-temporal chaos to continuous-wave operation is shown in Fig. 9.5. However, the criteria in respect to how to choose ω_2 are not as obvious as in the case of the twin-stripe laser. The mode to be suppressed varies with pump current and diffusion strength, and in some cases the best results are obtained when ω_2 lies between two transverse mode frequencies. The fact that stabilization is still possible could be plausible if the successive transverse modes are nonlinearly coupled in a cascade-like fashion. Then damping of particular modes could cause the suppression of the whole cascade of modes.

9.5 Spatially Structured Delayed Optical Feedback Control

Due to the great complexity and shortness of relevant spatial and temporal scales an application of schemes from the field of chaos control [14; 15] to a high-power semiconductor laser displaying complex spatio-temporal chaos is

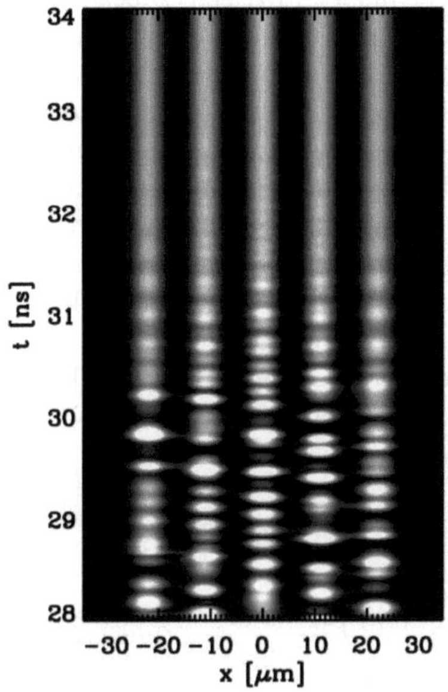

Fig. 9.5. Spatio-temporal dynamics of the five-stripe laser for $J = 80\,\mathrm{mA}$, $\gamma_{\mathrm{R}} = 3 \cdot 10^{-2}$, $\tau = 0.0877\,\mathrm{ns}$ and $\Phi = 0.06$ where the time delay and feedback phase are switched from zero to the above values at $t = 30\,\mathrm{ns}$

not straight forward. Particularly because of the small time scales involved in semiconductor laser dynamics an all-optical control scheme like delayed-feedback control is the method of choice [2]. However, naive application of a delayed optical feedback (DOF) control or stabilization method to the semiconductor laser even tends to increase spatio-temporal complexity in spatially distributed systems [2; 7]. On the other hand, we have seen above that with careful choice of the feedback parameters obtained from a complex eigenmode analysis, DOF may successfully be employed for stabilization of the typical spatio-temporal chaos in semiconductor laser arrays. In the broad-area laser, however, in addition to time-delayed feedback, spatial and spectral degrees of freedom have to be simultaneously stabilized by designing an appropriate control set-up.

In order to shed light on the fundamental interplay of the internal spatio-temporal dynamics in the active layer of a high-power laser diode and the spatially inhomogeneous delayed optical feedback both aspects have to be considered simultaneously in a theoretical description. Thus to account for the microscopic processes which act in concert with the macroscopic spatio-temporal interactions we will base our investigation on the Maxwell–Bloch equations and incorporate the influence of spatially structured delayed optical feedback (SDOF).

The *semiconductor laser Maxwell–Bloch delay-equations* consist of Maxwell's wave equations for the counterpropagating (in positive (+) and negative (−) z directions) optical fields $\boldsymbol{E}^{\pm}(\boldsymbol{r},t)$ in which the effect of structured delayed optical feedback is included and an ambipolar transport equation for the charge carrier density $N(\boldsymbol{r},t)$, where $\boldsymbol{r}=(x,z)$ indicates the longitudinal light field propagation direction z and the transverse direction x. This coupled system is in turn self-consistently coupled to spatially inhomogeneous semiconductor Bloch equations for the Wigner distributions $f^{\mathrm{e,h}}(k,\boldsymbol{r},t)$ of electrons (e) and holes (h) as well as the interband polarizations $p_k^{\pm}=p^{\pm}(k,\boldsymbol{r},t)$. The coupling between the microscopic spatio-spectral dynamics and the macroscopic propagation of the light field is mediated by the macroscopic nonlinear polarizations $\boldsymbol{P}^{\pm}=V^{-1}\sum_{\mathbf{k}}\boldsymbol{d}_k p^{\pm}(k,\boldsymbol{r},t)$, which in Maxwell's wave equation

$$\begin{aligned}\pm\frac{\partial}{\partial z}\boldsymbol{E}^{\pm}(\boldsymbol{r},t)+\frac{n_1}{c}\frac{\partial}{\partial t}\boldsymbol{E}^{\pm}(\boldsymbol{r},t)&=\frac{\mathrm{i}}{2}\frac{1}{K_z}\frac{\partial^2}{\partial x^2}\boldsymbol{E}^{\pm}(\boldsymbol{r},t)\\&-\left(\frac{\alpha}{2}+\mathrm{i}\eta\right)\boldsymbol{E}^{\pm}(\boldsymbol{r},t)+\frac{\mathrm{i}}{2}\frac{\Gamma}{n_1^2\epsilon_0 L}\boldsymbol{P}^{\pm}(\boldsymbol{r},t)\\&+\frac{\kappa}{\tau_{\mathrm{r}}}\boldsymbol{E}^{\pm}\left(x\frac{\tilde{r}}{\sqrt{\left(\frac{\mathrm{w}}{2}\right)^2+(L_{\mathrm{e}}+\tilde{r})^2}},L,t-\tau\right)\end{aligned}\qquad(9.11)$$

are the source of the optical fields. In the feedback term the resonator round trip times of the internal and external resonators are τ_{r} and τ, respectively. The back-coupling strength is denoted by $\kappa=(1-R_0)\sqrt{\frac{R_1}{R_0}}$ with R_0 and R_1 being the reflectivity of the front laser facet and of the external mirror, respectively. The spatially structured feedback as realized in the external resonator configuration, schematically displayed in Fig. 9.6, is characterized by the radius of curvature $\tilde{r}=0.5$ mm and the length of the external resonator $L_{\mathrm{e}}=2370$ μm; $\mathrm{w}=100$ μm is the transverse width of the emitter. In (9.11) K_z denotes the wavenumber of the propagating fields, n_1 the refractive index of the active layer, L the length of the structure, and $\alpha=\alpha(T_1)$ the linear absorption coefficient. The parameter η includes transverse (x) and vertical (y) variations of the refractive index due to the waveguide structure, and

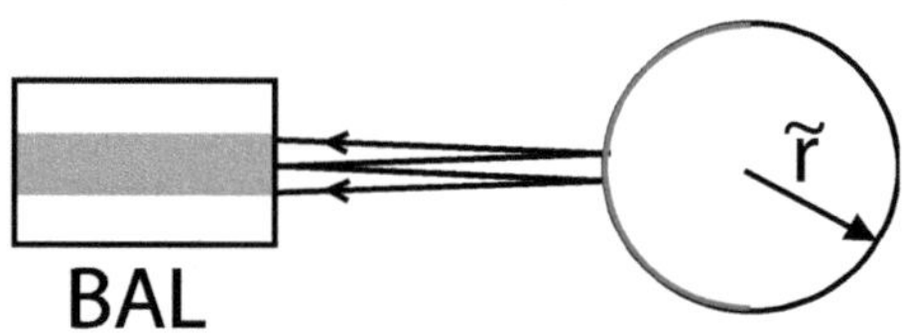

Fig. 9.6. Schematic geometry of the spatially structured delayed optical feedback stabilization scheme realized in the form of an unstable external resonator

the wave-guiding properties are described by the confinement factor Γ. The optical properties additionally depend on the local density of charge carriers, whose dynamics are governed by a carrier transport equation (see Chap. 2). In addition to the parameters describing the broad-area laser itself the principle components of an SDOF set-up are included in a quasi-two-dimensional approximation [1] in the microscopic simulations. They are, in particular, delayed coherent optical feedback from a mirror at a distance L_e (with respect to the outcoupling facet of the laser, cf. Fig. 9.6), the spatial structure of the feedback as a result of the mirror-curvature with a radius $\tilde{r}$, the reflectivities R_0 and R_1 of the laser and the external mirror, respectively, as well as the applied electrical pump current. With $R_0 \approx 10^{-5}$ and $R_1 \approx 98\%$ in the experimental realization compared to $R_0 = 10^{-4}$ and $R_1 = 70\%$ assumed slightly larger (smaller) in the simulations, the reflection coefficients chosen here are in a regime where the feedback strength is sufficient to control the laser output and at the same time retain the assumptions of the delayed feedback model [16].

Figures 9.7–9.9 display representative examples of the spatio-temporal dynamics and the temporally averaged spatial profiles of (a) the nearfield intensity, (b) the farfield intensity, and (c) the corresponding density of charge carriers of the broad-area laser during its initial 3 ns after start-up. The electrical pump current ($J = 1.5J_{\text{thr}}$) is applied at $t = 0$. While Fig. 9.7 for comparison shows the broad-area laser in the free-running condition (i.e. without the delayed optical feedback), Fig. 9.8 and Fig. 9.9 visualize the influence of optical feedback "turned on" at $t = 1.2$ ns. In the free-running condition, the formation of filaments and onset of their migration are clearly visible in Fig. 9.7. The processes which lead to this peculiar filamentation behavior are of microscopic nature and in the broad-area laser depend only to a certain degree on the particular lateral boundary conditions, e.g. the imposed wave-guiding structure. Rather, they are a direct consequence of a concert of processes. Due to the fast microscopic Coulomb scattering processes the relaxation time of the carrier density is larger than that of the optical field and the interband polarization. This leads to a localization of filaments of high intensity in wave-guiding channels formed by the carrier density: the carrier density is locally depleted by a filament of high intensity. An optical filament is thus located in a region of low gain (low carrier density) and relatively high refractive index. By the process of gain guiding the filament thus provides itself with the dielectric waveguide which is necessary for its support during propagation in the laser cavity. Due to the length of the (strongly asymmetric) internal cavity of the broad-area laser, however, the filaments are longitudinally inhomogeneous. The result is a wave-like reflection of the filaments. Due to the spatially non-uniform transport of carriers the filament front becomes longitudinally and laterally more and more inhomogeneous. At the same time, with uniform injection of charge carriers the local carrier density outside the filament is not depleted by stimulated emission and con-

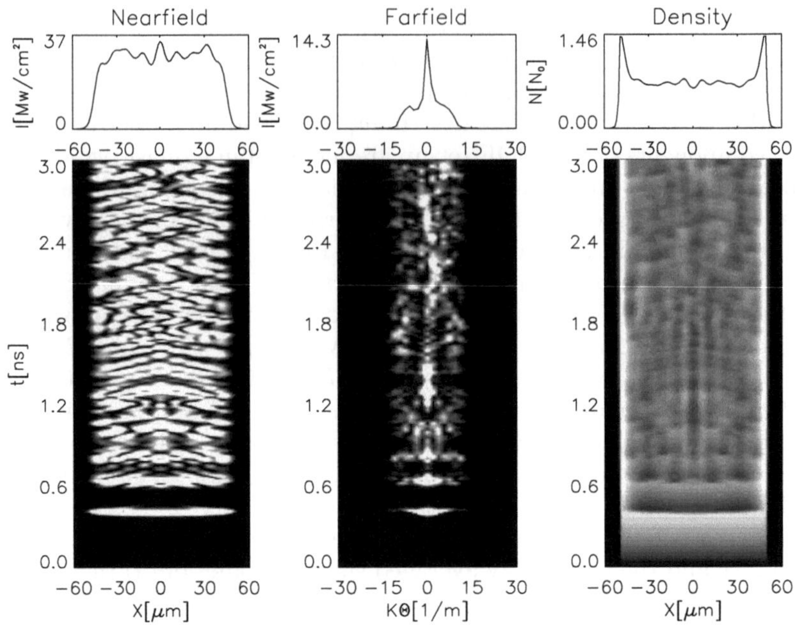

Fig. 9.7. start-up spatio-temporal dynamics of a free-running broad-area semiconductor laser. *Left column*: temporal evolvement of the nearfield intensity and time-averaged nearfield intensity profile. *Middle column*: corresponding dynamics of the Fraunhofer farfield with time averaged farfield profile with optical wavenumber K and farfield angle θ. *Right column*: spatio-temporal distribution of the charge carrier density and temporally averaged profile

sequently rises quickly to levels above the threshold charge carrier density. A new optical filament is thus created. Every new filament thus interacts with the previous filament via the medium nonlinearly, thereby destabilizing it. The result is a vividly irregular and fundamentally chaotic longitudinal and transverse interaction of optical filaments. Also, the gain guiding processes inside the laser cavity ensure that by sustaining relatively stable high values of the density at the edge of the laser stripe the optical field creates its own optical waveguide, i.e. an effective waveguide is formed. The combination of a transverse modulational instability with the propagation of the filaments as well as the carrier transport then leads to a symmetry breaking and the start of filament migration on a time scale of about 500 ps [17]. As a consequence, the temporally averaged profiles of the nearfield intensity and the carrier density appear asymmetric while the farfield both shows a strong angular spread in addition to the asymmetry. The spatio-temporal farfield trace reveals that this is due to irregular sub-ns beam-fanning.

In the situation portrayed in Fig. 9.8 the delayed optical feedback is turned on at t=1.2 ns. In Fig. 9.8 the external mirror is flat (no spatial structure),

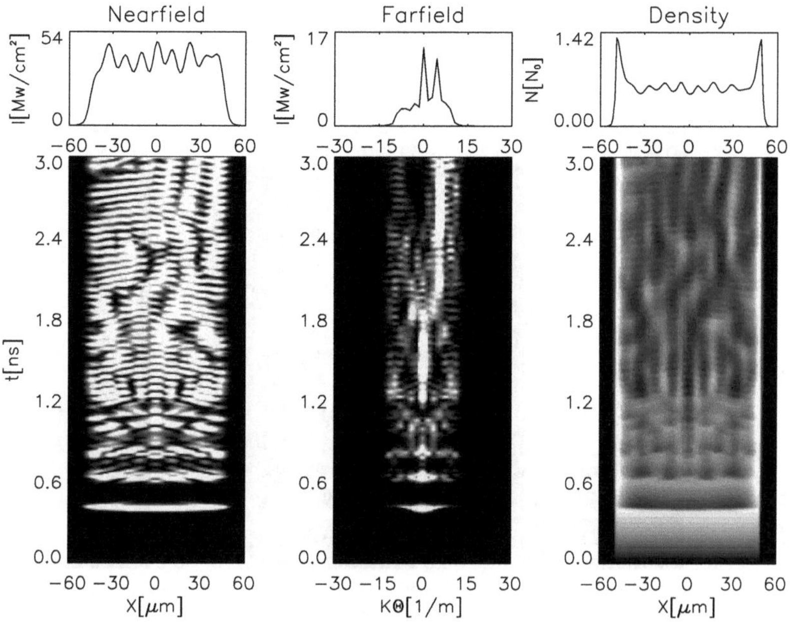

Fig. 9.8. Broad-area semiconductor laser with delayed optical feedback (applied at t = 1.2 ns) from a flat, weakly reflecting ($R \approx 0.01$) external mirror. The length of the external resonator is set to $L_e = 30$ μm, corresponding to a round trip time $\tau \approx 20$ ps

of low reflectivity ($R \approx 0.01$) and placed at a distance $L_e \approx 3000$ μm. As a consequence of this "mild" interference the filaments persist. Due to the regular feedback at regular intervals they are oscillatory stabilized while the farfield shows a more strongly pronounced angular jump compared to the free-running case. An increase of the reflectivity of the external mirror from $R \approx 0.01$ to $R \approx 0.5$ (with all other parameters unchanged) stabilizes the high-frequency oscillation and prevents the symmetry breaking. In Fig. 9.9 the spatio-temporal nearfield intensity trace appears with a fishbone pattern. The time averaged farfield is characterized by a single lobe with increased hight riding on a broader shoulder which represents the contribution of higher optical wavenumbers. It is these higher wavenumbers that have to be suppressed for a control scheme to be effective. Variation of the feedback parameters has shown that flat feedback, however, either tends to stabilize the fishbone oscillations or re-introduce the migrating nearfield filaments. It is not until there is additional spatial structuring of the external feedback mirror that the stabilization of a single lobed farfield becomes possible.

The left, middle, and right columns of the bottom row of Fig. 9.10 show the start-up spatio-temporal dynamics of the nearfield intensity, farfield intensity and charge carrier density, respectively, for the case of a broad-area

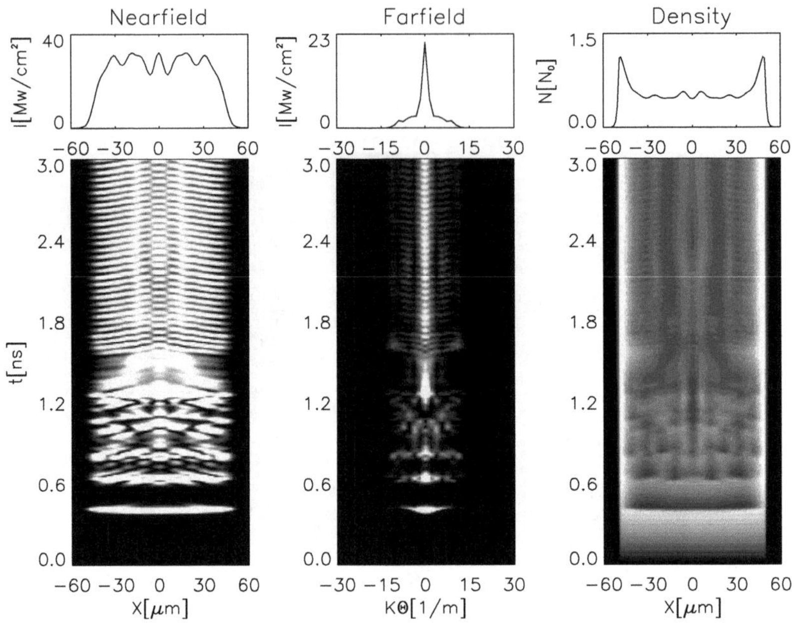

Fig. 9.9. Broad-area semiconductor laser with strong ($R \approx 0.5$), unstructured (flat mirror) delayed optical feedback

laser with SDOF. The top row depicts the corresponding time traces in the center ($x = 0$) of the laser stripe and the middle row displays the spatial field and density profiles at $t = 500$ ps. The laser is electrically pumped at twice threshold current. Note that during the initial 200 ps in the free-running condition (bottom row, $t < 200$ ps) the reflectivity at the outcoupling mirror has artificially been raised to $R_0 = 0.1$ for the broad-area laser to reach its threshold condition. Upon application of SDOF at $t = 200$ ps it is reduced to $R_0 = 10^{-4}$.

The general influence of SDOF is most clearly observed in the spatio-temporal nearfield and farfield intensity traces. After application (switch-on) of the SDOF the filaments in the nearfield are pinned, the wide farfield narrows considerably and the charge carrier density is depleted homogeneously in space. This clearly demonstrates the positive influence of appropriately tailored SDOF leading to successful stabilization of the spatially migrating intensity filaments. Note that the fast pulse-like peaks visible in the intensity profiles in the top row are a fingerprint of delayed feedback in the external resonator. Similar to the relaxation oscillations observed in a free-running semiconductor laser they gradually decrease in amplitude, however, on a more prolongated time scale. After a few nanoseconds they relax to a both spatially and temporally stable intensity profile. We further note that for visualization purposes and in order to highlight the transition from the free- running laser

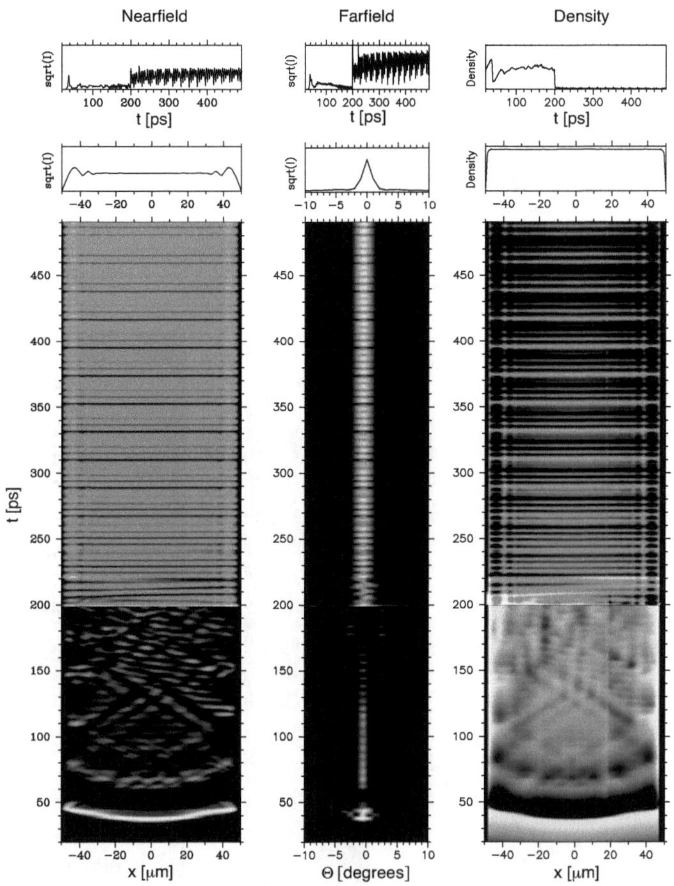

Fig. 9.10. Formation and dynamics of optical filaments in the initially free-running ($t = 200$ ps) start-up dynamics of a broad-area semiconductor laser ($w = 100$ µm, $J = 2$ A) and application of structured delayed optical feedback at $t = 200$ ps. *Left column*: spatio-temporal evolvement of the nearfield intensity, its time averaged profile at $t = 500$ ps and dynamics in the center ($x = 0$). *Middle column*: corresponding dynamics of the Fraunhofer farfield. *Right column*: spatio-temporal distribution of the charge carrier density

to the application of SDOF the vertical scale of the time trace of the density (middle row in Fig. 9.10) is considerably expanded in the vicinity of this transition. Upon application of SDOF at $t = 200$ ps the density is reduced but does not drop to zero as can be seen from the spatio-temporal evolution of the density. This reduction is due to the stronger field returned by the SDOF configuration.

While in Fig. 9.10 we have applied the SDOF after the initial start-up sequence of 200 ps, Fig. 9.11 shows the corresponding situation where SDOF (and the reduced reflectivity at the output mirror) is present right from the start in correspondence to the experiments. In contrast to Fig. 9.10 where

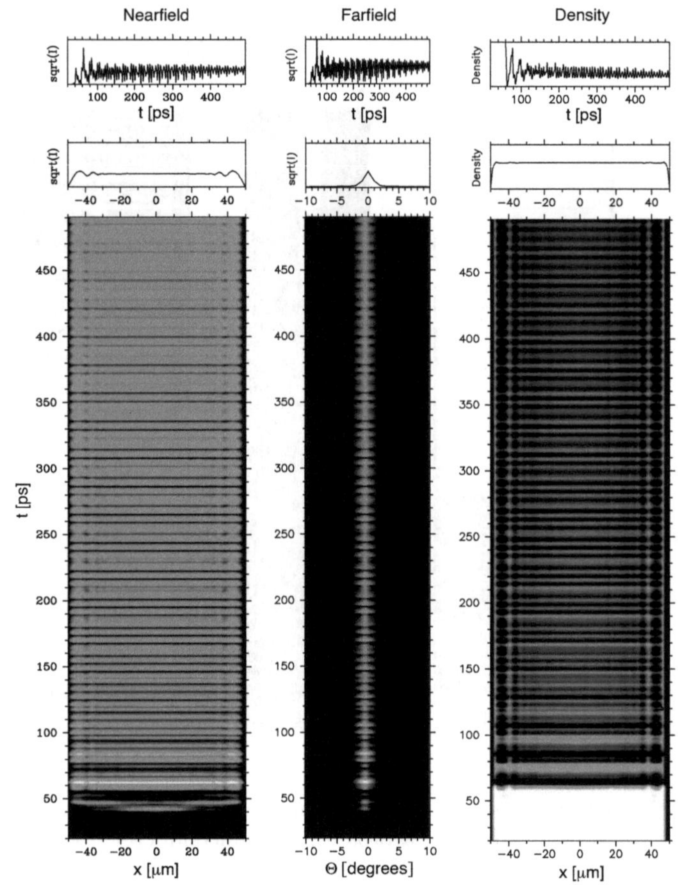

Fig. 9.11. Structured delayed optical feedback applied from the start

the relaxation oscillations led to break-up into multiple filaments, the relaxation oscillations are now characterized by lateral homogeneity. The higher transverse modes that appear in the free-running and flat-feedback cases as irregular and periodic high-frequency oscillations, respectively, are effectively suppressed in the SDOF configuration. Ideally, the temporal coherence of the output signal should be equally well maintained. When the feedback time (corresponding to the length of the external resonator) is adjusted to the internal propagation length the filament front experiences a stabilizing coherent push just in time to prevent a destabilizing instability as in the case of the free-running broad-area laser (Fig. 9.7). Figure 9.11 confirms that in spite of the competing dynamics of the internal laser and the feedback, which under certain conditions may induce coherence collapse [18] the spatio-temporal behavior of the broad-area laser with SDOF also relaxes to the spatially and temporally stable emission characterized by a narrow farfield.

9.6 Conclusion

In this chapter we have discussed the influence of delayed optical feedback on the spatio-temporal dynamics of spatially extended (coupled, broad-area, etc.) semiconductor lasers. The theory is based on phenomenological and microscopically founded extensions of the transverse Lang–Kobayashi equations. We show that the generally destabilizing influence of delayed optical feedback may be constructively employed for control of ultrafast instabilities and filamentation in coupled stripe and broad-area lasers. With the characteristic spectra pertaining to the stabilizing and destabilizing modes of laser arrays (which have been obtained by complex eigenmode analysis) a control scheme is described which allows supressing the spatio-temporal instabilities in laser arrays. For control of the continuous spatio-temporal filamentation dynamics in broad-area lasers, however, the controlling time delayed feedback has to be adequately spatially structured (filtered).

References

1. O. Hess and T. Kuhn. Spatio-temporal dynamics of semiconductor lasers: Theory, modeling and analysis. *Prog. Quant. Electr.*, 20:85–179, 1996.
2. C. Simmendinger and O. Hess. Controlling delay-induced chaotic behavior of a semiconductor laser with optical feedback. *Physics Lett. A*, 216:97–105, 1997.
3. M. E. Bleich, D. Hochheiser, J. V. Moloney, and J. E. S. Socolar. Controlling extended systems with spatially filtered, time-delayed feedback. *Phys. Rev. E*, 55:2119–2126, 1997.
4. C. Simmendinger, M. Münkel, and O. Hess. Controlling complex temporal and spatio-temporal dynamics in semiconductor lasers. *Chaos, Solitons & Fractals*, 10:851–864, 1999.
5. H. Aachihara, O. Hess, R. Indik, and J. V. Moloney. Semiconductor laser array dynamics: Numerical simulations on multi-stripe index-guided lasers. *J. Opt. Soc. Am. B*, 10:496–506, 1993.
6. I. Fischer, O. Hess, W. Elsäßer, and E. Göbel. Complex spatio-temporal dynamics in the nearfield of a broad-area semiconductor laser. *Europhys. Lett.*, 35:579–584, 1996.
7. M. Münkel, F. Kaiser, and O. Hess. Spatio-temporal dynamics of multi-stripe semiconductor lasers with delayed optical feedback. *Physics. Lett. A*, 222:67–75, 1996.
8. R. Lang and K. Kobayashi. External optical feedback effects on semiconductor injection laser properties. *IEEE J. Quant. Electr.*,, 16:347, 1980.

9. O. Hess. Spatio-temporal complexity in multi-stripe and broad-area semiconductor lasers. *Chaos, Solitons & Fractals*, 4:1597–1618, 1994.
10. D. Merbach, O. Hess, H. Herzel, and E. Schöll. Injection-induced bifurcations of transverse spatiotemporal patterns in semiconductor laser arrays. *Phys. Rev. E*, 52:1571–1578, 1995.
11. M. Münkel, F. Kaiser, and O. Hess. Stabilization of spatiotemporally chaotic semiconductor laser arrays by means of delayed optical feedback. *Phys. Rev. E*, 56:3868–3875, 1997.
12. H. G. Winful and S. S. Wang. Stability of phase locking in coupled semiconductor laser arrays. *Appl. Phys. Lett.*, 53:1894–1896, 1988.
13. S. S. Wang and H. G. Winful. Dynamics of phase-locked semiconductor laser arrays. *Appl. Phys. Lett.*, 52:1774–1776, 1988.
14. D. Auerbach, C. Grebogi, E. Ott, and J. Yorke. Controlling chaos in high dimensional systems,. *Phys. Rev. Lett.*, 69:3479–3482, 1992.
15. K. Pyragas. Continuous control of chaos, by self-controlling feedback. *Physics Lett. A*, 170:421–428, 1992.
16. K. Petermann. External optical feedback phenomena in semiconductor lasers. *IEEE. J. Selec. Top. Quant. Electr.*, 1:480, 1995.
17. I. Fischer, O. Hess, W. Elsäßer, and E. Göbel. Complex spatio-temporal dynamics in the nearfield of a broad-area semiconductor laser. *Europhys. Lett.*, 35:579–584, 1996.
18. D. Lenstra, B. H. Verbeck, and A. J. Den Boef. Coherence collapse in single-mode semiconductor laser due to optical feedback. *IEEE J. Quant. Eectr.*, QE-21:674–679, 1985.

10 Spatially Inhomogeneous Luminescence in Semiconductor Lasers

In this chapter we discuss the theoretical basis for describing spatially inhomogeneous spontaneous and amplified spontaneous emission (luminescence) in semiconductor lasers. The theory builds on and adopts the concepts of quantum Maxwell–Bloch equations (QMBEs) [1] describing the spatio-temporal dynamics of photoluminescence of excited semiconductor lasers. The QMBEs are derived from fully quantum mechanical operator dynamics describing the interaction of the light field with the quantum states of the electrons and the holes near the bandgap. Field–field correlations and field–dipole correlations quantum noise effects that cause spontaneous emission and amplified spontaneous emission are taken into account in a self-consistent way. On the basis of the QMBEs luminescence equations are derived that represent a starting point for analysis of spatially inhomogeneous light emission characteristics and coherence properties of complex laser structures such as for example VCSELs. After setting up a general quantum-kinetic description we will apply the formalism to quantum well lasers. The chapter concludes with an adaption of the theory to describe optically pumped semiconductor lasers in external cavities.

10.1 Introduction

In recent years increasing miniaturisation and improved cavity design has lead to a growing influence of quantum mechanical properties of light fields on the spatio-temporal dynamics and emission properties of modern semiconductor laser structures. A theoretical analysis of quantum effects thus is of high relevance for profound understanding of the physical properties of small or microstructured semiconductor lasers. Generally the underlying physical processes that are responsible for the generation of light affect the interplay between spontaneous and induced radiation and consequently the spatial and spectral emission properties. Sufficiently above the laser threshold the spatio-temporal dynamics of semiconductor lasers can be simulated successfully on the basis of semiclassical Maxwell–Bloch equations without explicitly including fundamental quantum effects in the light field (cf. Chap. 2). The classical treatment of the light field is justified by the high intensity of the laser light. However, the incoherent noise required by the uncertainty principle in both the electrical dipole of the semiconductor medium and the light field itself is of significant importance when several cavity modes compete or when the laser is close to threshold. The spatial coherence of spontaneous emission and amplified spontaneous emission is even more important in devices close to threshold or devices with a light field output dominated by spontaneous emission such as superluminescent diodes and ultra-low threshold semiconductor lasers [2; 3]. Ultra-low threshold lasers may actually operate in a regime of negative gain where spontaneous emission is the only source of radiation [2]. A description of the light field emitted by such devices therefore requires an explicit description of the spatial coherence in spontaneous emission as well. In order to understand the physical origin of the specific emission characteristics of a semiconductor laser it is necessary to investigate the fundamental processes that create and consequently determine the physical properties of light such as e.g. its (spatial and temporal) coherence properties, spatio-spectral beam characterstics, polarization fluctuations and temporal stability. These originate from the quantum-optical nature of light and its spatio-temporal coupling to the charge carrier plasma in a laser structure.

The history of spontaneous and stimulated emission processes starts with Einstein's rate description from which one can derive characteristic expressions for spontaneous as well as induced emission and absorption. The resulting transition propabilities can easily be applied to lasers and lead to respective laser rate equations for the description of simple semiconductor lasers. However, in these phenomenological models the spatio-temporal coherence of spontaneous emission is not taken into account. Photon rate equations for multi-mode operation of semiconductor lasers show that the spontaneous emission terms may contribute significantly to the spectral characteristics of the light field emitted by the laser [4; 5]. Such models assume a fixed mode structure determined entirely by the empty cavity. This assumption does not apply to gain guided lasers and to unstable resonators, however [6; 7; 8]. In these cases it is therefore desirable to explicitly describe the spatial coherence of spontaneous emission. Recent investigations [1] show that spontaneous emission processes can often be interpreted in terms of quantum fluctuations in light field and carrier density. A realistic description of luminescence consequently requires consideration of quantum fluctuations in both the light field and the carrier density of the semiconductor. An approach to the consistent inclusion of the quantum noise properties of the light field in the dynamics of semiconductor laser diodes using nonequilibrium Green's functions has been presented in [9; 10; 11].

In recent theories local luminescence equations describing the response of a semiconductor material to a δ-like singular excitation have been derived [12; 13]. That theory includes dynamic carrier scattering and dephasing on a very fundamental level and clearly proves the importance of many-body effects and higher order correlation terms for the realistic description of line shapes, gain curves and threshold charactistics of semiconductor lasers. Microscopic quantum theories consisting of a closed set of quantum correlation functions for excitons, phonons and photons [14] allow a detailed investigation of the build-up of excitonic occupation and its decay via secondary emission. Another approach [15] uses the Bethe–Salpeter equation for the polarization to calculate the luminescence spectrum for nonthermal carrier distributions.

In order to simulate the spatio-temporal dynamics of multi-mode operation, of lasers near threshold, low threshold lasers, superluminescent diodes and microstructured laser devices we formulate an alternative approach to the problem of spontaneous emission and amplified spontaneous emission on the basis of a Maxwell–Bloch description. By including the spatio-temporal dynamics of the interband dipole, non-Markovian terms are avoided. The spatial dependence of the light fields (i.e. light diffraction and counterpropagation effects) as well as the nonlinear dynamic coupling of the active charge carrier plasma to the light field dynamics thereby will explicitly be taken into account. The basis of the theoretical description and numerical simulations is spatio-temporally resolved luminescence equations that are microscopically coupled to Maxwell's wave equation. The luminescence equations will be derived from a fully quantum kinetic many-body description. They can be applied to the description of the light generating processes with explicit consideration of the quantum nature of light. In particular, we will microscopically include an optical pump process via suitable boundary conditions and consider an external resonator with delayed optical feedback equations.

In Sect. 10.2 the quantum dynamics of the interactions between the light field and the carrier system is formulated in terms of Wigner distributions for the carriers and of spatially continuous amplitudes for the light field. The equations are formulated for both bulk material and for quantum wells including the effects of anisotropic coupling to the polarization components of the light field. Section 10.3 summarizes the effects of the dynamics of the electron–hole system in the semiconductor material. The light field dynamics are introduced in Sect. 10.4. By quantizing Maxwell's equation the coupling constant g_0 introduced in Sect. 10.2 is expressed in terms of the interband dipole matrix element. The complete set of quantum Maxwell–Bloch luminescence equations is presented in Sect. 10.5. Based on this general formulation, specific approximate versions for quantum well edge-emitting and vertical-cavity surface-emitting lasers are derived. Extensions for inclusion of e.g. optical excitation or the influence of an external resonator are discussed in Sect. 10.6. Section 10.7 concludes this chapter.

10.2 Dynamics of the Light–Carrier Interaction

The fundamental properties of the light–matter interactions that are responsible for the dynamics of the light emission require a quantum theoretical description and have to be derived from operator algebra. We will thus derive the luminescence equations from the fundamental quantum kinetic description. On a microscopic basis they include and describe the interactions and correlations of the amplitudes of the light fields and the dipoles excited in the medium that cause and mediate the (spatial and temporal) coherence of spontaneous and stimulated emission.

The starting point for our description of quantum noise effects is the dynamics of quantum mechanical operators of the field and carrier system. Since the operator dynamics of the carrier system have been investigated in the context of Maxwell–Bloch equations before [16] and the light field equations correspond exactly to the classical Maxwell equations (see e.g. Chap. 2), in this chapter we will focus on the light–matter interaction and on the modifications required for simulation of a specific semiconductor laser system.

10.2.1 Hamiltonian Dynamics of Densities and Fields

A fundamental approach to investigating the physics of light generation is to start from the many-body Hamiltonian including Coulomb interactions in the charge carrier plasma as well as carrier–phonon interactions and to derive luminescence equations on the basis of the density matrix approach.

In the following we will describe the active semiconductor medium in terms of an isotropic two band model where, for the case of the holes, a suitably averaged effective mass is taken [16]. Generalization to more bands is straightforward. In order to include all relevant light generation processes we have to take into account not only the expectation values of the fields and carriers (electron and holes), but also the field–field and field–dipole correlations. In terms of the local annihilation operators for photons ($\hat{b}_R$), electrons ($\hat{c}_R$), and holes ($\hat{d}_R$), the Hamiltonian of the light–carrier interaction can be written as

$$\hat{H}^{\mathrm{cL}} = \hbar g_0 \sum_R \left(\hat{b}_R^\dagger \hat{c}_R \hat{d}_R + \hat{b}_R \hat{c}_R^\dagger \hat{d}_R^\dagger \right) , \tag{10.1}$$

where g_0 denotes the coupling constant characterizing the interaction between the carriers and the photons. The operator dynamics associated with this Hamiltonian is then given by

$$\left.\frac{\partial}{\partial t}\hat{b}_R\right|_{\mathrm{cL}} = -\mathrm{i}g_0 \hat{c}_R \hat{d}_R \tag{10.2}$$

$$\left.\frac{\partial}{\partial t}\hat{c}_R \hat{d}_{R'}\right|_{\mathrm{cL}} = \mathrm{i}g_0 \left(\hat{b}_R \hat{d}_R^\dagger \hat{d}_{R'} + \hat{b}_{R'} \hat{c}_{R'}^\dagger \hat{c}_R - \hat{b}_R \delta_{R,R'} \right) \tag{10.3}$$

$$\left.\frac{\partial}{\partial t}\hat{c}_R^\dagger \hat{c}_{R'}\right|_{\mathrm{cL}} = -\mathrm{i}g_0 \left(\hat{b}_{R'} \hat{c}_R^\dagger \hat{d}_{R'}^\dagger - \hat{b}_R^\dagger \hat{c}_{R'} \hat{d}_R \right) \tag{10.4}$$

$$\left.\frac{\partial}{\partial t}\hat{d}_R^\dagger \hat{d}_{R'}\right|_{\mathrm{cL}} = -\mathrm{i}g_0 \left(\hat{b}_{R'} \hat{c}_{R'}^\dagger \hat{d}_R^\dagger - \hat{b}_R^\dagger \hat{c}_R \hat{d}_{R'} \right) . \tag{10.5}$$

The discrete positions R correspond to the lattice sites of the Bravais lattice describing the semiconductor crystal. Each lattice site actually represents the spatial volume ν_0 of the Wigner–Seitz cell of the lattice. For the zincblende crystal structure this volume is equal to one quarter of the cubed lattice constant. In the case of AlGaAs structures the lattice constant is about 5.65×10^{-10}m and $\nu_0 \approx 4.5\times10^{-29}\mathrm{m}^3$ [17]. The photon annihilation operator $\hat{b}_R$ therefore describes the annihilation of a photon within a volume ν_0. Here, we will focus on the light–carrier interaction and the quantum noise contributions responsible for spontaneous emission. For that purpose, we extend the semiclassical description by including not only the expectation values of the field and dipole operators, $\langle\hat{b}_R\rangle$ and $\langle\hat{c}_R\hat{d}_{R'}\rangle$, respectively, but also the field–field correlations $\langle\hat{b}^\dagger{}_R\hat{b}_{R'}\rangle$ and the field–dipole correlation $\langle\hat{b}^\dagger_R\hat{c}_{R'}\hat{d}_{R''}\rangle$. The factorized equations of motion then read

$$\frac{\partial}{\partial t}\langle\hat{b}^\dagger_R\hat{b}_{R'}\rangle\bigg|_{\mathrm{cL}} = -\mathrm{i}g_0\left(\langle\hat{b}^\dagger_R\hat{c}_{R'}\hat{d}_{R'}\rangle - \langle\hat{b}^\dagger_{R'}\hat{c}_R\hat{d}_R\rangle^*\right) \tag{10.6}$$

$$\begin{aligned}\frac{\partial}{\partial t}\langle\hat{b}^\dagger_R\hat{c}_{R'}\hat{d}_{R''}\rangle\bigg|_{\mathrm{cL}} &= \mathrm{i}g_0\left(\langle\hat{b}^\dagger_R\hat{b}_{R'}\hat{d}^\dagger_{R'}\hat{d}_{R''}\rangle + \langle\hat{b}^\dagger_R\hat{b}_{R''}\hat{c}^\dagger_{R''}\hat{c}_{R'}\rangle\right.\\ &\quad\left.-\langle\hat{b}^\dagger_R\hat{b}_{R'}\rangle\delta_{R',R''}\right) + \mathrm{i}g_0\langle\hat{c}^\dagger_R\hat{c}_{R'}\hat{d}^\dagger_R\hat{d}_{R''}\rangle\end{aligned} \tag{10.7}$$

$$\frac{\partial}{\partial t}\langle\hat{c}^\dagger_R\hat{c}_{R'}\rangle\bigg|_{\mathrm{cL}} = \mathrm{i}g_0\left(\langle\hat{b}^\dagger_R\hat{c}_{R'}\hat{d}_R\rangle - \langle\hat{b}^\dagger_{R'}\hat{c}_R\hat{d}_{R'}\rangle^*\right) \tag{10.8}$$

$$\frac{\partial}{\partial t}\langle\hat{d}^\dagger_R\hat{d}_{R'}\rangle\bigg|_{\mathrm{cL}} = \mathrm{i}g_0\left(\langle\hat{b}^\dagger_R\hat{c}_R\hat{d}_{R'}\rangle - \langle\hat{b}^\dagger_{R'}\hat{c}_{R'}\hat{d}_R\rangle^*\right). \tag{10.9}$$

Note that this set of equations already represents a closed description of the field dynamics. If, as in many experimental configurations, the absolute phase of the light field and dipole operators may be considered unknown, these equations are sufficient for a description of the light–carrier interaction. However, when two time correlations are of interest or in the case of coherent excitation by injection of an external (resonoant or off-resonant) laser or by re-injection of emitted light after passage of an external resonator it may also be necessary to additionally consider the dynamics of the field and dipole expectation values, i.e.,

$$\frac{\partial}{\partial t}\langle\hat{b}_R\rangle\bigg|_{\mathrm{cL}} = -\mathrm{i}g_0\langle\hat{c}_R\hat{d}_R\rangle \tag{10.10}$$

$$\frac{\partial}{\partial t}\langle\hat{c}_R\hat{d}_{R'}\rangle\bigg|_{\mathrm{cL}} = \mathrm{i}g_0\left(\langle\hat{b}_R\rangle\langle\hat{d}^\dagger_R\hat{d}_{R'}\rangle + \langle\hat{b}_{R'}\rangle\langle\hat{c}^\dagger_{R'}\hat{c}_R\rangle - \langle\hat{b}_R\rangle\delta_{R,R'}\right). \tag{10.11}$$

10.2.2 Physical Background of the Factorization

In the following we will briefly discuss the implications of the factorization performed in the derivation of (10.6–10.9). In order to formulate the dynamics of the light–matter interaction without including higher order correlations, three terms have been factorized in the time derivative of the field–dipole correlation (10.7). The factorizations are

$$\langle \hat{b}_R^\dagger \hat{b}_{R''} \hat{c}_{R''}^\dagger \hat{c}_{R'} \rangle \approx \langle \hat{b}_R^\dagger \hat{b}_{R''} \rangle \langle \hat{c}_{R''}^\dagger \hat{c}_{R'} \rangle \tag{10.12}$$

$$\langle \hat{b}_R^\dagger \hat{b}_{R'} \hat{d}_{R'}^\dagger \hat{d}_{R''} \rangle \approx \langle \hat{b}_R^\dagger \hat{b}_{R'} \rangle \langle \hat{d}_{R'}^\dagger \hat{d}_{R''} \rangle \tag{10.13}$$

$$\langle \hat{c}_R^\dagger \hat{c}_{R'} \hat{d}_R^\dagger \hat{d}_{R''} \rangle \approx \langle \hat{c}_R^\dagger \hat{c}_{R'} \rangle \langle \hat{d}_R^\dagger \hat{d}_{R''} \rangle + \langle \hat{d}_{R'}^\dagger \hat{c}_{R'\prime}^\dagger \rangle \langle \hat{c}_R \hat{d}_R \rangle . \tag{10.14}$$

No additional factorizations are necessary in the density dynamics of photons, electrons and holes. The three factorizations are based on the assumption of statistical independence between the densities of photons, electrons and holes. Since we are not considering fluctuations in the particle densities, this is a necessary assumption.

Equations (10.12) and (10.13) separate the photon density from the carrier densities. These terms represent stimulated emission processes. Therefore, a correlation of the photon density with the carrier densities would lead to a modified stimulated emission rate. Below threshold this effect is small because the amplified spontaneous emission is distributed over many modes such that the local correlations between photon and carrier densities are weak. Above threshold the photon number fluctuations in the lasing mode cause relaxation oscillations. The photon number fluctuations are nearly ninety degrees out of phase with the carrier number fluctuations. Therefore, the time averaged correlation is still negligible.

For a simplification of the dipole–dipole correlation term $\langle \hat{c}_R^\dagger \hat{d}_{R'}^\dagger \hat{c}_{R''} \hat{d}_{R'''} \rangle$ we neglect in the possible factorizations all terms that are not directly correlated to the relevant physical quantities of the carriers (electrons and holes) and dipoles. Equation (10.14) thus consists of two terms. The first term represents spontaneous emission caused by the simultaneous presence of electrons and holes in the same location. Although it is reasonable to assume that the high rate of scattering at high carrier densities effectively reduces all electron–hole correlations to zero, it is important to note that the interband dipole $\langle \hat{c}_R \hat{d}_{R'} \rangle$ implies a phase correlation between the electrons and the holes. The second term of (10.14) gives a contribution that can be attributed to dipole–dipole interactions. Consequently, it consists of a product of dipole operators $\langle \hat{c}_R \hat{d}_{R'} \rangle^* \langle \hat{c}_{R''} \hat{d}_{R'''} \rangle$. Using the factorization the dynamics of the field–field and the field–dipole correlations are then identical to the dynamics of the products of the fields and dipoles.

Generally, spontaneous emission must always arise from random phase fluctuations. In principal, these are given by both the product of electron and hole densities and the product of phase dependent dipoles. In semiconductor lasers the phase dipole relaxes quickly due to scattering events, whereas the carrier densities are (in lowest order) preserved during scattering. Therefore

$$\langle \hat{c}_R \hat{d}_{R'} \rangle^* \langle \hat{c}_{R''} \hat{d}_{R'''} \rangle \ll \langle \hat{c}_R^\dagger \hat{c}_{R''} \rangle \langle \hat{d}_{R'}^\dagger \hat{d}_{R'''} \rangle \tag{10.15}$$

is usually a good assumption in semiconductor systems. However, this assumption does fail in the case of low carrier densities and high dipole inducing fields. As we will discuss later, this is the case e.g. in an optically pumped semiconductor laser. In order to keep our description on a very general level in the following we will include both the electron–hole interaction as well as the dipole–dipole interaction of the equation for the dipole–dipole correlation. Consequently, both may be important sources of spontaneous emission. The influence and magnitude of the individual contributions thereby are strongly determined by the configuration and boundary conditions of the particular semiconductor laser system considered.

10.2.3 Wigner Function Formulation

In order to connect the light–carrier interaction to the highly dissipative carrier transport equations, it is practical to transform the carrier and dipole densities using Wigner transformations [18]. Replacing the discrete density matrices by continuous ones obtained by polynomial interpolation allows e.g. for numerical purposes an arbitrary choice of the discretization scales, which generally will be much larger than a lattice constant. Analytically, it permits application of differential operators. Physically, the particle densities are smooth functions over distances of several lattice constants. A coherence length shorter than e.g. ten lattice constants would require k states with $\mid k \mid$ of at least one twentieth of the Brilloin zone diameter. In typical laser devices, however, the electrons and holes all accumulate near the fundamental gap at $k = 0$. Therefore, it is simply a matter of convenience to define the continuous densities such that

$$\rho^{\mathrm{e}}(r = R, r' = R') = \frac{1}{\nu_0}\langle \hat{c}_R^{\dagger}\hat{c}_{R'}\rangle \tag{10.16}$$

$$\rho^{\mathrm{h}}(r = R, r' = R') = \frac{1}{\nu_0}\langle \hat{d}_R^{\dagger}\hat{d}_{R'}\rangle \tag{10.17}$$

$$\rho^{\mathrm{dipole}}(r = R, r' = R') = \frac{1}{\nu_0}\langle \hat{c}_R\hat{d}_{R'}\rangle . \tag{10.18}$$

These continuous functions may then be transformed into Wigner functions by

$$f^{\mathrm{e}}(r,k) = \int \mathrm{d}^3r'\mathrm{e}^{-ikr'}\rho^{\mathrm{e}}\left(r - \frac{r'}{2}, r + \frac{r'}{2}\right) \tag{10.19}$$

$$f^{\mathrm{h}}(r,k) = \int \mathrm{d}^3r'\mathrm{e}^{-ikr'}\rho^{\mathrm{h}}\left(r - \frac{r'}{2}, r + \frac{r'}{2}\right) \tag{10.20}$$

$$p(r,k) = \int \mathrm{d}^3r'\mathrm{e}^{-ikr'}\rho^{\mathrm{dipole}}\left(r - \frac{r'}{2}, r + \frac{r'}{2}\right) . \tag{10.21}$$

The normalization of these Wigner functions has been chosen in such a way that a value of one represents the maximal phase space density possible for fermions, that is, one particle per state. Since the density of states in the six dimensional phase space given by r and k is $1/8\pi^3$, a factor of $1/8\pi^3$

will appear whenever actual carrier densities need to be obtained from the Wigner functions. However, the normalization in terms of the maximal possible phase space density is convenient because it represents the probability that a quantum state in a given region of phase space is occupied. Therefore, the Wigner distribution corresponding to the thermal equilibrium of a given particle density is directly given by the Fermi function.

To deal with the light field dynamics in the same manner, the field and field–field correlation variables must also be defined on a continuous length scale. In order to obtain photon densities, we define

$$\rho^{\mathcal{E}}(r = R) = \frac{1}{\sqrt{\nu_0}} \langle \hat{b}_R \rangle \tag{10.22}$$

$$\rho^{I}(r = R; r' = R') = \frac{1}{\nu_0} \langle \hat{b}^\dagger_R \hat{b}_{R'} \rangle . \tag{10.23}$$

Transformation to Wigner functions leads to

$$\mathcal{E}(r, q) = \int \mathrm{d}^3 r' \mathrm{e}^{-\mathrm{i}qr'} \rho^{\mathcal{E}} \left(r - \frac{r'}{2}, r + \frac{r'}{2} \right) \tag{10.24}$$

$$I(r; r', q) = \int \mathrm{d}^3 r' \mathrm{e}^{-\mathrm{i}qr'} \rho^{\mathcal{I}} \left(r - \frac{r'}{2}, r + \frac{r'}{2} \right) . \tag{10.25}$$

Finally, the dipole–field correlation must be defined accordingly such that

$$\Theta^{\mathrm{corr.}}(r = R; r' = R', r'' = R'') = \frac{1}{\sqrt{\nu_0^3}} \langle \hat{b}^\dagger_R \hat{c}_{R'} \hat{d}_{R''} \rangle \tag{10.26}$$

$$C(r; r', k) = \int \mathrm{d}^3 r'' \mathrm{e}^{-\mathrm{i}kr''} \Theta^{\mathrm{corr.}} \left(r; r' - \frac{r''}{2}, r' + \frac{r''}{2} \right) . \tag{10.27}$$

With these new definitions, the light–carrier interaction dynamics can now be expressed in a form that considers both the position and the momentum of the electrons and holes. The dynamics of emission and absorption now reads

$$\begin{aligned} \left. \frac{\partial}{\partial t} I(r; r', q) \right|_{\mathrm{cL}} &= -\mathrm{i} g_0 \frac{\sqrt{\nu_0}}{8\pi^3} \int \mathrm{d}^3 x \int \mathrm{d}^3 k \, \mathrm{e}^{\mathrm{i}kx} \\ &\quad \times \left(C \left(r + x; r, -q + \frac{k}{2} \right) - C^* \left(r + x; r, -q + \frac{k}{2} \right) \right) \end{aligned} \tag{10.28}$$

$$\begin{aligned} \left. \frac{\partial}{\partial t} C(r; r', k) \right|_{\mathrm{cL}} &= \mathrm{i} g_0 \sqrt{\nu_0} \, \frac{1}{8\pi^3} \int \mathrm{d}^3 x \int \mathrm{d}^3 q \, \mathrm{e}^{\mathrm{i}qx} \\ &\quad \times \left(f^{\mathrm{e}} \left(r', k + \frac{q}{2} \right) + f^{\mathrm{h}} \left(r', -k + \frac{q}{2} \right) - 1 \right) \\ &\quad \times I(r; r' + x, q) \\ &\quad + \mathrm{i} g_0 \sqrt{\nu_0} \, \frac{1}{8\pi^3} \int \mathrm{d}^3 q \, \mathrm{e}^{\mathrm{i}q(r - r')} \\ &\quad \times \left(f^{\mathrm{e}} \left(\frac{r + r'}{2}, k + \frac{q}{2} \right) f^{\mathrm{h}} \left(\frac{r + r'}{2}, -k + \frac{q}{2} \right) \right. \end{aligned}$$

$$+ \int \mathrm{d}^3k' \, p_q^* \left(\frac{r+r'}{2}, k \right) p_q \left(\frac{r+r'}{2}, k' \right) \Big) \Big) \tag{10.29}$$

$$\left. \frac{\partial}{\partial t} f^{\mathrm{e}}(r,k) \right|_{\mathrm{cL}} = \mathrm{i} g_0 \sqrt{\nu_0} \, \frac{1}{8\pi^3} \int \mathrm{d}^3x \int \mathrm{d}^3q \, \mathrm{e}^{\mathrm{i}qx} \tag{10.30}$$

$$\times \left(C \left(r+x; r, k + \frac{q}{2} \right) - C^* \left(r+x; r, k + \frac{q}{2} \right) \right)$$

$$\left. \frac{\partial}{\partial t} f^{\mathrm{h}}(r,k) \right|_{\mathrm{cL}} = \mathrm{i} g_0 \sqrt{\nu_0} \, \frac{1}{8\pi^3} \int \mathrm{d}^3x \int \mathrm{d}^3q \, \mathrm{e}^{\mathrm{i}qx} \tag{10.31}$$

$$\times \left(C \left(r+x; r, -k + \frac{q}{2} \right) - C^* \left(r+x; r, -k + \frac{q}{2} \right) \right)$$

$$\left. \frac{\partial}{\partial t} \mathcal{E}(r,q) \right|_{\mathrm{cL}} = -\mathrm{i} g_0 \frac{\sqrt{\nu_0}}{8\pi^3} \int \mathrm{d}^3k \, p_q(r,k) \tag{10.32}$$

$$\left. \frac{\partial}{\partial t} p_q(r,k) \right|_{\mathrm{cL}} = \mathrm{i} g_0 \sqrt{\nu_0} \, \frac{1}{8\pi^3} \int \mathrm{d}^3x \tag{10.33}$$

$$\times \left(f^{\mathrm{e}} \left(r, k + \frac{q}{2} \right) + f^{\mathrm{h}} \left(r, -k + \frac{q}{2} \right) - 1 \right) \mathcal{E}(r+x, q).$$

10.2.4 Local Approximation

The integrals over x and q represent seemingly non-local effects introduced by the transformation into Wigner functions. This property of the Wigner transformation retains the coherent effects in the carrier system. For the interaction of the carriers with the light field it ensures momentum conservation by introducing a non-local phase correlation in the dipole field corresponding to the total momentum of the electron and hole concentrations involved. Effectively, the integral over q converts the momentum part of the Wigner distributions into a coherence length. This coherence length then reappears in the spatial structure of the dipole field and the electromagnetic field generated by the carrier distribution. However, the coherence length in the carrier system is usually much shorter than the optical wavelength. In the Wigner functions describing the light–carrier interaction it can therefore be approximated by a spatial delta function. Here, we do this by noting that

$$\frac{1}{8\pi^3} \int \mathrm{d}^3q \, \mathrm{e}^{\mathrm{i}qx} = \delta(x). \tag{10.34}$$

If the effects of the momentum shift q in the Wigner functions is neglected the integrals may then be solved, yielding only local interactions between the carrier system and the light field:

$$\left. \frac{\partial}{\partial t} I(r; r', k) \right|_{\mathrm{cL}} = -\mathrm{i} g_0 \frac{\sqrt{\nu_0}}{8\pi^3} \left(C \left(r; r', q + \frac{k}{2} \right) \right.$$

$$\left. - C^* \left(r'; r, q - \frac{k}{2} \right) \right) \tag{10.35}$$

$$
\begin{aligned}
\left.\frac{\partial}{\partial t} C\left(r ; r^{\prime}, k\right)\right|_{\mathrm{cL}} = & \mathrm{i} g_0 \sqrt{\nu_0}\left(f^{\mathrm{e}}\left(r^{\prime}, k\right)+f^{\mathrm{h}}\left(r^{\prime},-k\right)-1\right) I\left(r ; r^{\prime}, k\right) \\
& +\mathrm{i} g_0 \sqrt{\nu_0}\, \delta\left(r-r^{\prime}\right)\Big(f^{\mathrm{e}}(r, k) f^{\mathrm{h}}(r,-k) \\
& +p^*\left(\frac{r+r^{\prime}}{2}, k\right) p\left(\frac{r+r^{\prime}}{2}, k\right)\Big)
\end{aligned} \tag{10.36}
$$

$$
\left.\frac{\partial}{\partial t} f^{\mathrm{e}}(r, k)\right|_{\mathrm{cL}} = \mathrm{i} g_0 \sqrt{\nu_0}\left(C(r ; r, k)-C^*(r ; r, k)\right) \tag{10.37}
$$

$$
\left.\frac{\partial}{\partial t} f^{\mathrm{h}}(r, k)\right|_{\mathrm{cL}} = \mathrm{i} g_0 \sqrt{\nu_0}\left(C(r ; r,-k)-C^*(r ; r,-k)\right) \tag{10.38}
$$

$$
\left.\frac{\partial}{\partial t} \mathcal{E}(r, k)\right|_{\mathrm{cL}} = -\mathrm{i} g_0 \frac{\sqrt{\nu_0}}{8 \pi^3}\, p(r, k) \tag{10.39}
$$

$$
\left.\frac{\partial}{\partial t} p(r, k)\right|_{\mathrm{cL}} = \mathrm{i} g_0 \sqrt{\nu_0}\left(f^{\mathrm{e}}(r, k)+f^{\mathrm{h}}(r,-k)-1\right) \mathcal{E}(r, k) . \tag{10.40}
$$

These equations now provide a compact description of the light–carrier interaction in a three dimensional semiconductor medium, including the light generating processes such as an incoherent quantum noise term which is the source of spontaneous emission. Following [13] we will denote the set of equations (10.35), that are the basis for our analysis of luminescence in semiconductor media the *semiconductor luminescence equations.*

10.2.5 Light–Carrier Interaction for Quantum Wells

Similar equations may also be formulated for a quantum well structure by replacing the phase space density of $1/8\pi^3$ with $1/4\pi^2$, reducing the spatial coordinates of the carrier system to two dimensions, and introducing a delta function for the coordinate perpendicular to the quantum well at the points where field coordinates correspond to dipole coordinates. Of course, the electromagnetic field remains three dimensional, even though the dipole it originates from is confined to two dimensions. In particular, the correlation $C(r; r', k)$ has both a three dimensional coordinate r and a two dimensional coordinate r'. It is therefore useful to distinguish the two dimensional and the three dimensional coordinates. In the following, the two dimensional carrier coordinates will be marked with the index $\|$. Note that in some cases both r and $r_\|$ appear in the equations. In those cases, the in-plane coordinates r_x and r_y are equal, while the perpendicular coordinate r_z must be equal to the quantum well coordinate z_0. The equations for the interaction of the three dimensional light field with the two dimensional electron–hole system in a single quantum well subband then read

$$\left.\frac{\partial}{\partial t} I(r;r',k_{\|})\right|_{\mathrm{cL}} = -\mathrm{i}g_0 \frac{\sqrt{\nu_0}}{4\pi^2}\left(C(r;r'_{\|},k_{\|})\delta(r'_z - z_0)\right.$$
$$\left. - C^*(r';r_{\|},k_{\|})\delta(r_z - z_0)\right) \tag{10.41}$$

$$\left.\frac{\partial}{\partial t} C(r;r'_{\|},k_{\|})\right|_{\mathrm{cL}} = \mathrm{i}g_0\sqrt{\nu_0}\ \left(f^{\mathrm{e}}(r'_{\|},k_{\|}) + f^{\mathrm{h}}(r'_{\|},-k_{\|}) - 1\right)$$
$$I(r;r',k_{\|})_{r'_z = z_0}$$
$$+\mathrm{i}g_0\sqrt{\nu_0}\ \delta(r_{\|} - r'_{\|})\delta(r_z - z_0)\Bigg(f^{\mathrm{e}}(r_{\|},k_{\|}) f^{\mathrm{h}}(r_{\|},-k_{\|})$$
$$+p^*\left(\frac{r+r'}{2},k_{\|}\right)p\left(\frac{r+r'}{2},k'_{\|}\right)\Bigg) \tag{10.42}$$

$$\left.\frac{\partial}{\partial t} f^{\mathrm{e}}(r_{\|},k_{\|})\right|_{\mathrm{cL}} = \mathrm{i}g_0\sqrt{\nu_0}\ \Big(C\left(r;r_{\|},k_{\|}\right)_{r_z=z_0}$$
$$-C^*\left(r;r_{\|},k_{\|}\right)_{r_z=z_0}\Big) \tag{10.43}$$

$$\left.\frac{\partial}{\partial t} f^{\mathrm{h}}(r_{\|},k_{\|})\right|_{\mathrm{cL}} = \mathrm{i}g_0\sqrt{\nu_0}\ \Big(C\left(r;r_{\|},-k_{\|}\right)_{r_z=z_0}$$
$$-C^*\left(r;r_{\|},-k_{\|}\right)_{r_z=z_0}\Big) \tag{10.44}$$

$$\left.\frac{\partial}{\partial t} \mathcal{E}(r,k_{\|})\right|_{\mathrm{cL}} = -\mathrm{i}g_0 \frac{\sqrt{\nu_0}}{4\pi^2} p(r_{\|},k_{\|})\delta(r_z - z_0) \tag{10.45}$$

$$\left.\frac{\partial}{\partial t} p(r_{\|},k_{\|})\right|_{\mathrm{cL}} = \mathrm{i}g_0\sqrt{\nu_0}\ \left(f^{\mathrm{e}}(r_{\|},k_{\|}) + f^{\mathrm{h}}(r_{\|},-k_{\|}) - 1\right)\mathcal{E}(r,k_{\|})_{r_z=z_0}. \tag{10.46}$$

Note that the value of g_0 will usually be slightly lower than the bulk value because the overlap of the spatial wavefunctions of the electrons and the holes in the lowest subbands is less than one. The equations derived above represent the interaction of a single conduction band and a single valence band with a single scalar light field. Neither the spin degeneracy of the carriers nor the polarization of the light field has been considered.

10.2.6 Spin Degeneracy and Light Field Polarization

Since the geometry of light field emission is highly dependent on polarization effects such effects should also be taken into account in the framework of this theory. The basic interaction between a single conduction band, a single valence band and a single light field polarization are accurately represented by (10.35–10.40) and (10.41–10.46). By adding the contributions of separate transitions, any many band system may be described based on these equations. In semiconductor quantum wells the situation is considerably simplified if only the lowest subbands are considered. Then there are only two completely separate transitions involving circular light field polarizations coupled to a single one of the two electron and hole bands. The quantum well structure does not interact with light fields which are linearly polarized in the direction perpendicular to the plane of the quantum well. The equations for quantum wells are therefore supplemented by adding an index of + or − to each variable.

The situation in the bulk system is much more involved. The transitions occur between the two fold degenerate spin 1/2 system of the electrons and the four fold degenerate spin 3/2 system of the holes. All three polarization directions of the light field are equally possible, connecting each of the electron bands with three of the four hole bands. However, since the effective mass of the two heavy–hole bands is much larger than the effective mass of the light–holes (e.g. by a factor of eight in GaAs), only a small fraction of the holes will be in the light–hole bands (about 6% in GaAs for equilibrium distributions). Consequently, the carrier subsystem can again be separated into two pairs of bands. However, the light field polarization emitted by the electron heavy–hole transitions in bulk material is circularly polarized with respect to the relative momentum $2k$ of the electron and the heavy–hole. Since usually there is no strong directional anisotropy in the k-space distribution of the carriers, it can be assumed that one third of the k-space volume contributes to each polarization direction and the equations may be formulated accordingly.

In the following we will assume that the Wigner distributions of the two pairs of bands considered can be formulated with luminescence equations for each polarization direction (σ^+,σ^-). As we will see in the following section the two carrier-subsystems interact with each other via carrier–carrier as well as carrier–phonon scattering. Inclusion of the polarization in our equations allows explicit consideration of hole burning effects in the spin and polarization dynamics which may occur in vertical-cavity surface-emitting lasers [19; 20; 21].

The complete set of dynamical equations can now be formulated by adding the carrier dynamics and the linear part of Maxwell's equations to the light–carrier interaction.

10.3 Carrier Kinetics

The microscopic carrier dynamics can be derived from the full many-body Hamiltonians $H^{\mathrm{e,h}}_{\mathrm{c-c}}$ and $H^{\mathrm{e,h}}_{\mathrm{c-ph}}$ for the carrier–carrier and carrier–phonon interaction respectively.

Using the Heisenberg equation one can set up a hierarchy of equations of motion for the expectation values of commutators describing carrier–carrier and carrier–phonon interaction processes. Evaluation of the commutators and factorization leads to Boltzmann equations for the dynamic carrier scattering processes. They consist of characteristic in and out scattering rates with respect to a carrier state of carrier wavenumber k. For a simplification of the notation in the following we will express the k-dependence of the physical quantities in terms of a suffix and omit the indication of the (r, r')-dependence. The change in carrier distributions, $\frac{\partial}{\partial t} f_k^{\mathrm{e,h}}\big|^{\mathrm{sc}}$, and in the dipole, $\frac{\partial}{\partial t} p_k\big|^{\mathrm{sc}}$, caused by carrier–carrier and carrier–phonon interactions can be expressed as [13]

$$
\begin{aligned}
\frac{\partial}{\partial t} f_k^{\mathrm{e,h}}\Big|^{\mathrm{sc}} &= -\Gamma_k^{\mathrm{e,h,out}} f_k^{\mathrm{e,h}} + \Gamma_k^{\mathrm{e,h,in}} (1 - f_k^{\mathrm{e,h}}) \\
\frac{\partial}{\partial t} p_k\Big|^{\mathrm{sc}} &= \left[-\frac{1}{2} \left(\Gamma_k^{\mathrm{e,out}} + \Gamma_k^{\mathrm{e,in}} + \Gamma_k^{\mathrm{h,out}} + \Gamma_k^{\mathrm{h,in}} \right) - \mathrm{i}\Omega_k \right] p_k \\
\frac{\partial}{\partial t} C_k &= \left[\left(\Gamma_k^{\mathrm{e,out}} + \Gamma_k^{\mathrm{e,in}} + \Gamma_k^{\mathrm{h,out}} + \Gamma_k^{\mathrm{h,in}} \right) - \mathrm{i}\Omega_k \right] C_k
\end{aligned}
\tag{10.47}
$$

where $\Gamma_k^{\mathrm{e,h,out}}$ and $\Gamma_k^{\mathrm{e,h,in}}$ are the out and in-scattering rates for the carriers (electrons and holes) respectively.

For the carrier–carrier interaction these rates read

$$
\begin{aligned}
\Gamma_k^{\mathrm{e,out}} &= \sum_{k',q} \sum_{\sigma=\sigma^+,\sigma^-} \frac{2\pi}{\hbar} \delta\left(E_k^{\mathrm{e}} - E_{k-q}^{\mathrm{e}} + E_{k'}^{\mathrm{e,h}} - E_{k'+q}^{\mathrm{e,h}} \right) |V_q|^2 \\
&\quad \times f_{k'\sigma}^{\mathrm{e,h}} \left(1 - f_{k'+q\,\sigma}^{\mathrm{e,h}}\right) \left(1 - f_{k-q\,\sigma}^{\mathrm{e}}\right) \\
\Gamma_k^{\mathrm{e,in}} &= \sum_{k',q} \sum_{\sigma=\sigma^+,\sigma^-} \frac{2\pi}{\hbar} \delta\left(-E_k^{\mathrm{e}} + E_{k-q}^{\mathrm{e}} - E_{k'}^{\mathrm{e,h}} + E_{k'+q}^{\mathrm{e,h}} \right) |V_q|^2 \\
&\quad \times \left(1 - f_{k'\sigma}^{\mathrm{e,h}}\right) f_{k'+q\,\sigma}^{\mathrm{e,h}} f_{k-q\,\sigma}^{\mathrm{e}} .
\end{aligned}
\tag{10.48}
$$

The respective rates of the carriers with the phonons are given by

$$
\begin{aligned}
\Gamma_k^{\mathrm{e,out}} &= \sum_{q} \sum_{\pm} \sum_{\sigma=\sigma^+,\sigma^-} \frac{2\pi}{\hbar} \delta\left(E_k^{\mathrm{e}} - E_{k-q}^{\mathrm{e}} \pm \hbar\omega_{\mathrm{LO}} \right) |\gamma_q|^2 \\
&\quad \times f_{k'\sigma}^{\mathrm{e,h}} \left(n_q + \frac{1}{2} \mp \frac{1}{2}\right) \left(1 - f_{k-q\,\sigma}^{\mathrm{e}}\right) \\
\Gamma_k^{\mathrm{e},in} &= \sum_{q} \sum_{\pm} \sum_{\sigma=\sigma^+,\sigma^-} \frac{2\pi}{\hbar} \delta\left(-E_k^{\mathrm{e}} + E_{k-q}^{\mathrm{e}} \pm \hbar\omega_{\mathrm{LO}} \right) |\gamma_q|^2 \\
&\quad \times f_{k'\sigma}^{\mathrm{e,h}} \left(n_q + \frac{1}{2} \mp \frac{1}{2}\right) f_{k-q\,\sigma}^{\mathrm{e}} .
\end{aligned}
\tag{10.49}
$$

Assuming that the phonons are in thermal equilibrium the phonon population function n_q can be described by a thermal Bose function. $E^{\mathrm{e,h}}$ is the kinetic carrier frequency (including Coulomb renormalization) with respect to the bandgap frequency ω_0. The momentum dependent frequency Ω_k is given by

$$
\Omega_k = \varepsilon_k - \omega_0 \tag{10.50}
$$

where ε_k is the kinetic energy of the electron–hole pair with the Coulomb renormalization.

The real physical phase oscillations of $p_k(r)$ would include an additional phase factor of $\exp[-i\omega_0 t]$. However, the only physical effect of this oscillation is to establish resonance with the corresponding frequency range in the electromagnetic field, the dynamics of which we consider next.

10.4 Maxwell's Equation

The Heisenberg equations of motion describing the operator dynamics of the electromagnetic field operators are identical to the classical Maxwell's equations. In terms of the electromagnetic field $E(r)$ and the dipole densities $P(r)$, the equation reads

$$\nabla \times (\nabla \times E(r)) + \frac{\epsilon_r}{c^2}\frac{\partial^2}{\partial t^2}\left(E(r) + \frac{1}{\epsilon_r\epsilon_0}P(r)\right) = 0, \tag{10.51}$$

where ϵ_0 and c are the dielectric constant and the speed of light in vacuum, respectively, and ϵ_r is the dielectric constant in the background semiconductor medium.

Maxwell's equation describes the light field dynamics for all frequencies. Since we are only interested in frequencies near the bandgap frequency ω_0, it is useful to separate the phase factor of $\exp[-i\omega_0 t]$, defining $E(r) = \exp[-i\omega_0 t]E_0(r)$. Now $E_0(r)$ can be considered to vary slowly in time relative to $\exp[-i\omega_0 t]$. Therefore, the time derivatives may be approximated by

$$\exp[i\omega_0 t]\frac{\partial^2}{\partial t^2}\exp\left[-i\omega_0 t\right]E_0(r) \approx -\omega_0^2 E_0(r) - 2i\omega_0\frac{\partial}{\partial t}E_0(r). \tag{10.52}$$

Similarly, $P_0(r)$ may be defined such that $P(r) = \exp[-i\omega_0 t]P_0(r)$. The approximation used here may even be of zero order since we are primarily interested in the dynamics of the electromagnetic field:

$$\exp\left[i\omega_0 t\right]\frac{\partial^2}{\partial t^2}\exp\left[-i\omega_0 t\right]P_0(r) \approx -\omega_0^2 P_0(r). \tag{10.53}$$

The temporal evolution of the electromagnetic field now reads

$$\frac{\partial}{\partial t}E_0(r) = -i\frac{\omega_0}{2k_0^2\epsilon_r}\left(\nabla \times (\nabla \times E_0(r)) - \epsilon_r k_0^2 E_0(r)\right) - i\frac{\omega_0}{2\epsilon_r\epsilon_0}P_0(r), \tag{10.54}$$

where $k_0 = \omega_0/c$ is the vacuum wavevector length corresponding to ω_0. In (10.54) the field dynamics is described in terms of electromagnetic units, that is the fields represent forces acting on charges. To switch scales to the photon densities represented by $\mathcal{E}(r)$, energy densities have to be considered. Since the energy of each photon will be close to the bandgap energy $\hbar\omega_0$, the energy density of the electromagnetic field is given by

$$\hbar\omega_0\mathcal{E}^*(r)\mathcal{E}(r) = \frac{\epsilon_r\epsilon_0}{2}E_0^*(r)E_0^*(r). \tag{10.55}$$

Therefore, the field may be expressed as photon density amplitude using

$$E_0(r) = \sqrt{\frac{2\hbar\omega_0}{\epsilon_r\epsilon_0}} \int_k \mathcal{E}(r,k). \tag{10.56}$$

The dipole density $P_0(r)$ may be expressed in terms of $\rho^{\text{dipole}}(r,r)$ and $p(r,k)$ by noting that the density $\rho^{\text{dipole}}(r,r)$ is the dipole density in units of one half the atomic dipole given by the interband dipole matrix element d_{cv} at $k=0$. The factor of one-half is a logical consequence of the property that $\langle \hat{c}_R \hat{d}_{R'} \rangle \leq 1/2$. Thus, a fully polarized lattice would have a dipole density of $\rho^{\text{dipole}}(r, r=1)/(2\nu_0)$ which must correspond to $P(r) = d_{\text{cv}}/\nu_0$. Note that d_{cv} contains an arbitrary phase factor depending only on the definition of the states used for its determination. For convenience, we assume a definition of phases such that d_{cv} is real. The dipole density $P_0(r)$ may then be written as

$$\begin{aligned} P(r) &= d_{\text{cv}}\rho^{\text{dipole}} \\ &= \frac{1 \mid d_{\text{cv}} \mid}{8\pi^3} \sum_{\sigma=\sigma^+,\sigma^-} \int \mathrm{d}^3k \; p(r,k). \end{aligned} \tag{10.57}$$

Written in terms of $\mathcal{E}(r,k)$ and $p(r,k)$, the complete field dynamics now reads

$$\begin{aligned} \frac{\partial}{\partial t}\mathcal{E}(r,k) &= -\mathrm{i}\frac{\omega_0}{2k_0^2\epsilon_r}\left(\nabla\times\left(\nabla\times\mathcal{E}\left(r,k\right)\right) - \epsilon_r k_0^2 \mathcal{E}\left(r,k\right)\right) \\ &\quad -\mathrm{i}\sqrt{\frac{\omega_0}{2\hbar\epsilon_r\epsilon_0}}\frac{\mid d_{\text{cv}} \mid}{8\pi^3} \sum_{\sigma=\sigma^+,\sigma^-} p(r,k) \\ &= -\mathrm{i}\frac{\omega_0}{2k_0^2\epsilon_r}\left(\nabla\times\left(\nabla\times\mathcal{E}\left(r,k\right)\right) - \epsilon_r k_0^2 \mathcal{E}\left(r,k\right)\right) \\ &\quad -\mathrm{i}g_0\frac{\sqrt{\nu_0}}{8\pi^3} \sum_{\sigma=\sigma^+,\sigma^-} p(r,k). \end{aligned} \tag{10.58}$$

The coupling frequency g_0 introduced in (10.1) may be expressed in terms of the dipole matrix element d_{cv}:

$$g_0 = \sqrt{\frac{\omega_0}{2\hbar\epsilon_r\epsilon_0\nu_0}} \mid d_{\text{cv}} \mid. \tag{10.59}$$

With this equation, the operator dynamics of the light field operator $\hat{b}_R$ corresponds to the field dynamics of the Maxwell–Bloch equations for classical fields. By applying the linear propagation dynamics of the field to $I_{ij}(r;r',k)$ and $C_{ij}(r;r',k)$ as well, it is now possible to formulate a complete set of quantum Maxwell–Bloch luminescence equations.

10.5 Quantum Maxwell–Bloch Luminescence Equations

10.5.1 Three Dimensional Quantum Maxwell–Bloch Luminescence Equations for Quantum Wells

On the basis of the discussion in the previous sections the quantum Maxwell–Bloch luminescence equations can be set up for a given (bulk or quantum well) semiconductor medium. In recent years the quantum well laser structure has become basic laser structure for both edge-emitting and surface-emitting lasers. In the following we will thus restrict the formulation of the equations to a quantum well structure. Details of the application to bulk media may be found in [1]. Using the terminology of (10.41–10.46), the quantum Maxwell–Bloch luminescence equations for a multi quantum well structure with Q quantum wells read

$$\left.\frac{\partial}{\partial t} f^{\mathrm{e}}(r_{\|}, k_{\|})\right|_{\mathrm{cL}} = \mathrm{i} g_0 \sqrt{\nu_0}\ \left(C(r; r_{\|}, k_{\|}) - C^*(r; r_{\|}, k_{\|})\right) \tag{10.60}$$

$$\left.\frac{\partial}{\partial t} f^{\mathrm{h}}(r_{\|}, k_{\|})\right|_{\mathrm{cL}} = \mathrm{i} g_0 \sqrt{\nu_0}\ \left(C(r_{\|}; r_{\|}, -k_{\|}) - C^*(r_{\|}; r_{\|}, -k_{\|})\right) \tag{10.61}$$

$$\begin{aligned}
\frac{\partial}{\partial t} C_{ij_{\|}}(r; r'_{\|}, k_{\|}) = & \left[-\left(\Gamma_{k_{\|}}^{\mathrm{e,out}} + \Gamma_{k_{\|}}^{\mathrm{e,in}} + \Gamma_{k_{\|}}^{\mathrm{h,out}} + \Gamma_{k_{\|}}^{\mathrm{h,in}}\right) - \mathrm{i}\Omega(k_{\|})\right] \\
& \times C_{ij_{\|}}(r; r'_{\|}, k_{\|}) \\
& -\mathrm{i}\frac{\omega_0}{2k_0^2}\left(\sum_k \frac{\partial}{\partial r_k}\epsilon_{\mathrm{r}}^{-1}\frac{\partial}{\partial r_k} C_{ij_{\|}}(r; r'_{\|}, k_{\|})\right. \\
& \left. - \sum_k \frac{\partial}{\partial r_i}\epsilon_{\mathrm{r}}^{-1}\frac{\partial}{\partial r_k} C_{kj_{\|}}(r; r'_{\|}, k_{\|}) + k_0^2 C_{ij_{\|}}(r; r'_{\|}, k_{\|})\right) \\
& + \mathrm{i} g_0 Q \sqrt{\nu_0}\ \left(f^{\mathrm{e}}\left(r'_{\|}, k_{\|}\right) + f^{\mathrm{h}}\left(r'_{\|}, k_{\|}\right) - 1\right) \\
& \times U_{ij}(r; r', k_{\|})_{r'_z = z_0} \\
& + \mathrm{i} g_0 Q \sqrt{\nu_0}\ \delta(r_{\|} - r'_{\|})\delta(r_z - z_0)\delta_{ij_{\|}} \\
& \times\left(f^{\mathrm{e}}\left(r_{\|}, k_{\|}\right)\cdot f^{\mathrm{h}}\left(r_{\|}, k_{\|}\right)\right. \\
& \left. + p^*\left(r_{\|}, k_{\|}\right)\cdot p\left(r_{\|}, k_{\|}\right)\right)
\end{aligned} \tag{10.62}$$

$$\begin{aligned}
\frac{\partial}{\partial t} I_{ij}(r; r', k_{\|}) = & -\mathrm{i}\frac{\omega_0}{2k_0^2}\sum_k\left(\frac{\partial}{\partial r_k}\epsilon_{\mathrm{r}}^{-1}\frac{\partial}{\partial r_k} - \frac{\partial}{\partial r'_k}\epsilon_{\mathrm{r}}^{-1}\frac{\partial}{\partial r'_k}\right) I_{ij}(r; r', k_{\|}) \\
& + \mathrm{i}\frac{\omega_0}{2k_0^2}\sum_k\left(\frac{\partial}{\partial r_i}\epsilon_{\mathrm{r}}^{-1}\frac{\partial}{\partial r_k} I_{kj}(r; r', k_{\|})\right. \\
& \left. - \frac{\partial}{\partial r'_j}\epsilon_{\mathrm{r}}^{-1}\frac{\partial}{\partial r'_k} I_{ik}(r; r', k_{\|})\right)
\end{aligned}$$

$$-\mathrm{i}g_0\frac{\sqrt{\nu_0}}{4\pi^2}\left(C_{ij_\parallel}(r;r'_\parallel,k_\parallel)\delta(r'_z-z_0)\right.$$
$$\left.-C^*_{ji_\parallel}(r';r_\parallel,k_\parallel)\delta(r_z-z_0)\right)$$

$$U_{ij}(r;r',k_\parallel)=I_{ij}(r;r',k_\parallel)+\int_{k'}V_{k,k'}I_{ij}(r;r',k'_\parallel) \tag{10.63}$$

$$\frac{\partial}{\partial t}p_{i_\parallel}(r_\parallel,k_\parallel)=\left[-\left(\Gamma^{\mathrm{e,out}}_{k_\parallel}+\Gamma^{\mathrm{e,in}}_{k_\parallel}+\Gamma^{\mathrm{h,out}}_{k_\parallel}+\Gamma^{\mathrm{h,in}}_{k_\parallel}\right)-\mathrm{i}\Omega(k_\parallel)\right]$$
$$\times p_{i_\parallel}(r_\parallel,k_\parallel)$$
$$+\mathrm{i}g_0Q\sqrt{\nu_0}\left(f^{\mathrm{e}}\left(r_\parallel,k_\parallel\right)+f^{\mathrm{h}}\left(r_\parallel,k_\parallel\right)-1\right)$$
$$\times\left(\mathcal{E}_i(r,k_\parallel)_{r_z=z_0}+\int_{k'}V_{k,k'}\mathcal{E}_i(r,k'_\parallel)_{r_z=z_0}\right) \tag{10.64}$$

$$\frac{\partial}{\partial t}\mathcal{E}_i(r,k_\parallel)=\mathrm{i}\frac{\omega_0}{2k_0^2}\left(\sum_k\frac{\partial}{\partial r_k}\epsilon_{\mathrm{r}}^{-1}\frac{\partial}{\partial r_k}\mathcal{E}_i(r,k_\parallel)\right.$$
$$\left.-\frac{\partial}{\partial r_i}\epsilon_{\mathrm{r}}^{-1}\frac{\partial}{\partial r_k}\mathcal{E}_k(r,k_\parallel)+k_0^2\mathcal{E}_i(r,k_\parallel)\right)$$
$$-\mathrm{i}g_0\frac{\sqrt{\nu_0}}{4\pi^2}p_{i_\parallel}(r_\parallel,k_\parallel)\delta(r_z-z_0). \tag{10.65}$$

Many particle effects such as bandgap renormalization or the Coulomb enhancement lead to corrections in the carrier energies and carrier potential [12]. The corrections in carrier energies are considered via the momentum dependent frequency $\Omega(k)$. For a consideration of the carrier potential modified by many-body effects we have introduced the local field $U(r;r',k_\parallel)_{r'_z=z_0}$ in (10.60) which consists of the field–field correlations and a term describing the renormalization due to the Coulomb interactions. (Note that the same correction have been made in $\mathcal{E}_i(r,k_\parallel)_{r_z=z_0}$.)

A specific laser structure and design of the active area may be considered with a spatially varying ϵ_{r}. We note that different classes of laser diodes may be distinguished by the location of reflective and non-reflective edges. If the reflective surface is perpendicular to the plane of the quantum wells, the laser is an edge-emitter. If the reflectivity is very high on the surface planes parallel to the quantum wells, the laser is a vertical-cavity surface-emitting laser (VCSEL). The effect and structure of an external cavity will be considered later by extending the equation for $\mathcal{E}$ to a delayed optical feedback wave equation.

10.5.2 Two Dimensional Quantum Maxwell–Bloch Luminescence Equations for VCSELs

Due to its spatial symmetry and small resonator dimensions the vertical-cavity surface-emitting laser represents a system in which the fundamental quantum mechanical processes strongly affect the complex carrier and light field dynamics. For reasons of momentum conservation the transitions involved in the lasing process can be attributed to either right or left circular polarized light, which may interact via flucutation and scattering processes thereby causing a nonlinear complex polarization dynamics of the emitted light. Thereby it is the interplay of spontaneous and induced emission, in particular, that affects the luminescence and consequently the emission characteristics of the laser. In combination with spatial flucutations and material inhomogenities they influence the radiation recombination of electron–hole pairs as well as the coupling of transitions belonging to right and left circular polarized light. The boundary conditions given by the gain of the optically excited medium and the cavity then determine the build-up and decay of spatial and temporal coherence of the generated light.

In a VCSEL the light field propagating perpendicular to the quantum well structure is strongly confined by highly reflective mirrors above and below the quantum wells. Therefore both the possible polarization directions and the spatial dynamics remain two dimensional. Only the z direction may be eliminated by averaging over a single longitudinal mode. Coupling terms between the polarization directions should be taken into account, even if they are small. For VCSELs, the assumptions read

$$\mathcal{E}(r,k) \approx \tilde{\mathcal{E}}(r_{\|},k_{\|})\xi(r_z) \tag{10.66}$$

$$\begin{aligned}\frac{\partial}{\partial r_z}\mathcal{E}_z(r,k) \approx &-\frac{\partial}{\partial r_x}\left(\mathcal{E}_x(r,k)+\frac{2g_0\sqrt{\nu_0}}{4\pi\omega_0}\delta(r_z-z_0)p_x(r_{\|},k_{\|})\right)\\ &-\frac{\partial}{\partial r_y}\left(\mathcal{E}_y(r,k)+\frac{2g_0\sqrt{\nu_0}}{4\pi\omega_0}\delta(r_z-z_0)p_y(r_{\|},k_{\|})\right).\end{aligned} \tag{10.67}$$

The latter condition takes into account the divergence of $\mathcal{E}(r)$ caused by the polarization $\mathbf{p}_{\|}(r_{\|},k_{\|})$. This is an important contribution to the quantum Maxwell–Bloch equations since it coherently couples orthogonal polarizations. The properties of the envelope function $\xi(r_z)$ are defined as

$$\int \mathrm{d}r_z \mid \xi(r_z) \mid^2 = 1 \tag{10.68}$$

$$\frac{\partial^2}{\partial r_z^2}\xi(r_z) \approx -k_0^2\xi(r_z). \tag{10.69}$$

Since all coordinates are two dimensional, the index $\|$ (that has indicated the two dimensional coordinates in the quantum well equations) will be omitted. Instead, the two dimensional variables are marked with a tilde. All indices and coordinates of such variables are only defined in two dimensions. The variables are defined as

$$\tilde{C}_{ij}(r;r',k) = \int \mathrm{d}r_z\xi(r_z)\xi^*(r'_z=z_0)C_{ij_{\|}}(r;r'_{\|},k_{\|}) \tag{10.70}$$

$$\tilde{I}_{ij}(r;r',k) = \int \mathrm{d}r_z\, \mathrm{d}r'_z\xi(r_z)\xi^*(r'_z)I_{ij}(r;r',k) \tag{10.71}$$

$$\tilde{p}_i(r,k) = \xi^*(r_z = z_0) p_{i_\parallel}(r_\parallel, k_\parallel) \tag{10.72}$$

$$\tilde{\mathcal{E}}_i(r,k) = \int \mathrm{d}r_z \xi^*(r_z)\mathcal{E}_i(r,k). \tag{10.73}$$

Note that the dipole variable $p_{i_\parallel}(r_\parallel, k_\parallel)$ is rescaled by the field density of the envelope function $\xi(r_z)$ at $r_z = z_0$. The dipole given by $\tilde{p}_i(r,k)$ is therefore the average dipole density in the whole cavity, not just in the quantum wells. Combining the carrier–carrier with the carrier–light interaction and adding terms describing the carrier injection and nonradiative recombination we can formulate a set of equations for the desription of luminescence in vertical-cavity surface-emitting semiconductor lasers.

$$\begin{aligned}
\frac{\partial}{\partial t}\tilde{f}^{\mathrm{e}}_\sigma(r,k) &= \tilde{\Lambda}^{\mathrm{e}}(r) - \gamma_{\mathrm{nr}}\tilde{f}^{\mathrm{e}}_\sigma(r,k) \\
&\quad + \mathrm{i}g_0\frac{\sqrt{\nu_0}}{4\pi^2}\left(\sum_i \left(\tilde{C}_{ii}(r;r,k) - \tilde{C}^*_{ii}(r;r,k)\right)\right. \\
&\quad + \sum_{ij}\frac{1}{k_0^2\epsilon_\mathrm{r}}\int \mathrm{d}^2r'\delta(r-r') \\
&\quad \times \left.\left(\frac{\partial^2}{\partial r'_i r'_j}\tilde{C}_{ij}(r;r',k) - \frac{\partial^2}{\partial r_i r_j}\tilde{C}^*_{ij}(r';r,k)\right)\right)
\end{aligned} \tag{10.74}$$

$$\begin{aligned}
\frac{\partial}{\partial t}\tilde{f}^{\mathrm{h}}_\sigma(r,k) &= \tilde{\Lambda}^{\mathrm{h}}(r) - \gamma_{\mathrm{nr}}\tilde{f}^{h}(r,k) \\
&\quad + \mathrm{i}g_0\frac{\sqrt{\nu_0}}{4\pi^2}\left(\sum_i \left(\tilde{C}_{ii}(r;r,k) - \tilde{C}^*_{ii}(r;r,k)\right)\right. \\
&\quad + \sum_{ij}\frac{1}{k_0^2\epsilon_\mathrm{r}}\int \mathrm{d}^2r'\delta(r-r') \\
&\quad \times \left.\left(\frac{\partial^2}{\partial r'_i r'_j}\tilde{C}_{ij}(r;r',k) - \frac{\partial^2}{\partial r_i r_j}\tilde{C}^*_{ij}(r';r,k)\right)\right)
\end{aligned}$$

$$\begin{aligned}
\frac{\partial}{\partial t}\tilde{C}_{ij}(r;r',k) &= \left[-\left(\Gamma^{\mathrm{e,out}}_{k_\parallel} + \Gamma^{\mathrm{e,in}}_{k} + \Gamma^{\mathrm{h,out}}_{k} + \Gamma^{\mathrm{h,in}}_{k}\right) - \mathrm{i}\Omega(k)\right] \\
&\quad \times\tilde{C}_{ij}(r;r',k) \\
&\quad -(\kappa_i - \mathrm{i}\delta\omega_i)\tilde{C}_{ij}(r;r',k) - \mathrm{i}\frac{\omega_0}{2k_0^2\epsilon_\mathrm{r}}\Delta_r\tilde{C}_{ij}(r;r',k) \\
&\quad + \mathrm{i}g_0\sigma\sqrt{\nu_0}\,\left(f^{\mathrm{e}}(r',k)) + f^{\mathrm{h}}(r',k) - 1\right)\tilde{U}_{ij}(r;r') \\
&\quad + \mathrm{i}g_0\sigma\sqrt{\nu_0}\,\delta(r-r')\delta_{ij}\left(f^{\mathrm{e}}(r,k)\cdot f^{\mathrm{h}}(r,k)\right. \\
&\quad \left. + p(r,k)\cdot p^*(r,k)\right)
\end{aligned} \tag{10.75}$$

$$\tilde{U}_{ij}(r;r',k_{\|}) = \tilde{I}_{ij}(r;r',k_{\|}) + \int_{k'} V_{k,k'}\tilde{I}_{ij}(r;r',k'_{\|}) \tag{10.76}$$

$$\begin{aligned}\frac{\partial}{\partial t}\tilde{I}_{ij}(r;r',k) = &-(\kappa_i+\kappa_j)\tilde{I}_{ij}(r;r',k) - \mathrm{i}\frac{\omega_0}{2k_0^2\epsilon_\mathrm{r}}\left(\Delta_r - \Delta_{r'}\right)\tilde{I}_{ij}(r;r',k)\\ &-\mathrm{i}g_0\frac{\sqrt{\nu_0}}{4\pi^2}\left(\tilde{C}_{ij}(r;r',k) - \tilde{C}^*_{ji}(r';r,k)\right.\\ &\left.+\frac{1}{k_0^2\epsilon_\mathrm{r}}\left(\frac{\partial^2}{\partial r'_j r'_k}\tilde{C}_{ik}(r;r',k) - \frac{\partial^2}{\partial r_i r_k}\tilde{C}^*_{jk}(r';r,k)\right)\right)\end{aligned} \tag{10.77}$$

$$\begin{aligned}\frac{\partial}{\partial t}\tilde{p}_i(r,k) = &\left[-\left(\Gamma_k^{\mathrm{e,out}} + \Gamma_k^{\mathrm{e,in}} + \Gamma_k^{\mathrm{h,out}} + \Gamma_k^{\mathrm{h,in}}\right) - \mathrm{i}\Omega(k)\right]\tilde{p}_i(r,k)\\ &+\mathrm{i}g_0\sigma\sqrt{\nu_0}\left(f^\mathrm{e}(r,k) + f^\mathrm{h}(r,k) - 1\right)\\ &\times\left(\tilde{\mathcal{E}}_i(r,k) + \int_{k'} V_{k,k'}\tilde{\mathcal{E}}_i(r,k')\right)\end{aligned} \tag{10.78}$$

$$\begin{aligned}\frac{\partial}{\partial t}\tilde{\mathcal{E}}_i(r,k) = &-(\kappa_i + \mathrm{i}\delta\omega_i)\tilde{\mathcal{E}}_i(r) + \mathrm{i}\frac{\omega_0}{2k_0^2\epsilon_\mathrm{r}}\Delta\tilde{\mathcal{E}}_i(r,k)\\ &-\mathrm{i}g_0\frac{\sqrt{\nu_0}}{4\pi^2}\left(\tilde{p}_i(r,k) + \sum_j \frac{1}{k_0^2\epsilon_\mathrm{r}}\frac{\partial^2}{\partial r_i r_j}\tilde{p}_j(r,k)\right),\end{aligned} \tag{10.79}$$

with σ being the confinement factor along the z direction,

$$\sigma = Q \mid \xi(r_z = z_0) \mid^2 . \tag{10.80}$$

κ is the cavity loss rate. In practice one experimentally observes a certain polarization stability; it is taken into account by using slightly different reflectivities for the x and $y-$ polarizations. The cavity loss rate is therefore given by κ_x and κ_y. Experimental results [22] suggest that $(\kappa_y - \kappa_x)/(\kappa_x) \approx 10^{-3} - 10^{-2}$. A birefringence of $\delta\omega_{x/y}$ is also included to denote the difference between the bandgap frequency and the longitudinal frequencies of the confined light field for the two polarization directions. The birefringence $\delta\omega_x - \delta\omega_y$ is usually in the GHz range.

In (10.74) we have introduced the pump rates $\Lambda^{\mathrm{e,h}}(r)$ describing the carrier injection into the respective carrier states of the active semiconductor medium and a rate for nonradiative recombination $\gamma_{\mathrm{nr}} f^\mathrm{e}(r_{\|}, k_{\|})$ that cannot be neglected in typical semiconductor lasers. The field $\tilde{U}_{ij}(r;r')$ includes the bandgap renormalization due to many particle effects [12].

Equations (10.74) represent a starting point for the study of spatial polarization patterns and fluctuations in VCSELs. For more realistic models, it may also be desirable to include a spatial dependence of the birefringence and the dichroism. Also, nonlinear effects may be introduced, e.g. by separating the carrier densities for right and left circular polarization [19].

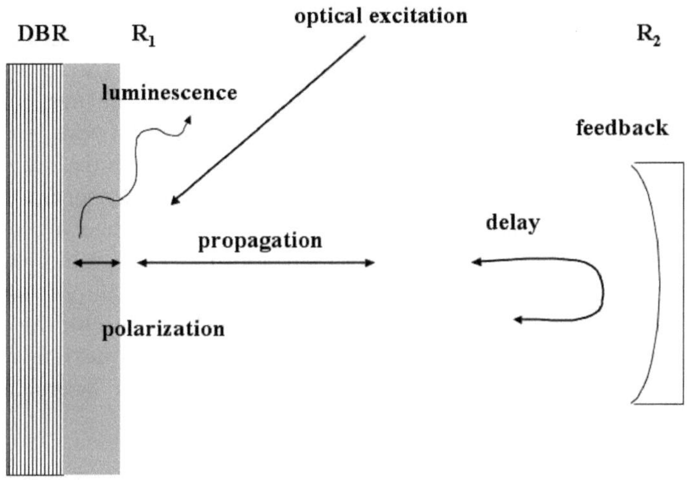

Fig. 10.1. Schematic configuration of an optically pumped vertical-extended-cavity surface-emitting laser

10.6 Optically Excited Semiconductor Lasers

In order to simulate the physical properties of a real semiconductor laser device the luminescence equations discussed above have to be modified. In particular, the laser configuration, the epitaxial design and the pump process play a decisive role for the emission properties and consequently have to be considered appropriately. In the Bloch equations the epitaxial structure can be taken into account via material parameters for effective masses and dipole matrix elements. In the optical wave equations the wave-guiding properties are modelled with appropriate spatially resolved effective indices. The laser configuration and the pump process on the other hand can be included using suitable boundary conditions for the light fields and the carrier distributions.

Recent experimental investigations have shown that by intense optical pumping of a semiconductor laser structure (typically a vertical-cavity surface-emitting laser structure) placed in an external resonator a compact laser source may be created with added wavelength flexibility and good beam quality. A sketch of a typical configuration is depicted in Fig. 10.1. The carrier dynamics in the active layer is via the interband polarization coupled to the spatio-temporal dynamics of the light fields. These are, in turn, coupled to the light field dynamics in the external resonator. The ultrafast carrier interactions directly linked with the (comparatively slow) round trip time of the optical field in the active medium and in the external resonator thereby lead to a complex interplay of physical processes that span a large temporal regime ranging from fs up to a few hundred ps. This dynamic interplay of (laser) internal light matter interactions with external conditions given by e.g. an optical pump laser (with specific frequency, beam profile and duration of the excitation) or a specific external resonator design determines the spatio-temporal dynamics and the emission properties of the laser source.

The underlying physical effects are of fundamental nature, yet directly relevant to the device physics. Among them are spontaneous and induced emission processes, quantum fluctuations, spatial hole burning, amplitude and phase of the optical feedback signal (defined by the external resonator), spatial and spectral filtering and nonlinearities, carrier heating induced by radiative recombination and via the optical pump process, nonlinear spectral gain and refractive index dispersion. To underline the importance of this innovative laser system in the following we will thus present the respective extensions that have to be made to the luminescence equations to provide a realistic description of vertical-cavity surface-emitting lasers in external resonators.

10.6.1 Luminescence and Delayed Optical Feedback by an External Resonator

In Chap. 9 we have discussed the influence of delayed optical feedback on the spatio-temporal dynamics of semiconductor lasers. Several theoretical models have been derived and numerical simulations have been performed (see e.g. [23], see also Chap. 9) that explicitly take into account the influence of delayed optical feedback on the spatio-temporal dynamics in edge-emitting lasers. This includes both the modeling of unwanted (but in experiments often observed) instabilities induced by the back-coupling mechanism [24] and the application of an external resonator for a stabilization and control of the emission properties of high-power edge-emitting lasers [25]. In our discussion in Chap. 9 we have implicitly assumed that the lasers themselves are operated significantly above threshold such that stimulated emission dominates the lasing processes. Consequently the theory and simulations have been based e.g. on the (coherent) Maxwell–Bloch delay equations. There the delayed feedback is represented by a separate delay term in the wave equations that simulates the temporally delayed re-injection of the optical signal. We have seen that in principle a large variety of nonlinear phenomena may occur that depend on the chosen set of parameters. The influence of an external cavity on the spatio-temporal dynamics of broad-area semiconductor lasers may result in both the control of instabilities and spatio-temporally chaotic behavior.

Recently, innovative optically pumped semiconductor laser systems are considered that typically consist of an optically excited laser gain medium placed inside an external resonator (see Fig. 10.1). The configuration of an optically pumped semiconductor laser directly uses the spatio-temporal interaction between the light re-injected after a round trip in the external resonator and the spontaneously emitted light field distribution within the laser. Thus we have to take into account both (amplified) spontaneous and stimulated emission as well as their interplay. In our theoretical description of delayed optical feedback systems we will thus resort to using the luminescence equations in which the effect of structured delayed optical feedback is self-consistently included. Consequently the theory represents a microscopically founded and quantum-kinetic extension of the transverse Lang–Kobayashi Model discussed in Chap. 9.

In order to consider the spatio-temporal interaction between laser-internal light and light that has passed an external resonator we introduce in the luminescence equations (10.74) delay-terms that simulate the interaction between the light field at the output facet of the laser and the temporally delayed field contribution that has passed the external resonator. In terms of the expectation values of the spectral photon density amplitude $\mathcal{E}(r,k)$ which is via $E_0(r) = \sqrt{\frac{2\hbar\omega_0}{\epsilon_r\epsilon_0}} \int_k \mathcal{E}(r,k)$ correlated to the light field we can assume that an additional contribution $\mathcal{E}_{\text{delay}}(r,q)$ enters the luminescence equations describing the fraction of emitted light re-entering the laser via its output facet (i.e. at $z = L$). Both the amplitude and the phase of this field thereby are defined by the round trip time of the external resonator and the reflectivity of the external mirror. Keeping in mind that the photon amplitude is defined via the expectation value of the photon creation operator, $\langle \hat{b}_R \rangle$, in a similar way we can modify the equation for $I(r;r',q)$ and introduce the field–field correlation $I_{\text{delay}}(r;r',q)$ between the light field at the output facet of the laser and the temporally delayed field contribution that has passed the external resonator. We thus obtain the following delay terms for the description of the effect caused by the re-injected light at one of the laser's facets ($\tilde{r} = (x,y,L)$),

$$\mathcal{E}_{\text{delay}}(\tilde{r},q) = \int \mathrm{d}^3\tilde{r}' \mathrm{e}^{-\mathrm{i}q\tilde{r}'} \rho^{\mathcal{E}}_{\text{delay}}\left(\tilde{r} - \frac{\tilde{r}'}{2}, \tilde{r} + \frac{\tilde{r}'}{2}\right) \tag{10.81}$$

$$I_{\text{delay}}(\tilde{r};\tilde{r}' = L,q) = \int \mathrm{d}^3\tilde{r}' \mathrm{e}^{-\mathrm{i}q\tilde{r}'} \rho^{\mathcal{I}}_{\text{delay}}\left(\tilde{r} - \frac{\tilde{r}'}{2}, \tilde{r} + \frac{\tilde{r}'}{2}\right), \tag{10.82}$$

with

$$\rho^{\mathcal{E}}_{\text{delay}}(r = R) = \frac{1}{\sqrt{\nu_0}} \gamma_{\text{delay}} \mathrm{e}^{\mathrm{i}\Phi} \langle \hat{b}_R \rangle \tag{10.83}$$

$$\begin{aligned}\rho^{I}_{\text{delay}}(r = R; r' = R') &= \frac{1}{\nu_0} \langle \hat{b}^\dagger_R \gamma_{\text{delay}} \mathrm{e}^{\mathrm{i}\Phi} \hat{b}_{R'} \rangle \\ &= \frac{1}{\nu_0} \gamma_{\text{delay}} \mathrm{e}^{\mathrm{i}\Phi} \langle \hat{b}^\dagger_R \hat{b}_{R'} \rangle .\end{aligned} \tag{10.84}$$

For the description of VCSEL the z coordinate can be eliminated so that no differentation between variables within the laser and at the facets is required, i.e. $r = \tilde{r}$. In (10.83) γ_{delay} describes the strength of the back coupling defined by the reflectivity of the external resonator. The modified equation for the field–field correlation then reads

$$
\begin{aligned}
\frac{\partial}{\partial t}\tilde{I}_{ij}(r;r',k) = &-(\kappa_i+\kappa_j)\tilde{I}_{ij}(r;r',k) - \mathrm{i}\frac{\omega_0}{2k_0^2\epsilon_\mathrm{r}}\left(\Delta_r - \Delta_{r'}\right)\tilde{I}_{ij}(r;r',k) \\
&-\mathrm{i}g_0\frac{\sqrt{\nu_0}}{4\pi^2}\Bigg(\tilde{C}_{ij}(r;r',k) - \tilde{C}^*_{ji}(r';r,k) \\
&+\frac{1}{k_0^2\epsilon_\mathrm{r}}\left(\frac{\partial^2}{\partial r'_j r'_k}\tilde{C}_{ik}(r;r',k) - \frac{\partial^2}{\partial r_i r_k}\tilde{C}^*_{jk}(r';r,k)\right)\Bigg) \\
&+\frac{c}{2nL}\gamma_\mathrm{delay}(r)\mathrm{e}^{\mathrm{i}\Phi}\tilde{I}_{ij}(r;r',k)
\end{aligned}
\qquad (10.85)
$$

$$
\begin{aligned}
\frac{\partial}{\partial t}\tilde{\mathcal{E}}_i(r,k) = &-(\kappa_i + \mathrm{i}\delta\omega_i)\tilde{\mathcal{E}}_i(r) + \mathrm{i}\frac{\omega_0}{2k_0^2\epsilon_\mathrm{r}}\Delta\tilde{\mathcal{E}}_i(r,k) \\
&-\mathrm{i}g_0\frac{\sqrt{\nu_0}}{4\pi^2}\left(\tilde{p}_i(r,k) + \sum_j \frac{1}{k_0^2\epsilon_\mathrm{r}}\frac{\partial^2}{\partial r_i r_j}\tilde{p}_j(r,k)\right) \\
&+\frac{c}{2nL}\gamma_\mathrm{delay}(r)\mathrm{e}^{\mathrm{i}\Phi}\mathcal{E}(r,q,t-\tau_\mathrm{cav})
\end{aligned}
\qquad (10.86)
$$

In (10.85) delayed optical feedback is represented by the feedback parameters $\gamma_\mathrm{delay}(r)$ (feedback strength), τ_cav (delay time), and Φ (feedback phase), respectively. Note that $\gamma_\mathrm{delay}(\mathbf{r})$ may also describe a spatially structured delayed optical feedback (cf. Chap. 9). The feedback strength is given by $\kappa = (1-R_0)\sqrt{\frac{R_1}{R_0}}$ with R_0 and R_1 being the reflectivity of the front laser facet and of the external mirror, respectively.

The application of (10.85) on a VCSEL in an external cavity allows calculation of spatio-temporal luminescence properties. In particular, this includes the resonances between the modes of the external cavity, the modes of the microcavity formed by the geometry of the laser and the dipole dynamics of the medium. As a result, (10.85) may be used to analyze the origin of a cavity-induced increase or suppression of spontaneous emission. Since spontaneous emission determines the phase coherence in the cavity, this dynamic interplay finally affects the intensity and spatial emission characteristics in the far field of a laser. We recall that the delayed optical feedback luminescence equations (10.85) model the interaction between the resonator-internal light field distributions with the re-injected light having passed the external resonator in a self-consistent way. The explicit consideration of the temporal, spatial and spectral dependence in the equations allows simulation of arbitrary resonator configurations, inclusion of spatial or spectral selective feedback mechanisms and inclusion of nonlinear resonator-internal elements.

So far we have considered the control of laser emission by self-injection of parts of the emitted light via an external resonator. In a VCSEL the control of spatio-spectral laser properties can be further increased if the active area is optically pumped by an external light source. As a final extension to our luminescence equations in the following we will thus describe and discuss the possibility of inclusion of an optical pump mechanism.

10.6.2 Spatio-Temporal Dynamics of Optically Pumped Semicondutor Lasers

The emission properties of semiconductor lasers with large extension of the active area typically show a complex filamentation behavior that arises from the spatio-temporal coupling and interplay of light propagation, diffraction, carrier diffusion and microscopic carrier scattering processes. The transfer of the concept of optical pumping (that so far has successfully been applied to solid state lasers) to the semiconductor laser allows a combination of power scaling involved in the high gain of semiconductor laser devices and high beam quality provided by direct optical excitation. Preliminary experimental investigations show that this concept is indeed very promising for advanced laser technologies and applications.

In the situation of optically pumped semiconductor laser devices the injection of a light field high above the bandgap generates a (partially incoherent) excited electron–hole plasma which then leads to a hierarchy of carrier relaxation processes within the bands followed by (low momentum) radiative carrier recombination that can be observed as (photo) luminescence.

In principle there are two ways to include this pump process in the luminescence equations. One method is to use heated carrier distributions for simulation of high energy carrier accumulation generated by a pump pulse. This can be done by assuming e.g. a δ or (to include the finite bandwidth of the pump pulse) Lorenzian shaped carrier generation around a given wave vector k. The superposition of this high energy carrier accumulation and the quasi-equilibrium carrier distribution given by the respective Fermi–Dirac functions (where one has to insert the plasma temperatures and chemical potentials of electrons and holes) can then be used for initialization of the carrier distributions f_k^{e} and f_k^{h} in the luminescence equations (10.74). The carrier relaxation within the bands and the radiative recombination described by the luminescence equations then automatically lead to a conversion of the carrier inversion (created by the optical excitation of the semiconductor medium) into the fundamental mode at a wavelength given by the resonator of the semiconductor laser. This method is appropriate for the study of the *ultrafast response* of the *carrier system* in an excited semiconductor medium.

As an alternative one can introduce the spatial, spectral and temporal properties of an optical pump laser via an additional field entering the luminescence equations. This method allows direct inclusion and analysis of *transverse light field dynamics* effects induced by e.g. a particular spatial profile, a specific transverse phase distribution (defining the focusing into the medium) or coherence properties of an optical pump laser. The derivation of the respective modifications of the luminescence equations are analogues to the derivation of the delay terms for the description of external resonators.

However, for simulation of an optical pump process, the amplitude ($\tilde{I}_{\mathrm{inj}}$) and phase (Φ_{inj}) of the additional injection terms are given by the amplitude and phase distribution of the injected field, respectively (and not by a delay line and an external mirror as in the case of the external resonator configuration). In particular, the spatial (i.e. spatial beam shape), spectral (i.e. central frequency, band width, line shape) and temporal (i.e. duration of the excitation) properties can directly be expressed via the r, k and t dependence of $\tilde{I}_{\mathrm{inj}}$. For example, a light pulse of duration τ_{pulse}, Gaussian beam profile (spatial center r_0, beam width w_0), central wavenumber k_{inj} and Lorentzian shaped spectrum may be expressed as

$$\tilde{I}_{\mathrm{inj}}(r;r',k) = \tilde{I}^0_{\mathrm{inj}} \mathrm{e}^{-\left(\frac{(r-r_0)}{w_0}\right)^2} \mathrm{e}^{-(t-\tau_{\mathrm{pulse}})} \frac{\Gamma^2}{\Gamma^2 + (k-k_{\mathrm{inj}})} \tag{10.87}$$

The resulting luminescence equations for optically pumped surface emitting lasers finally read

$$\begin{aligned}\frac{\partial}{\partial t}\tilde{I}_{ij}(r;r',k) = &-(\kappa_i+\kappa_j)\tilde{I}_{ij}(r;r',k) - \mathrm{i}\frac{\omega_0}{2k_0^2\epsilon_{\mathrm{r}}}\left(\Delta_r - \Delta_{r'}\right)\tilde{I}_{ij}(r;r',k) \\ &-\mathrm{i}g_0\frac{\sqrt{\nu_0}}{4\pi^2}\left(\tilde{C}_{ij}(r;r',k) - \tilde{C}^*_{ji}(r';r,k)\right. \\ &\left.+\frac{1}{k_0^2\epsilon_{\mathrm{r}}}\left(\frac{\partial^2}{\partial r'_j r'_k}\tilde{C}_{ik}(r;r',k) - \frac{\partial^2}{\partial r_i r_k}\tilde{C}^*_{jk}(r';r,k)\right)\right) \\ &+\tilde{I}_{\mathrm{inj}}(r;r',k) \\ &+\frac{c}{2nL}\mathrm{e}^{\mathrm{i}\Phi_{\mathrm{inj}}}\tilde{I}_{ij}(r;r',k)\end{aligned} \tag{10.88}$$

$$\begin{aligned}\frac{\partial}{\partial t}\tilde{\mathcal{E}}_i(r,k) = &-(\kappa_i + \mathrm{i}\delta\omega_i)\tilde{\mathcal{E}}_i(r) + \mathrm{i}\frac{\omega_0}{2k_0^2\epsilon_{\mathrm{r}}}\Delta\tilde{\mathcal{E}}_i(r,k) \\ &-\mathrm{i}g_0\frac{\sqrt{\nu_0}}{4\pi^2}\left(\tilde{p}_i(r,k) + \sum_j \frac{1}{k_0^2\epsilon_{\mathrm{r}}}\frac{\partial^2}{\partial r_i r_j}\tilde{p}_j(r,k)\right) \\ &+\frac{c}{2nL}\gamma_{\mathrm{delay}}(r)\mathrm{e}^{\mathrm{i}\Phi}\mathcal{E}(r,q,t-\tau_{\mathrm{cav}}) \\ &+\mathcal{E}_{\mathrm{inj}}(r,q,t)\,.\end{aligned} \tag{10.89}$$

10.7 Conclusions

The quantum Maxwell–Bloch luminescence equations (QMBEs) for spatially inhomogeneous semiconductor lasers discussed in this chapter take into account the quantum mechanical nature of the light field as well as that of the carrier system. The only approximation used in the derivation of the intensity and correlation dynamics is the one of statistical independence between the two carrier systems and the light field. In the QMBEs the effects of coherent spatio-temporal quantum fluctuations (that are generally not considered in the semiclassical Maxwell–Bloch equations for semiconductor laser devices) have thus been taken into account.

The spontaneous emission term appears side by side with the gain and absorption terms in the dynamics of the field–dipole correlation. In this way the spatial coherence of spontaneous emission and amplified spontaneous emission is consistently described by the quantum Maxwell–Bloch equations. In general, the quantum Maxwell–Bloch equations derived for quantum well structures and vertical-cavity surface-emitting lasers provide a starting point for a detailed analysis of spatial coherence patterns in diverse semiconductor laser geometries such as broad-area or ultra-low threshold lasers. It combines quantum kinetics of hot-carrier relaxation and quantum theory of spontaneous emission. In future investigations theoretical studies may allow for analysis of hot luminescence being a sensitive measure of microscopic carrier scattering processes.

To underline the importance of semiconductor luminescence in advanced semiconductor lasers, we have extended the theory of semiconductor photo luminescence to model the spatio-temporal dynamics of vertical-cavity surface-emitting lasers in external resonators. They are of high importance for profound understanding of the physical processes within these systems and for detailed laser and resonator design.

References

1. H. F. Hofmann and O. Hess. Quantum Maxwell-Bloch equations for spatially inhomogenous semiconductor lasers. *Phs. Rev. A*, 59:2342–2358, 1999.
2. G. Björk, A. Karlsson, and Y. Yamamoto. Definition of a laser threshold. *Phys. Rev. A*, 50:1675–1680, 1994.
3. Y. Yamamoto and R.E. Slusher. Optical processes in microcavities. *Physics Today*, June 1993:66, 1993.
4. K. J. Ebeling. *Integrated Optoelectronics*. Springer-Verlag, Berlin, 1993.
5. T. Lee, C.A. Burrus, J.A. Copeland, A.G. Dentai, and D. Marcuse. Short cavity ingaasp: Dependence of mode spectra and single-longitudinal-mode power on cavity length. *IEEE J. Quantum Electron.*, QE-18:1101, 1982.
6. K. Petermann. Calculated spontaneous emission factor for double-heterostructure injection lasers with gain-induced waveguiding. *IEEE J. Quantum Electron.*, QE-15:566–570, 1979.
7. A. Yariv and S. Margalit. On spontaneous emission into guided modes with curved wavefronts. *IEEE J. Quantum Electron.*, QE-18:1831, 1982.
8. A. E. Siegmann. *Coherence and Quantum Optics VII*, chapter Dynamics of High-Power Diode Lasers. Plenum, New York, 1996.
9. F.Jahnke, K.Henneberger, W.Schäfer, and S.W. Koch. Transient nonequilibrium and many-body effects in semiconductor microcavity lasers. *J. Opt. Soc. Am. B*, 10:2394–2405, 1993.

10. F.Jahnke and S.W. Koch. Many-body theory for semiconductor microcavity lasers. *Phys. Rev. A*, 52:1712–1727, 1995.
11. K. Henneberger and S.W. Koch. Quantum kinetics of semiconductor light emission and lasing. *Phys. Rev. Lett.*, 76:182–1823, 1996.
12. M. Kira, F. Jahnke, and S. W. Koch. Microscopic theory of excitonic signatures in semiconductor photoluminescence. *Phys. Rev. Lett.*, 81:3263–3266, 1998.
13. M. Kira, F. Jahnke, W. Hoyer, and S. W. Koch. Quantum theory of spontaneous emission and coherent effects in semiconductor microstructures. *Prog. Quantum Electronics*, 23:189–279, 1999.
14. A. Thränhardt, S. Kuckenburg, A. Knorr, T. Meier, and S. W. Koch. Quantum theory of phonon-assisted exciton formation and luminescence in semiconductor quantum wells. *Phys. Rev. B*, 62:2706–2720, 2000.
15. K. Hannewald, S. Glutsch, and F. Bechstedt. Theory of photoluminescence in semiconductors. *Phys. Rev. B*, 81:4519–4525, 2000.
16. O. Hess and T. Kuhn. Maxwell-Bloch equations for spatially inhomogeneous semiconductor lasers I: Theoretical description. *Phys. Rev. A*, 54:3347–3359, 1996.
17. P. Y. Yu and M. Cardona. *Fundamentals of Semiconductors.* Springer-Verlag, Berlin, 1996.
18. E. Wigner. On the quantum correction for thermodynamic equilibrium. *Phys. Rev.*, 40:749–759, 1932.
19. M. San Miguel, Q. Feng, and J.V. Moloney. Light-polarization dynamics in surface-emitting semiconductor lasers. *Phys. Rev. A*, 52:1728–1739, 1995.
20. H. F. Hofmann and O. Hess. Quantum noise and polarization fluctuations in vertical-cavity surface-emitting lasers. *Phys. Rev. A*, 56:868–876, 1997.
21. H. van der Lem and D. Lenstra. Saturation-induced frequency shift in the noise spectrum of a birefringent vertical-cavity surface-emitting laser. *Opt. Lett.*, 22:1698–1700, 1997.
22. A. K. J. van Doorn, M.P. van Exter, A. M. van der Lee, and J.P. Woerdman. Coupled-mode description for the polarization state of a vertical-cavity semiconductor laser. *Phys. Rev. A*, 55:1473–1484, 1997.
23. C. Simmendinger, M. Mnkel, and O. Hess. Controlling complex temporal and spatio-temporal dynamics in semiconductor lasers'. *Chaos, Solitons and Fractals*, 10:851–864, 1999.
24. G. H. M. van Tartwijk and D. Lenstra. Semiconductor lasers with optical injection and feed-back. *Qantum Semiclass. Opt.*, 7:87–143, 1995.
25. C. Simmendinger, D. Preier, and O. Hess. Stabilization of chaotic spatiotemporal filamentation in large broad area lasers by spatially structured optical feedback. *Optics Express*, 5:4854, 1999.

Index

Springer Tracts in Modern Physics

149 **High-Resolution X-Ray Scattering from Thin Films and Multilayers**
By V. Holý, U. Pietsch, and T. Baumbach 1999. 148 figs. XI, 256 pages

150 **QCD at HERA**
The Hadronic Final State in Deep Inelastic Scattering
By M. Kuhlen 1999. 99 figs. X, 172 pages

151 **Atomic Simulation of Electrooptic and Magnetooptic Oxide Materials**
By H. Donnerberg 1999. 45 figs. VIII, 205 pages

152 **Thermocapillary Convection in Models of Crystal Growth**
By H. Kuhlmann 1999. 101 figs. XVIII, 224 pages

153 **Neutral Kaons**
By R. Beluševi 1999. 67 figs. XII, 183 pages

154 **Applied RHEED**
Reflection High-Energy Electron Diffraction During Crystal Growth
By W. Braun 1999. 150 figs. IX, 222 pages

155 **High-Temperature-Superconductor Thin Films at Microwave Frequencies**
By M. Hein 1999. 134 figs. XIV, 395 pages

156 **Growth Processes and Surface Phase Equilibria in Molecular Beam Epitaxy**
By N.N. Ledentsov 1999. 17 figs. VIII, 84 pages

157 **Deposition of Diamond-Like Superhard Materials**
By W. Kulisch 1999. 60 figs. X, 191 pages

158 **Nonlinear Optics of Random Media**
Fractal Composites and Metal-Dielectric Films
By V.M. Shalaev 2000. 51 figs. XII, 158 pages

159 **Magnetic Dichroism in Core-Level Photoemission**
By K. Starke 2000. 64 figs. X, 136 pages

160 **Physics with Tau Leptons**
By A. Stahl 2000. 236 figs. VIII, 315 pages

161 **Semiclassical Theory of Mesoscopic Quantum Systems**
By K. Richter 2000. 50 figs. IX, 221 pages

162 **Electroweak Precision Tests at LEP**
By W. Hollik and G. Duckeck 2000. 60 figs. VIII, 161 pages

163 **Symmetries in Intermediate and High Energy Physics**
Ed. by A. Faessler, T.S. Kosmas, and G.K. Leontaris 2000. 96 figs. XVI, 316 pages

164 **Pattern Formation in Granular Materials**
By G.H. Ristow 2000. 83 figs. XIII, 161 pages

165 **Path Integral Quantization and Stochastic Quantization**
By M. Masujima 2000. 0 figs. XII, 282 pages

166 **Probing the Quantum Vacuum**
Pertubative Effective Action Approach in Quantum Electrodynamics and its Application
By W. Dittrich and H. Gies 2000. 16 figs. XI, 241 pages

167 **Photoelectric Properties and Applications of Low-Mobility Semiconductors**
By R. Könenkamp 2000. 57 figs. VIII, 100 pages

168 **Deep Inelastic Positron-Proton Scattering in the High-Momentum-Transfer Regime of HERA**
By U.F. Katz 2000. 96 figs. VIII, 237 pages

169 **Semiconductor Cavity Quantum Electrodynamics**
By Y. Yamamoto, T. Tassone, H. Cao 2000. 67 figs. VIII, 154 pages

170 **d–d Excitations in Transition-Metal Oxides**
A Spin-Polarized Electron Energy-Loss Spectroscopy (SPEELS) Study
By B. Fromme 2001. 53 figs. XII, 143 pages

Springer Tracts in Modern Physics

171 **High-T_c Superconductors for Magnet and Energy Technology**
By B. R. Lehndorff 2001. 139 figs. XII, 209 pages

172 **Dissipative Quantum Chaos and Decoherence**
By D. Braun 2001. 22 figs. XI, 132 pages

173 **Quantum Information**
An Introduction to Basic Theoretical Concepts and Experiments
By G. Alber, T. Beth, M. Horodecki, P. Horodecki, R. Horodecki, M. Rötteler, H. Weinfurter, R. Werner, and A. Zeilinger 2001. 60 figs. XI, 216 pages

174 **Superconductor/Semiconductor Junctions**
By Thomas Schäpers 2001. 91 figs. IX, 145 pages

175 **Ion-Induced Electron Emission from Crystalline Solids**
By Hiroshi Kudo 2002. 85 figs. IX, 161 pages

176 **Infrared Spectroscopy of Molecular Clusters**
An Introduction to Intermolecular Forces
By Martina Havenith 2002. 33 figs. VIII, 120 pages

177 **Applied Asymptotic Expansions in Momenta and Masses**
By Vladimir A. Smirnov 2002. 52 figs. IX, 263 pages

178 **Capillary Surfaces**
Shape – Stability – Dynamics, in Particular Under Weightlessness
By Dieter Langbein 2002. 182 figs. XVIII, 364 pages

179 **Anomalous X-ray Scattering for Materials Characterization**
Atomic-Scale Structure Determination
By Yoshio Waseda 2002. 132 figs. XIV, 214 pages

180 **Coverings of Discrete Quasiperiodic Sets**
Theory and Applications to Quasicrystals
Edited by P. Kramer and Z. Papadopolos 2002. 128 figs., XIV, 274 pages

181 **Emulsion Science**
Basic Principles. An Overview
By J. Bibette, F. Leal-Calderon, V. Schmitt, and P. Poulin 2002. 50 figs., IX, 140 pages

182 **Transmission Electron Microscopy of Semiconductor Nanostructures**
An Analysis of Composition and Strain State
By A. Rosenauer 2003. 136 figs., XII, 238 pages

183 **Transverse Patterns in Nonlinear Optical Resonators**
By K. Staliūnas, V. J. Sánchez-Morcillo 2003. 132 figs., XII, 226 pages

184 **Statistical Physics and Economics**
Concepts, Tools and Applications
By M. Schulz 2003. 54 figs., XII, 244 pages

185 **Electronic Defect States in Alkali Halides**
Effects of Interaction with Molecular Ions
By V. Dierolf 2003. 80 figs., XII, 196 pages

186 **Electron-Beam Interactions with Solids**
Application of the Monte Carlo Method to Electron Scattering Problems
By M. Dapor 2003. 27 figs., X, 110 pages

187 **High-Field Transport in Semiconductor Superlattices**
By K. Leo 2003. 164 figs.,XIV, 240 pages

188 **Transverse Pattern Formation in Photorefractive Optics**
By C. Denz, M. Schwab, and C. Weilnau 2003. 143 figs., XVIII, 331 pages

189 **Spatio-Temporal Dynamics and Quantum Fluctuations in Semiconductor Lasers**
By O. Hess, E. Gehrig 2003. 91 figs., XIV, 230 pages

MIX
Papier aus verantwortungsvollen Quellen
Paper from responsible sources
FSC® C105338

If you have any concerns about our products,
you can contact us on
ProductSafety@springernature.com

In case Publisher is established outside the EU,
the EU authorized representative is:
Springer Nature Customer Service Center GmbH
Europaplatz 3, 69115 Heidelberg, Germany

Printed by Libri Plureos GmbH
in Hamburg, Germany